D0712836

Protease Inhibitors in AIDS Therapy

Compliments of

INFECTIOUS DISEASE AND THERAPY

Series Editor

Burke A. Cunha

Winthrop-University Hospital
Mineola, and
State University of New York School of Medicine
Stony Brook, New York

1. Parasitic Infections in the Compromised Host, *edited by Peter D. Walzer and Robert M. Genta*
2. Nucleic Acid and Monoclonal Antibody Probes: Applications in Diagnostic Methodology, *edited by Bala Swaminathan and Gyan Prakash*
3. Opportunistic Infections in Patients with the Acquired Immunodeficiency Syndrome, *edited by Gifford Leoung and John Mills*
4. Acyclovir Therapy for Herpesvirus Infections, *edited by David A. Baker*
5. The New Generation of Quinolones, *edited by Clifford Siporin, Carl L. Heifetz, and John M. Domagala*
6. Methicillin-Resistant *Staphylococcus aureus*: Clinical Management and Laboratory Aspects, *edited by Mary T. Cafferkey*
7. Hepatitis B Vaccines in Clinical Practice, *edited by Ronald W. Ellis*
8. The New Macrolides, Azalides, and Streptogramins: Pharmacology and Clinical Applications, *edited by Harold C. Neu, Lowell S. Young, and Stephen H. Zinner*
9. Antimicrobial Therapy in the Elderly Patient, *edited by Thomas T. Yoshikawa and Dean C. Norman*
10. Viral Infections of the Gastrointestinal Tract: Second Edition, Revised and Expanded, *edited by Albert Z. Kapikian*
11. Development and Clinical Uses of Haemophilus b Conjugate Vaccines, *edited by Ronald W. Ellis and Dan M. Granoff*
12. *Pseudomonas aeruginosa* Infections and Treatment, *edited by Aldona L. Baltch and Raymond P. Smith*
13. Herpesvirus Infections, *edited by Ronald Glaser and James F. Jones*
14. Chronic Fatigue Syndrome, *edited by Stephen E. Straus*
15. Immunotherapy of Infections, *edited by K. Noel Masihi*
16. Diagnosis and Management of Bone Infections, *edited by Luis E. Jauregui*

17. Drug Transport in Antimicrobial and Anticancer Chemotherapy, *edited by Nafsika H. Georgopapadakou*
18. New Macrolides, Azalides, and Streptogramins in Clinical Practice, *edited by Harold C. Neu, Lowell S. Young, Stephen H. Zinner, and Jacques F. Acar*
19. Novel Therapeutic Strategies in the Treatment of Sepsis, *edited by David C. Morrison and John L. Ryan*
20. Catheter-Related Infections, *edited by Harald Seifert, Bernd Jansen, and Barry M. Farr*
21. Expanding Indications for the New Macrolides, Azalides, and Streptogramins, *edited by Stephen H. Zinner, Lowell S. Young, Jacques F. Acar, and Harold C. Neu*
22. Infectious Diseases in Critical Care Medicine, *edited by Burke A. Cunha*
23. New Considerations for Macrolides, Azalides, Streptogramins, and Ketolides, *edited by Stephen H. Zinner, Lowell S. Young, Jacques F. Acar, and Carmen Ortiz-Neu*
24. Tickborne Infectious Diseases: Diagnosis and Management, *edited by Burke A. Cunha*
25. Protease Inhibitors in AIDS Therapy, *edited by Richard C. Ogden and Charles W. Flexner*

Additional Volumes in Production

Antimicrobial Pharmacodynamics in Theory and Clinical Practice, *edited by Charles H. Nightingale, Takeo Murakawa, and Paul G. Ambrose*

Protease Inhibitors in AIDS Therapy

edited by

Richard C. Ogden

Agouron Pharmaceuticals, Inc.
San Diego, California

Charles W. Flexner

The Johns Hopkins University School of Medicine and
School of Hygiene and Public Health
Baltimore, Maryland

Marcel Dekker, Inc. New York • Basel

ISBN: 0-8247-0461-4

This book is printed on acid-free paper.

Headquarters
Marcel Dekker, Inc.
270 Madison Avenue, New York, NY 10016
tel: 212-696-9000; fax: 212-685-4540

Eastern Hemisphere Distribution
Marcel Dekker AG
Hutgasse 4, Postfach 812, CH-4001 Basel, Switzerland
tel: 41-61-261-8482; fax: 41-61-261-8896

World Wide Web
http://www.dekker.com

The publisher offers discounts on this book when ordered in bulk quantities. For more information, write to Special Sales/Professional Marketing at the headquarters address above.

Current printing (last digit):
10 9 8 7 6 5 4 3 2 1

PRINTED IN THE UNITED STATES OF AMERICA

Foreword

The history of antiretroviral therapy begins with zidovudine (AZT), which was initially studied in an AIDS therapeutic trial in 1986. Interim analysis of the unblinded data in September 1986 showed spectacular results—19 deaths among placebo recipients and a single death in the AZT recipients. The single exception in the AZT group was a patient who stopped taking AZT because of intolerance. The Data and Safety Monitoring Board stopped the study for ethical reasons, all participants in the trial received AZT, and the FDA provided an expedited review, which resulted in approval of the drug in March 1987. At that time, it appeared that we were well on the way to a cure. However, it was soon learned that any benefit from AZT was short-lived; in retrospect, this appears to reflect low potency combined with the rapid evolution of resistance. During the subsequent 8 years, four additional nucleoside analogs were introduced, combination therapy using two nucleoside analogs became standard, and AIDS clinicians debated the relative merits of drugs in this class while deaths ascribed to AIDS continued to rise. There is serious question that these drugs directed against reverse transcriptase had any substantial impact on the natural history of HIV infection either alone or in combination.

The second HIV enzyme to be targeted was aspartyl protease. These ''designer'' drugs posed substantial challenges for development due to problems with bioavailability, short half-life, and toxicity. The first FDA-approved member of the class, saquinavir (Invirase®), became available in late 1995, and this was quickly followed by the approval of ritonavir and indinavir. By mid-1996, the era of protease inhibitors was well engrained in medical practice, although the placement of these drugs and their use in various combinations was the subject of substantial confusion in the early months. Nevertheless, treatment strategies rapidly evolved to ''triple therapy,'' using two nucleosides and a protease inhibitor. More recently, use of double protease inhibitors to exploit the drug interaction of ritonavir to magnify the peak level or AUC of the companion agent has become common.

The introduction of protease inhibitors has had a profound impact on HIV care, arguably the most dramatic impact on any therapy for an important infectious disease since the introduction of penicillin. The continuous increase in HIV-related deaths reached its zenith in the United States in 1995, then decreased by 26% in 1996, and by an additional 47% in 1997. Multiple studies have documented the dramatic impact of this class of drugs on the morbidity and mortality of HIV infection. There was an equally dramatic decrease in the rate of HIV-related complications, including opportunistic infections, and the rate of hospitalizations for HIV-related medical problems. This change was one of the most dramatic seen in any disease over the past several decades. Some moribund patients had dramatic reversals resulting in robust health. Patients who were being prepared to die now had retraining to prepare to live. ''HIV burn-out,'' a common concern in HIV care programs, suddenly disappeared. These results rocketed through medicine, David Ho was named Man of the Year by *Time* magazine in 1996 for his work in this area, and protease inhibitors were identified as the scientific discovery of the year by *Science* in 1997.

The advances in therapy brought about by the introduction of protease inhibitors also had a down side. The initial projection that the new therapy would bring cure within 2 to 3 years was conceptually defeated by the demonstration of HIV latency, which suggests the improbability of cure even with maximum viral suppression for 60 years or longer. This disease, unlike most other infectious diseases, could not utilize antimicrobials and natural defenses to eliminate the pathogens. The treatment also became incredibly complex, leading to the widely held impression that HIV is no longer a ''primary care disease'' but one that requires substantial expertise by a new cadre of physicians—the HIV care providers—who generally define their expertise by the number of HIV-infected patients in their care. It was also challenging to the patients, who needed to adhere to the new regimens with a level of diligence that, by prior studies, is rarely achieved in any chronic disease. Moreover, the consequences of nonadherence were particularly punishing because of the evolution of resistance, first to the individual agents and then to the entire class. A mysterious group of side effects under the umbrage of ''lipodystrophy'' have arisen, with unclear pathophysiology and long-term consequences.

While protease inhibitors have revolutionized the field of HIV therapeutics, they have also introduced a variety of new challenges. The short history of this class has to be one of the most fascinating stories in medical therapeutics. It occupies a very central position in the history of HIV, which some have identified as the major epidemic of the twentieth century. Drs. Ogden and Flexner are to be commended for compiling this text on HIV protease inhibitors, which I believe to be the first in existence. This book should be useful to practitioners, patients, and those developing new classes of antiretroviral drugs. Let us hope that the successes documented in this book will continue accruing, to the benefit of all those infected with HIV.

John G. Bartlett

Preface

In the history of drug development, no target has attracted more attention in a shorter period of time than the protease from HIV-1. Within 5 years of the purification, characterization, and structure elucidation of the enzyme in 1988, tens of thousands of potential inhibitors had been synthesized and screened, and over 20 compounds had been tested in human subjects by 11 pharmaceutical and biotechnology companies. By 1996, treatment options for many people with AIDS changed significantly when three of these compounds were approved for prescription use by the U.S. Food and Drug Administration. Clinical trials showed that these compounds had an unprecedented capacity to suppress viral load, reduce disease progression, and prolong life. For many, this story represents the first real hope for effective treatment of an otherwise fatal infection. For others, this is the first solid success story at the beginning of a new era of molecular medicine. This book provides a timely and comprehensive look at the class of HIV protease inhibitors from a scientific and clinical perspective.

The human immunodeficiency virus was perhaps the first new pathogenic agent to be dissected at the structural and functional level by the tools of modern interdisciplinary science. Advances in recombinant DNA technology, rapid DNA sequence analysis, and detection and amplification of virus nucleic acid from tissue were initially used to characterize and categorize the virus. Improved techniques of structural biology and the design, synthesis, and screening of potential drugs have been brought to bear on the search for effective and durable therapy. The influence of publicly funded academic and corporate basic research continues to be felt in continuing studies of both drug discovery and HIV pathogenesis. This topic, as it relates to HIV protease, is covered in Chapter 1.

The accounts of the discovery and evaluation of novel, distinct HIV protease inhibitors by both large pharmaceutical firms and biotechnology start-up companies illustrate the range of strategies and techniques available for modern drug discovery. Chapters 2 through 7 include preliminary clinical studies during which such issues as oral bioavailability, pharmacokinetic profile, efficacy, resistance, and tolerability are first uncovered.

Chapter 8 is the first of five chapters that compare available data for all the marketed HIV protease inhibitors and those in late clinical development. It starts with a description of the pharmacological attributes of anti-HIV protease drug candidates, particularly critical in the absence of an in vivo animal model for AIDS. Of major importance to the clinician is an understanding of the pharmacokinetics, metabolism, and drug interactions of the protease inhibitors, topics covered in both Chapters 8 and 9. Chapter 10 gives a comparative account of clinical efficacy, followed by Chapter 11 on reported toxicities and Chapter 12 on the emergence and management of antiviral resistance.

Finally, Chapter 13 describes the necessary collaboration between drug developers and the regulatory authorities that led to the rapid approval of this class of agents, based largely on surrogate viral endpoints. The issues relating to long-term safety evaluation and full approval are also described.

We would like to acknowledge the hard work and dedication of all involved in the production of this book, especially the contributors and the editorial staff at Marcel Dekker, Inc.: Elyce Misher, Jinnie Kim, and the late Graham Garratt, under whose guidance this book was conceived.

We also acknowledge the help of our professional colleagues, families, friends, and those with HIV infection, who served as a constant source of inspiration for this project. C.F. would especially like to thank his wife, Carol, and children, James, Bill, Christopher, and Sara, who put up with his late nights, travel schedule, and occasional irritable moods, all side effects of participating in the development of this amazing class of drugs. He would also like to thank Laura Rocco for editorial assistance. R.O. would like to thank Regina Rooney for her love, support, and tolerance of his travels. He also acknowledges the support of his friends and colleagues at Agouron.

Many issues make HIV/AIDS a singular medical syndrome, in both the developed and developing worlds. The drugs described in this book—when used in combination with other antiretroviral agents—have extended and improved the lives of those living with the virus in many areas of the developed world. Their use has helped us understand the disease and to stimulate further AIDS research, the impact of which will be more global.

Richard C. Ogden
Charles W. Flexner

Contents

Contributors

C. T. Baker Medicinal Chemistry, Vertex Pharmaceuticals, Cambridge, Massachusetts

J. S. Boger Vertex Pharmaceuticals, Cambridge, Massachusetts

S. P. Chambers Vertex Pharmaceuticals, Cambridge, Massachusetts

R. Alan Chrusciel, Ph.D. Structural, Analytical, and Medicinal Chemistry Research, Pharmacia & Upjohn, Kalamazoo, Michigan

D. D. Deininger Vertex Pharmaceuticals, Cambridge, Massachusetts

Bruce D. Dorsey, Ph.D. Medicinal Chemistry, Merck Research Laboratories, West Point, Pennsylvania

Ian B. Duncan, Ph.D. Roche Discovery Welwyn, Welwyn Garden City, Hertfordshire, England

M. Dwyer Vertex Pharmaceuticals, Cambridge, Massachusetts

L. Elsayed Vertex Pharmaceuticals, Cambridge, Massachusetts

John W. Erickson, Ph.D. Director, Tibotec Institute for Therapeutics Research, Tibotec, Inc., Rockville, Maryland

Charles W. Flexner, M.D. Departments of Medicine, Pharmacology and Molecular Sciences, and International Health, The Johns Hopkins University School of Medicine, and School of Hygiene and Public Health, Baltimore, Maryland

J. Fulghum Vertex Pharmaceuticals, Cambridge, Massachusetts

Marshall J. Glesby, M.D., Ph.D. Department of Medicine, Weill Medical College of Cornell University, New York, New York

Roy M. Gulick, M.D., M.P.H. Department of Medicine, Weill Medical College of Cornell University, New York, New York

Dale J. Kempf Pharmaceutical Products Division, Abbott Laboratories, Abbott Park, Illinois

E. E. Kim Vertex Pharmaceuticals, Cambridge, Massachusetts

Daniel R. Kuritzkes, M.D. Division of Infectious Diseases, University of Colorado Health Sciences Center, Denver, Colorado

B. Li Vertex Pharmaceuticals, Cambridge, Massachusetts

D. J. Livingston Vertex Pharmaceuticals, Cambridge, Massachusetts

M. A. Murcko Vertex Pharmaceuticals, Cambridge, Massachusetts

Jeffrey Murray Division of Antiviral Drug Products, U.S. Food and Drug Administration, Rockville, Maryland

M. A. Navia Vertex Pharmaceuticals, Cambridge, Massachusetts

Judith A. Nicholas, Ph.D. Drug Discovery, MDS Pharma Services, Bothell, Washington

P. Novak Vertex Pharmaceuticals, Cambridge, Massachusetts

Richard C. Ogden Agouron Pharmaceuticals, San Diego, California

S. Pazhanisamy, Ph.D. Vertex Pharmaceuticals, Cambridge, Massachusetts

B. G. Rao, Ph.D. Vertex Pharmaceuticals, Cambridge, Massachusetts

Sally Redshaw, Ph.D. Department of Medicinal Chemistry, Roche Discovery Welwyn, Welwyn Garden City, Hertfordshire, England

Siegfried H. Reich, Ph.D. Department of Medicinal Chemistry, Agouron Pharmaceuticals, San Diego, California

C. Stuver Vertex Pharmaceuticals, Cambridge, Massachusetts

Suvit Thaisrivongs, Ph.D. Structural, Analytical, and Medicinal Chemistry Research, Pharmacia & Upjohn, Kalamazoo, Michigan

J. A. Thomson Vertex Pharmaceuticals, Cambridge, Massachusetts

R. D. Tung Vertex Pharmaceuticals, Cambridge, Massachusetts

Joseph P. Vacca, Ph.D. Department of Medicinal Chemistry, Merck Research Laboratories, West Point, Pennsylvania

Benjamin Young, M.D., Ph.D. Rose Medical Center, and Division of Infectious Diseases, University of Colorado Health Sciences Center, Denver, Colorado

Protease Inhibitors in AIDS Therapy

1
HIV-1 Protease as a Target for AIDS Therapy

John W. Erickson
Tibotec Institute for Therapeutics Research, Tibotec, Inc., Rockville, Maryland

I. INTRODUCTION

The story of the discovery and development of HIV-1 protease (HIV PR) inhibitors is a remarkable one in many ways. Perhaps first and foremost, it represents a watershed in the fight against AIDS, providing new hope and optimism in the face of widespread pessimism and despair. It represents the first major harvest of the commercial fruits of basic research on the molecular biology of viruses, which had been borne in the early 1970s. It represents the coming-of-age of rational drug design in an era when antiviral drug discovery had been dominated by screening, a method that seemed to be giving diminishing returns. Up until the introduction of the PR inhibitors in the 1990s, the most common strategy for antiviral drug discovery was based on trial-and-error screening of compounds using cell-based systems of viral replication. The current crop of nucleoside and non-nucleoside inhibitors of HIV-1 reverse transcriptase (HIV RT) were all discovered from antiviral screening efforts. In contrast, none of the PR inhibitors were discovered using cell-based screens. Their discovery depended largely on the skillful translation of knowledge on structure and mechanism into hypothesis-driven approaches to the design, synthesis, and evaluation of novel chemical entities, which ultimately proved to be more effective and less toxic, in many instances, than the nucleoside analogues. Thus, the PR inhibitors represent a radical new paradigm in antiviral drug discovery as well as a major success story for modern antiviral therapy.

There are currently five U.S. Food and Drug Administration (FDA)-approved PR inhibitors for clinical use—saquinavir, ritonavir, indinavir, nelfinavir, and amprenavir—and several others are undergoing clinical trials [e.g., tipranavir (PNU-140690), and ABT-378] or are in preclinical development. This is a remarkable achievement because owing to a combination of ignorance and skepticism, HIV PR was not seriously considered as a target for antiviral therapy until the late 1980s. The reason for the skepticism is easy to understand—viral protease inhibitors had not yet been invented. Fortunately, however, science and luck were, this time, on the side of the HIV-infected individual. Since their introduction into the AIDS pharmacopeia in the mid-1990s, protease inhibitors have become first-line antiviral agents for the control of HIV infection and are widely used in most highly active antiretroviral therapy (HAART) regimens that constitute the current standard of care (1). Most importantly, for the first time in the history of HIV infection, a class of drugs has been found that provides a dramatic and durable suppression of HIV replication and onset of symptoms.

The discovery and early development of the HIV PR inhibitors constitute important anecdotal chapters in the annals of drug discovery, and their stories are appropriately told in individual chapters in this book, written by those who were intimately involved in the drug discovery process. Unfortunately, the book on HIV PR inhibitors does not end with their regulatory approvals and availability to patients and physicians. The reason is that despite the success of protease inhibitors in AIDS therapy, their widespread use has led to the emergence of drug-resistant HIV variants, many of which are broadly cross-resistant (2,3). What is worse, a large number of these genetically distinct, but biologically selected, drug-resistant HIV strains represent unique infectious entities from a therapeutic viewpoint, and are being transmitted with increasing frequency (4). Drug resistance is not restricted to protease inhibitors, but is seen with all mechanistic classes of antiretroviral agents, including the reverse transcriptase inhibitors and the new fusion inhibitors. The drug-resistance problem poses new challenges for drug design as well as for the treatment of existing infections. This introductory chapter will discuss the key biological and structural features of HIV PR that define its usefulness as an antiviral target, and that also provide a basis for understanding viral mechanisms of drug resistance to protease inhibitors.

II. HIV AND AIDS

Fewer than 20 years ago, about 1981, acquired immunodeficiency syndrome (AIDS) was first recognized as a unique clinical syndrome, manifested by opportunistic infections or malignancies associated with an underlying defect of the immune system characterized by the progressive loss of CD4 helper T cells. Nearly always fatal unless treated, the worldwide prevalence of AIDS is close to 10 million cases, based on recent estimates from the 1998 AIDS meeting held in Geneva, Switzerland. The estimated number of HIV-infected individuals is much higher at over 30 million infections. Current estimates of global disease burden are consistent with earlier, 10-year-old projections of global prevalence at between 40 and 100 million HIV-infected individuals by the year 2000. It is important to realize that the incidence of AIDS a decade ago, when HIV PR inhibitor projects were in their early stages, was a small fraction of today's figure. Thus, the epidemiological forecast was a critical element that influenced market-based decisions about committing pharmaceutical research and development (R&D) resources to the discovery of antiviral agents for a disease that was at an otherwise early stage in its global pandemic. Other critical elements were scientific advances in our understanding of the molecular and structural biology of retroviruses, a field of research that had remained largely obscure to the pharmaceutical industry. Despite a great deal of fanfare in the 1960s and 1970s, retroviruses had never really been conclusively linked to a human disease. Nonetheless, much of our current understanding of HIV infection and, in particular, the identification and characterization of targets for antiviral drug design relied heavily on basic virology lessons learned from research on animal retroviruses before the discovery of HIV (for a comprehensive review of the molecular biology of retroviruses, see Ref. 5).

The discovery of HIV-1, and its association with AIDS in 1984, led to an explosion of research on the molecular virology of this infectious agent. In parallel, an intensive, global drug discovery effort was mounted to find a cure for this fatal disease. These two activities have combined to profoundly influence current thinking and strategies for new antiviral drug discovery. Efforts to inhibit HIV continue to represent one of the most active areas of antiviral research today, and nearly every aspect of the viral life cycle has become a target for antiviral drug discovery (6,7). Because of the success obtained with HIV PR inhibitors, modern antiviral strategies are turning more and more to target-based approaches for the design of safer, more specific, and effective drugs. What were the key factors that led to these successes? For answers, it may be instructive to review briefly the scientific developments and rationale that led up to the identification of HIV PR as a target for antiviral drug discovery.

III. ROLE OF HIV PR IN THE VIRAL LIFE CYCLE

The HIV genome, similar to all other retroviral genomes, is a single-stranded, positive-sense RNA molecule that is organized into three major coding elements: *gag*, *pol*, and *env* genes. The *gag* and *pol* gene products are translated from a single unspliced polycistronic mRNA that encodes both genes (Fig. 1). A stop codon in the unspliced RNA leads to the translation of a 55-kDa Gag polyprotein,

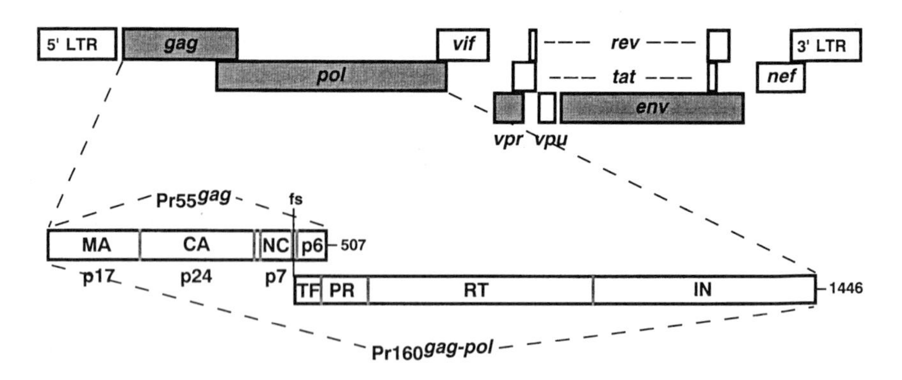

Figure 1 Genome organization and translational strategy for HIV: Structural (*gag, pol*, and *env*) genes are shaded; regulatory (*tat* and *rev*) and accessory (*vif, nef, vpr*, and *vpu*) genes are clear. Common to all retroviruses, the *gag* and *pol* gene products are translated on free ribosomes in the cytoplasm from newly synthesized unspliced viral RNA. Translation usually occurs through to a stop codon at the 3′-end of the *gag* gene, resulting in the structural polyprotein, $Pr55^{gag}$. About 5% of the time, ribosome frameshifting during translation of *gag* results in the synthesis of a Gag–Pol fusion protein, $Prl60^{gag-pol}$. The frameshift site (fs) is located upstream of the Gag p6 protein such that a transframe polypeptide, TF, is incorporated into Gag–Pol in place of p6. The functions of the p6 and TF proteins are unclear. The total number of amino acids contained by each polyprotein is indicated at the end of each molecule. See text for individual protein abbreviations. (From Ref. 8.)

Pr55gag, that contains sequences of the structural proteins of the virion—matrix (MA), capsid (CA), nucleocapsid (NC)—along with the peptides p2, p1, and p6 that are involved in the assembly and morphogenesis of mature capsids. The *pol* gene encodes the viral enzymes necessary for replication: protease (PR), reverse transcriptase (RT), and integrase (IN). These proteins are also translated as part of larger polyprotein precursor, Pr160gag–pol, which results from ribosomal frameshift and readthrough during translation of the *gag* gene. The frameshift site has been mapped to lie between the NC and p6 coding sequences. The *gag* and *gag–pol* gene products in mature virions are found in a ratio of 20:1 which represents the frequency of ribosomal frameshifting, about 5%. Thus, frameshifting is used as a regulatory mechanism to ensure that large numbers of the structural proteins of the virion are synthesized relative to the viral enzymes that are needed in only catalytic amounts. The N-termini of Pr55 and Pr160 both contain a covalently attached myristic acid moiety that is added cotranslationally and targets the polyproteins to the cellular membrane where virus assembly and budding takes place. The Pr55 precursor protein is believed to play a central role in directing virion assembly and RNA packaging, based on studies with other retroviruses that show that enveloped nucleoprotein core particles can form from Gag precursor proteins in the absence of *pol* and *env* gene products.

Proteolytic processing of Pr55gag and Pr160gag–pol during virus assembly and maturation is performed by the viral PR, which is itself encoded by the *pol* gene (8; and references therein). The *env* gene product, gp160, is processed into gp120 and gp41 by a cellular protease. The processing products of HIV PR include the *gag*-encoded structural proteins and peptides—MA, CA, NC, p1, p2, and p6—and the *pol* enzymes—RT, IN, and PR. All of these products are found in mature infectious viral particles and result from cleavages at unique amino acid sequences that span the N- and C-termini of the mature proteins (9) (Fig. 2). The sequences recognized by HIV PR are diverse, but certain general features can be recognized. Hydrophobic amino acids are preferred at the P1-P1′ residues that flank the scissile peptide bond; aliphatic and Glu/Gln residues are prevalent at P2′; aromatic residues are rarely found at P3′; and, small residues are preferred at P2. Several sequences contain an aromatic residue at P1, followed by a Pro

PROCESSING SITES FOR HIV-1 PROTEASE

Site	P4 P3 P2 P1 P1′ P2′ P3′ P4′
MA/CA	-Ser-Gln-Asn-**Tyr/Pro**-Ile-Val-Gln-
CA/p2	-Ala-Arg-Val-**Leu/Ala**-Glu-Ala-Met-
p2/NC	-Ala-Thr-Ile-**Met/Met**-Gln-Arg-Gly-
NC/p1	-Arg-Gln-Ala-**Asn/Phe**-Leu-Gly-Lys-
p1/p6	-Pro-Gly-Asn-**Phe/Leu**-Gln-Ser-Arg-
TF/PR	-Ser-Phe-Asn-**Phe/Pro**-Gln-Ile-Thr-
PR/RT	-Thr-Leu-Asn-**Phe/Pro**-Ile-Ser-Pro-
RT/IN	-Arg-Lys-Val-**Leu/Phe**-Leu-Asp-Gly-
RT (internal)	-Alu-Glu-Thr-**Phe/Tyr**-Val-Asp-Gly-

Figure 2 Cleavage site sequences in Gag and Gag–Pol polyproteins recognized by HIV PR. Cleavage occurs between residues in the P1/P1′ positions, indicated in bold and separated by a slash. RT (internal) represents a PR-mediated cleavage at the junction of the p51/RNase H domains which yields the active p66/p51 heterodimer found in isolated virus particles (64). The nomenclature of Schechter and Berger (65) is used to designate residue positions in the substrate sequence. (From Ref. 51.)

at P1′. Although all retroviral proteases appear to be structurally and functionally related, their cleavage site preferences vary widely. Efforts to predict cleavage sites for HIV PR have met with limited success (10,11) and our understanding of the basis of PR specificity is incomplete (12,13). However, identification of cleavage site sequences quickly led to the successful generation of a variety of synthetic substrates that facilitated the design of rapid and quantitative assays of PR activity (14,15).

The HIV PR has long been known to be toxic to cells, and this has prompted a search to identify cellular proteins that may be cleaved by HIV PR. Several investigators have shown that key proteins, such as NF-κB and certain cytoskeletal proteins, are cleaved in HIV infected cells (16). The possible involvement of PR in the early stages of retroviral replication was suggested initially on the basis of observations with equine infectious anemia virus (17) and later with HIV (18,19). However, recent data from several groups have demonstrated that PR inhibitors fail to block the synthesis of proviral DNA, its integration into cellular DNA, and transcription (20,21). Similar conclusions were reached using conditional lethal HIV-1 PR mutants as a probe (22). Thus, even if PR plays a role in early stages of infection, it is unlikely to be an essential one under normal circumstances.

In 1988, the late I. Segal and co-workers observed that deletion mutagenesis of the HIV PR gene resulted in the production of virus particles that had an immature morphology and were noninfectious (23). These results were confirmed by mutation of the active site aspartic acids and, subsequently, by chemical inhibition with PR inhibitors (24,25). This seminal experiment provided conclusive proof that HIV PR is essential for the life cycle of HIV and defined this enzyme as an important target for the design of specific antiviral agents for AIDS. A similar conclusion had been reached for the PR of murine leukemia virus in 1985 (26,27). However, the HIV PR studies provided the boost needed by many groups to launch drug discovery programs for this target.

IV. STRUCTURE AND MECHANISM OF HIV PR

Similar to all proteases, viral proteases can be classified according to their mechanism of action as serine, cysteine, aspartic, or metalloproteases. Identification of the mechanistic family that a viral protease belongs to is the key to predicting its structure and function, and may unlock strategies for inhibitor design that were previously developed for homologous members of the family. This concept was extremely valuable for the design of HIV PR inhibitors, which benefited from strategies that had been developed for designing potent and selective inhibitors of human renin, an aspartic protease involved in regulation of blood pressure by the renin–angiotensinogen system. Homology modeling and biochemical inhibition studies on retroviral proteases had led to the early hypothesis that these enzymes were related mechanistically to the aspartic protease family, typified by pepsin (28,29). The active site of these bilobed enzymes contains two aspartic acids, one from each lobe, that participate in the catalysis of peptide bond breakage. Crystal structures of aspartic proteases from cellular organisms revealed that the N- and C-domains associate to form an active site with approximate two-fold symmetry at the protein backbone level (30). These observations led to the suggestion that the cellular enzymes had evolved by a duplication event of a primordial aspartic protease gene (31). Because the sequence length of retroviral proteases is about one-third that of aspartic proteases, it was proposed that the former enzymes are composed of two identical subunits, each of which contributes a single aspartic acid to the active site (32). This hypothesis was verified by the crystal structure determination of Rous sarcoma virus (RSV) protease and, subsequently, of HIV PR by several laboratories (for review, see Ref. 33).

The HIV PR dimer consists of two identical, noncovalently associated subunits of 99 amino acid residues associated in a twofold (C-2) symmetric fashion (Fig. 3). The dimer is stabilized by a four-stranded antiparallel β-sheet formed by the interlocking N- and C-termini of each subunit. The active site of the enzyme is actually formed at the dimer interface and contains two conserved catalytic aspartic acid residues, one from each monomer. The substrate binding cleft is com-

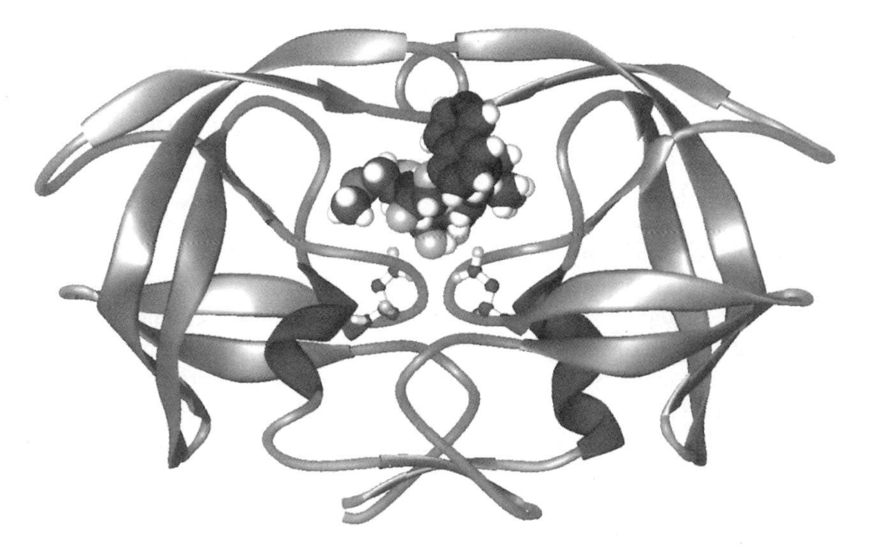

Figure 3 Ribbon diagram of HIV PR complexed with an inhibitor, KNI-272, taken from the crystal structure (66). Inhibitor is shown as a space-filling model and the two active site aspartic acids are drawn as ball-and-stick figures. (From Ref. 67.)

posed of equivalent residues from each subunit and is bound on one side by the active site aspartic acids, Asp25 and Asp125, and on the other by a pair of twofold related, antiparallel β-hairpin structures, or "flaps." The flaps make several direct interactions with the inhibitor, and probably undergo significant structural changes on binding. Molecular dynamics studies indicate that the flaps are highly flexible and must undergo large localized conformational changes during the binding and release of inhibitors and substrates (34).

To date, several hundred crystal structures have been solved for various HIV protease–inhibitor complexes—a testimony to the importance placed on structural information in the process of inhibitor design (reviewed in Refs. 33, 35–37). Structural comparison of the early peptidomimetic inhibitor complexes revealed certain common features. The enzyme contains several well-defined pockets, or subsites, in its active site region into which inhibitor side chains protrude, resulting in tight-binding interactions between enzyme and inhibitor (Fig. 4A). The inhibitor and enzyme make a pattern of complementary hydrogen bonds between their backbone atoms (see Fig. 4B). In some instances, these hydrogen bonds are mediated by bridging water molecules. A unique feature found in the structure of HIV PR–inhibitor complexes is the presence of a water molecule that forms bridging hydrogen bonds between the NH atoms of Ile50 and Ile150 in the two flaps, and the P2 and P1′ backbone carbonyl groups of the inhibitor. This water is close to the twofold axis of the enzyme and is distinct from the water molecule that has been identified in the active site of uncomplexed structures of aspartic proteinases, including HIV PR. The latter has been implicated in substrate catalysis. Because a similar pattern of hydrogen bonds are believed to be made for both substrates and peptidomimetic inhibitors, specificity was believed to reside in the pattern of largely nonpolar subsite interactions between inhibitor and enzyme side chain atoms.

V. INHIBITOR DESIGN

The approaches employed in the discovery phase of the HIV PR inhibitors can be classified on the basis of the rationale used for their design and optimization as being either substrate-based or structure-based. These distinctions are somewhat arbitrary because many inhibitors can fall into more than one category. However, with the possible exception of saquinavir, all of the HIV PR inhibitors currently in use or in the clinic have benefited from structure-based approaches in their early discovery phase.

A. Substrate-Based Approaches

The close structural and functional relations between the retroviral and cellular aspartic proteases, together with knowledge of the HIV PR cleavage site se-

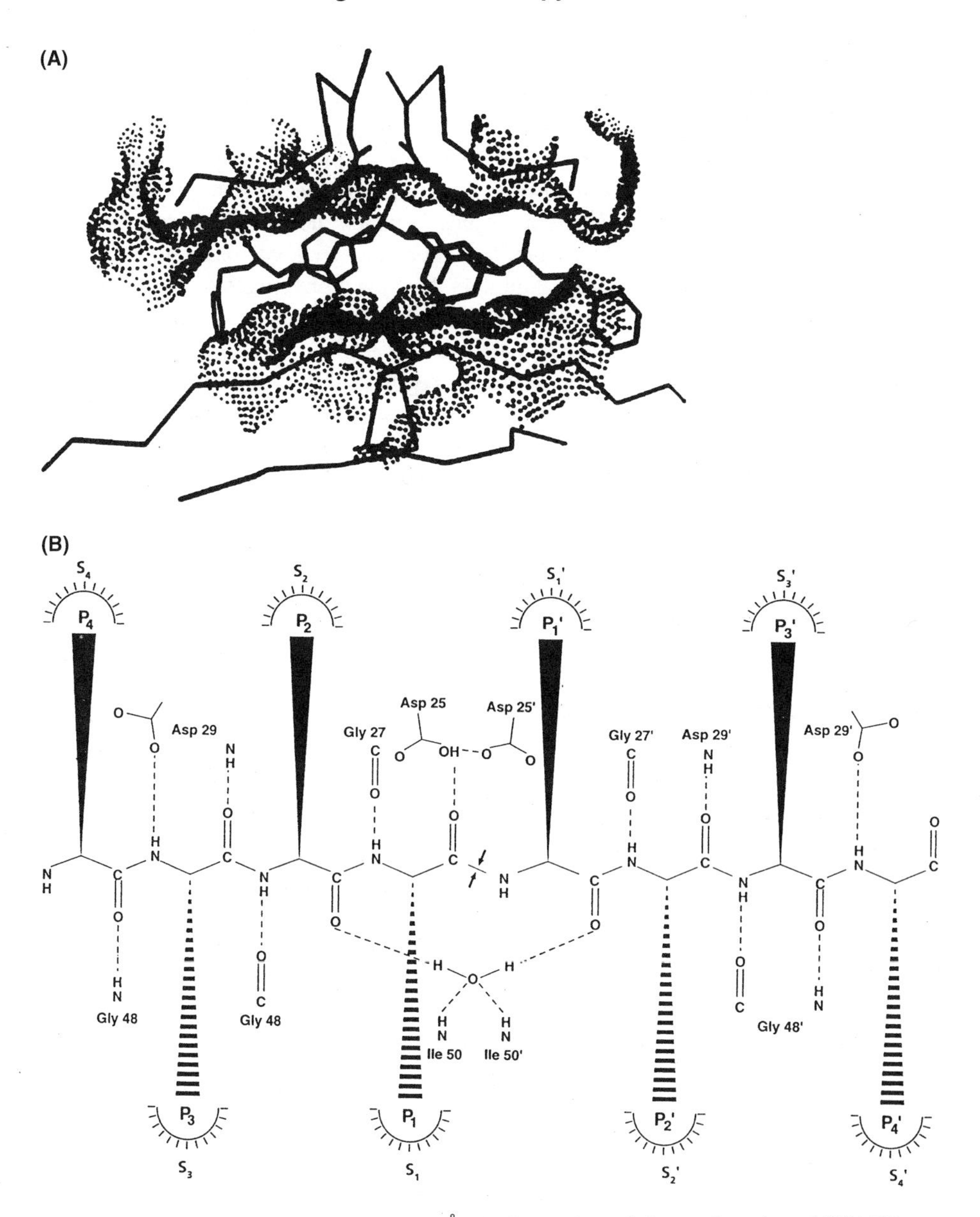

Figure 4 (A) View of a 10- to 15-Å–thick section of the active site of HIV PR complexed with a typical inhibitor. A solvent-accessible surface of the enzyme portion illustrates the high degree of structural complementarity between the inhibitor and active site region. (B) General scheme of active site interactions in HIV PR with a substrate-based inhibitor. Hydrogen bonds are drawn as dashed lines. (From: A, Ref. 68; B, Ref. 33.)

quences on the Gag and Gag–Pol polyproteins, immediately opened the avenue of substrate-based approaches that had been developed for designing peptidomimetic inhibitors of the aspartic protease, human renin. This enzyme was the focus of intensive drug discovery efforts in the 1980s aimed at the design of selective antihypertensive agents. Substrate-based peptidomimetic inhibitors for renin were essentially peptide substrate analogues in which the scissile peptide bond had been replaced by a noncleavable, transition-state analogue or isostere (Fig. 5). Many transition–state-based inhibitors had been investigated that would mimic the hydrated amide presumed to represent the intermediate in amide hydrolysis catalyzed by the two active site aspartic acids (38). A hydroxy group contained in nearly all HIV protease inhibitors is believed to mimic the transition-state diol structure and interact with the catalytic aspartates, residues 25 and 125. While much of the early investigation involved identifying suitable cores, the subsequent refinement of these structures to reduce their molecular weight and peptidic character, while still maintaining potency became the focus of intensive efforts by many research groups. An early success story from these efforts was that of saquinavir (Fig. 6) which was reported by the Hoffmann–La Roche group in 1990 (39) and was the first PR inhibitor to receive FDA approval for clinical treatment of AIDS in 1996.

B. Structure-Based Approaches

The mid-1980s witnessed the widespread introduction of protein crystallography and computational chemistry into mainstream pharmaceutical discovery research laboratories with the promise that ''rational drug design'' would revolutionize drug discovery. Although these disciplines often experienced initial difficulty in becoming effectively integrated with medicinal chemistry on the one hand, and molecular biology and biochemistry on the other, perhaps no other drug design target has more widely validated the investment in structure-based design than HIV PR. The crystal structure of HIV PR immediately provided a structural basis for the development of a new generation of inhibitors that did not need to rely strictly on substrate or peptide mimicry. Crystal structures were used to identify novel chemical templates as well as to optimize the interactions found in early leads from target-based screening of peptide or nonpeptide libraries. When the original lead was peptide-based, crystal structures were used to design nonpeptide groups to replace peptide moieties wherever possible. Hundreds of crystal structures of key HIV PR–inhibitor complexes have been determined during the past decade and used to rationalize structure–activity relations (SARs) at an atomic level. Virtually all of the groups working on HIV PR inhibitor design have been guided, to some extent, by structural data, although the exact manner in which medicinal chemistry and structural insights were blended together varies from example to example. These efforts have led to a structurally diverse compendium

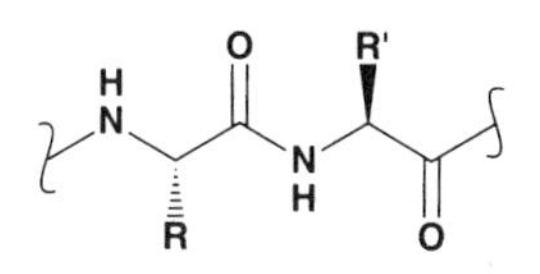

Peptide Bond

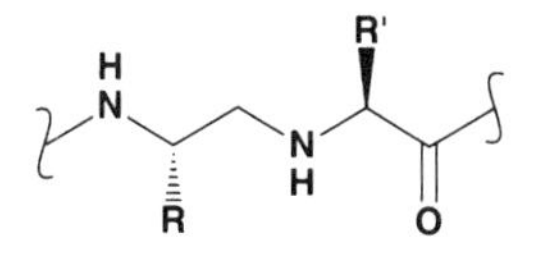

Reduced Amide

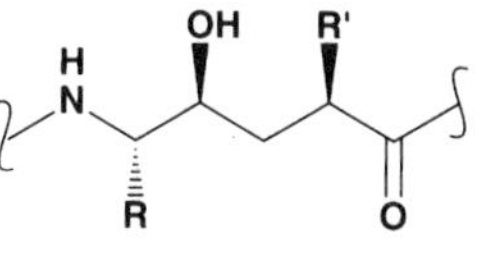

Hydroxyethylene

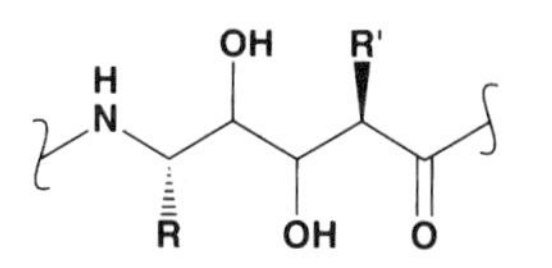

Dihydroxyethylene

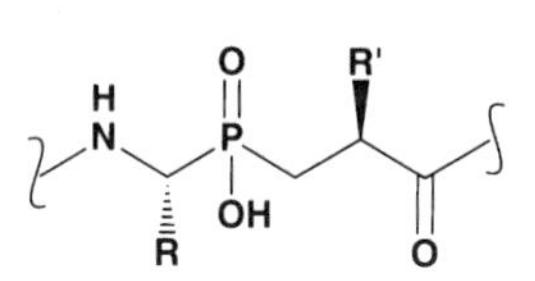

Phosphinate

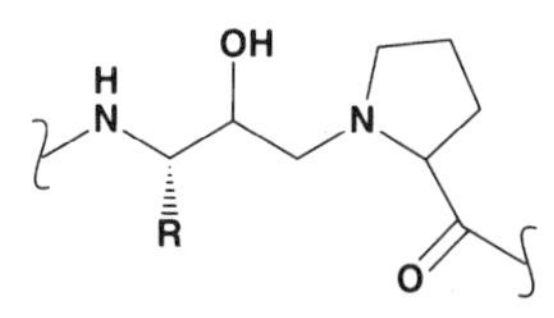

Hydroxyethylamine (R'-cyclic)

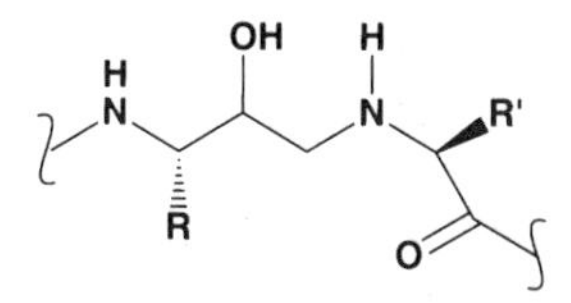

Hydroxyethylamine (R'-acyclic)

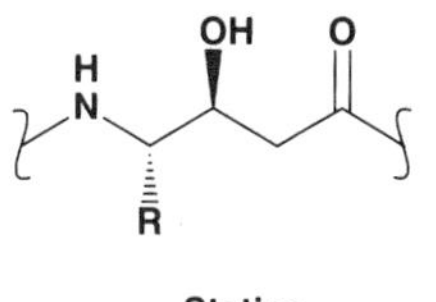

Statine

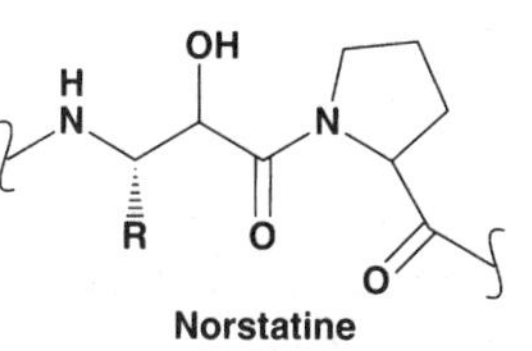

Norstatine

Figure 5 Chemical structures of transition-state isosteres, often referred to as cores or templates, used in the design of HIV PR inhibitors. A hydroxyethylamine (R′-cyclic) core was used in the design of saquinavir.

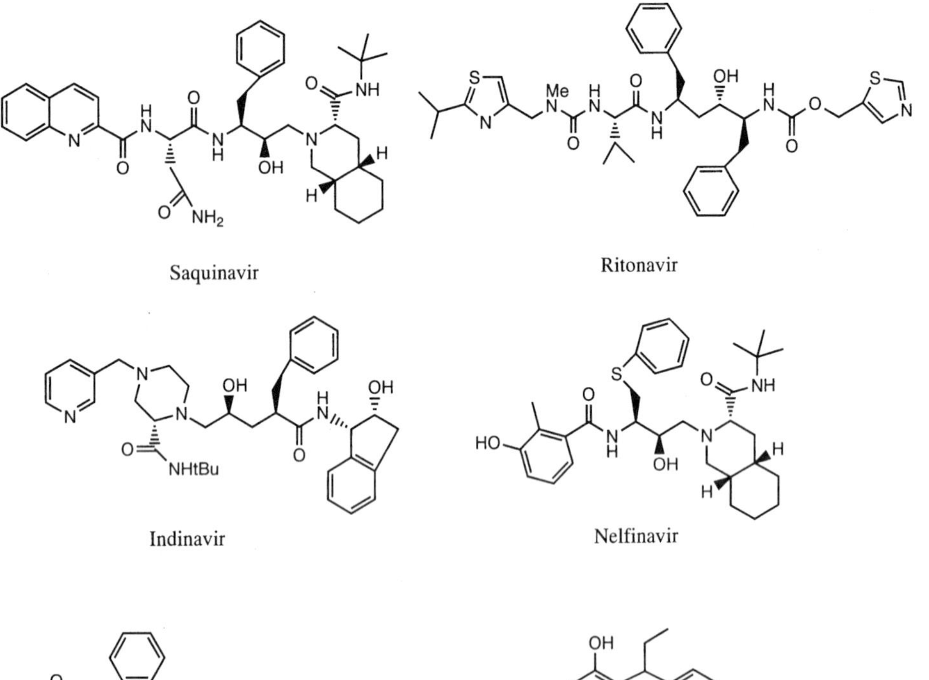

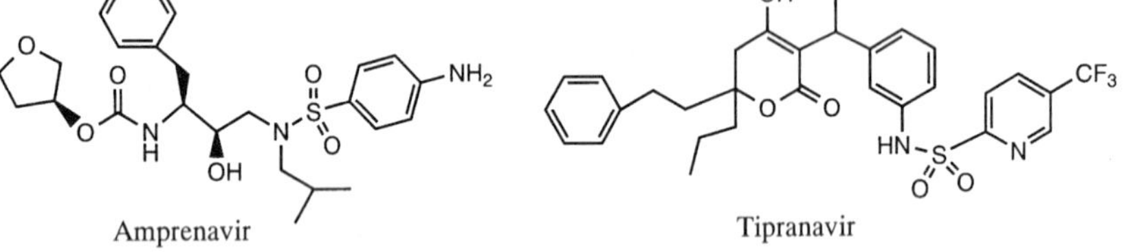

Figure 6 Chemical structures of selected HIV PR inhibitors.

of inhibitors that include nelfinavir (AG-1343), indinavir (MK-639), and amprenavir (VX-478) (see Fig. 6).

A powerful application of crystallography and modeling to the design of HIV PR inhibitors is exemplified by the work of the Parke–Davis and Upjohn groups, who independently identified 4-hydroxypyrone-containing compounds as novel, nonpeptidic leads during target-based screening of their chemical inventories (40,41). The initial leads were actually analogues of coumarin or warfarin that had marginal potency in the enzyme assay, with 50% inhibiting concentration (IC_{50}) values in the micromolar range, and had no detectable antiviral activity. Both groups used structure-based design methods to modify the cyclic templates of the initial leads and to add complimentary ‘‘side chains’’ that resulted in compounds with potent antiviral activity. The pyrones represent a novel template that is easily synthesized and readily optimizable for oral bioavailability. The dihydropyrone analogue, PNU-140690 (tipranavir; see Fig. 6), is currently undergoing clinical trials.

C. Symmetry-Based Approaches

One of the first examples of applying pure structural principles to the design of HIV PR inhibitors was that of the symmetry-based inhibitors originally described by the Abbott group. They postulated that HIV PR, being a symmetric dimer, might be inhibited by symmetric inhibitors that were complementary in a three-dimensional sense to the twofold symmetry of the enzyme (42,43). This approach led to the potent diamino diol and deshydroxy diol inhibitors, typified by the first clinical candidate of this series, A-77003. Further optimization of oral bioavailability, solubility, and plasma half-life led to the successful clinical development of ritonavir (ABT-538; see Fig. 6).

D. Selectivity

The retroviral proteases display considerable structural homology with the cellular aspartic proteases, despite the apparent lack of sequence homology. The structural similarities are most apparent at the backbone level, with the highest degree of structural correspondence in the active site region (44). The ease in obtaining high selectivity of inhibitors for HIV PR is due to the numerous differences in the subsites, and also to the fact that HIV PR has a symmetric active site region (42). The latter property causes HIV PR to stand apart from its cellular, single-chain counterparts. Indeed, it is rare to find tight-binding inhibitors of HIV PR that exhibit cross-specificity with human aspartic proteases.

E. From Inhibitors to Drugs

Knowledge of the structure and function of HIV PR and its relation to other aspartic proteases has led to the successful development of a wide variety of potent and chemically diverse inhibitors. Although the initial lead compounds were generated in various ways, the availability of crystal structures of these leads or of analogous compounds bound to the active site of HIV PR facilitated structure-based approaches to the optimization of interactions to increase potency. Once a certain level of potency against the enzyme was achieved, considerations such as improving antiviral potency, improving bioavailability (45), and reducing cost could be addressed within the known structural limitations provided by these crystal structures. The latter activity often dwarfed the former in terms of human and material resources required to achieve a suitable clinical, or even preclinical development candidate, as will be documented in subsequent chapters in this book. An indication of the magnitude of discovery activity in this field can be seen in the large number of cores or templates that have been used for elaboration of HIV protease inhibitors (for a recent review, see Ref. 4). It is difficult to overemphasize the enormous amount of medicinal chemistry and crystallography that has been performed over the last decade in support of the discovery and development of the PR inhibitors. These efforts also underscore the difficulty of transforming a potent PR inhibitor into an effective antiviral drug suitable for human use. To wit, the number of novel structural and chemical templates that are represented in the five FDA-approved drugs is a small subset of what has actually been attempted. Thus, it is perhaps more accurate to describe the early phase of drug discovery as inhibitor (or ligand) design as opposed to the commonly used phrase ''drug design.''

VI. PHARMACOKINETICS OF HIV PR INHIBITORS

Unlike the majority of RT inhibitors, most HIV PR inhibitors have serious pharmacokinetic limitations (46,47). Poor oral absorption, serum protein binding, liver enzyme metabolism, and other factors can all but eliminate the antiviral benefits of many potent protease inhibitors. Currently approved protease inhibitors need to be taken often and in large quantities to maintain effective antiviral concentrations in the blood. Of the currently available antiviral drugs for AIDS, protease inhibitors are among the most effective, but are costly and require difficult treatment regimens. As a result, the failure or drop-out rate of patients receiving protease inhibitor therapy tends to be relatively high, the compliance of treatment with protease inhibitors is likely to be poor, and the development of resistance to these highly effective drugs is a growing problem (48). The limited bioavailability and plasma half-life of many PR inhibitors are partly due to metabolism by cytochrome P450 enzymes. In contrast, ritonavir was actually a potent inhibitor of these enzymes. While the inhibition of metabolic liver enzymes is generally to be avoided in drug candidates, for ritonavir it allowed the compound to achieve sufficient blood levels to be effective clinically, and it received FDA approval.

A particularly vexing issue for the clinical development of HIV PR inhibitors has been serum protein binding. The clinical failures of two HIV PR inhibitors, SC-52151 and KNI-272, have been attributed largely to plasma protein binding, in particular to the α-acid glycoproteins (AAG) found in humans (49,50). The effect of protein binding on antiviral efficacy is now routinely studied and is commonly expressed as the fold increase in antiviral IC_{50} value in the presence of a given concentration of AAG. Thus, AAG binding can be considered as a kind of pharmacological resistance mechanism (51). Many groups actually use

AAG binding as a key selection criterion in the identification of new HIV PR drug candidates. Because all HIV PR inhibitors bind to AAG to a high degree (52), and AAG levels are known to vary on an individual patient basis, it is difficult to predict drug failure (let alone success) on the basis of in vitro protein binding assays per se. Quantitative therapeutic drug monitoring and physical chemical studies of protein binding under physiological conditions are needed if we are to accurately predict the influence of protein binding on clinical efficacy.

VII. RESISTANCE TO HIV PR INHIBITORS

Although many promising new anti-HIV drugs have been developed, their effectiveness has been hampered by the emergence of drug-resistant variants. Mutant viruses will emerge in the presence of antiviral agents whenever the balance of mutant virus replication is favorable (i.e., the mutant provides a selective advantage to the virus in the presence of drug). Biological selection of HIV mutants is particularly favorable owing to the nature of HIV infection, which is chronic, persistent, and characterized by a high steady-state rate of replication. Pharmacodynamic studies using potent HIV PR inhibitors have revealed that the combined half-life of plasma virus and virus-producing cells in the body is on the order of 2 days or less, with new virus being produced at a rate of 10^8–10^9 virions per day (53,54). These conditions, coupled with the error rate of HIV-1 reverse transcriptase, fewer than 1:10,000 bases (or about 1 per viral genome for every cycle of replication), favor rapid mutation and selection of drug-resistant virus (55). Mutant viruses can also be generated through errors in viral RNA synthesis, and recombination events. The rate at which a resistance virus will emerge is inversely related to the number of mutations (N) required to give the virus a selective growth advantage in the presence of the drug. Millions of copies of all possible single-site mutants will be generated in every new round of virus replication. For these reasons, monotherapy with antiretroviral drugs is now used only in very short efficacy trials of new drugs, for the likelihood of selecting resistant viruses is so high. In practice, drugs such as nevirapine, for which high-level resistance can be achieved with a single mutation relative to wild-type, should never be given to HIV-infected individuals at subtherapeutic doses. Clinical resistance to newly introduced anti-HIV agents has become the rule, and resistance to PR inhibitors is no exception (56,57). The widespread use of protease inhibitors has led to the emergence of drug-resistant HIV variants, many of which are widely cross-resistant (2,3,57). Recent reports of the selection and transmission of multidrug-resistant HIV strains that contain multiple PR and RT mutations in the *pol* gene and also in cleavage site sequences in the *gag* gene have raised new alarms in the AIDS community (2,58). An understanding of the biological and structural mechanisms of resistance to HIV PR inhibitors is necessary to predict optimal salvage therapies; indeed, the full implications of drug resistance for disease outcome have not yet been realized. A great deal of progress has been made at unraveling the structural, biochemical, and virological mechanisms of resistance to PR and RT inhibitors in light of the three-dimensional atomic structures of these proteins (56) (Fig. 7). An important goal of these studies is to gain insights that may lead to new strategies to combat resistance, and to eventually develop paradigms for structure-based strategies to combat resistance to anti-infectious disease agents (51).

Drug-resistant HIV strains can be considered to represent distinct infectious entities and pose new challenges for drug design as well as drug treatment of existing infections. Polymorphisms have been observed in 49 (49.5%) of the 99 amino acids of the HIV protease monomer (3,59). Of these, substitutions at 18 or more positions have been correlated with loss of responsiveness to protease inhibitor treatment. A recent statistical comparison of sequences from untreated versus treated patients indicates the actual number of resistance-associated mutational sites may be much higher (3). The particular sequence and pattern of mutations selected by protease inhibitors is believed to be somewhat drug-specific

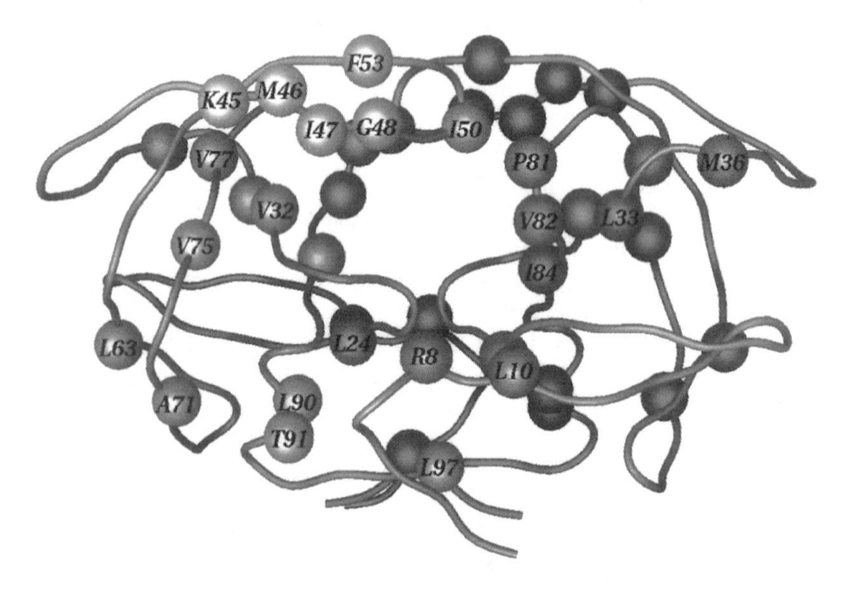

Figure 7 Mutations commonly associated with viral resistance to HIV PR inhibitors are shown as spheres located on a backbone representation of the HIV PR structure. Only one of each pair of symmetry-related amino acid positions is labeled for clarity. The twofold rotation axis is vertical.

and often patient-specific, but high-level resistance is typified by multiple mutations (often five or more) in the protease gene, which give rise to cross-resistance to all of the protease inhibitors (Fig. 8). As many as 19 different mutations in the protease gene have been observed in individual viral genomes from clinical isolates derived from patients who have developed high level cross resistance to protease inhibitors (M. P. de Bethune, personal communication). The apparently large number of mutational pathways that can lead to cross-resistant mutant viruses can make the choice of a protease inhibitor for salvage therapy seem hopelessly complex. Even the choice of protease inhibitor with which to initiate therapy can be a risky enterprise that may inadvertently select for an undesired resistance pathway. However, there is some light in this tunnel. Recent studies

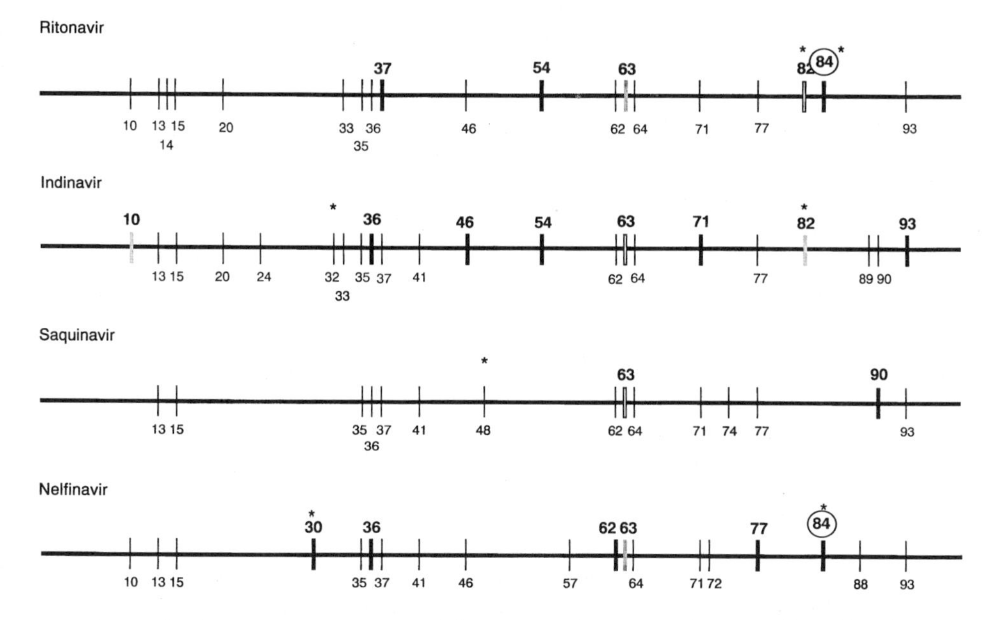

Figure 8 Resistance patterns for HIV protease inhibitors: Amino acid loci for HIV protease are designated by residue number. Loci where substitutions have been found in at least 10, 30, 50, and 75% of HIV-1-infected patients treated with protease inhibitors are designated by thin, thick, shaded, and open bars, respectively. Active site residues are denoted by asterisks. Residue 84 is circled to indicate that substitution frequencies were taken from in vitro-selected mutants; mutations at 84 are seen in fewer than 10% of clinical samples. (Clinical data courtesy of the Stanford HIV RT and Protease Sequence Database; R Shafer, D Jung, D Stevenson) (url: hivdb.stanford.edu/hiv/.)

suggest that prior phenotypic resistance profiling of a patient's HIV RT and PR genotype to the panel of current drugs improves treatment outcome.

VIII. FUTURE CHALLENGES FOR DRUG DESIGN AND TREATMENT

There are two major challenges that lie ahead for HIV PR inhibitor design. One concerns the issue of pharmacokinetics. As potent and bioavailable as these inhibitors are, they need to be even more potent and more bioavailable to maximize their effectiveness and usefulness. The daily requirement for PR inhibitors is rather massive, as it is in the range of 1 g or more of drug that must be taken two to three times daily. Side effects can be troublesome, if not debilitating, for most of these compounds. Indinavir therapy leads to painful recurrence of kidney stones in about 20% of individuals. Ritonavir has numerous drug interactions, as it is a highly potent inhibitor of a major liver enzyme, cytochrome P450 3A4, involved in drug metabolism. A syndrome of peripheral lipodystrophy (fat wasting), hyperlipidemia, and insulin resistance is emerging as a frequent side effect of long-term PR inhibitor therapy (60). Finally, these drugs tend to be costly to make. Coupled with the massive dosages required for maintenance of round-the-clock maximal viral suppression, the economics of PR inhibitor therapy is not a trivial consideration. Thus, a major challenge is the design of more potent, less costly, and more bioavailable inhibitors. The potency issue to some extent hinges on the theoretical limitation of potency that can be achieved for HIV PR inhibitors and how close are we to this limit? A related question is can we design inhibitors with a better plasma half-life and MW profile and thereby achieve significant improvements in bioavailability and ease of synthesis?

The second major challenge confronting the use of HIV PR inhibitors is the problem of how to effectively combat drug resistance. Given the unusual constraints on HIV PR, and our understanding of inhibitor binding and resistance at an atomic level of detail, it should, at a minimum, be possible to design inhibitors that target different mutants. These inhibitors could then be used together in multidrug therapy approaches. It may even be possible to design inhibitors that target multiple mutants simultaneously. The potency of saquinavir against multiple active site mutants suggests that this may be feasible. In certain instances, these second-line drugs might prompt the selection of wild-type revertants, that can then be treated with the original drug. Classic sequential therapy approaches have been unsuccessful in controlling infectious diseases and have given way to multidrug treatments. However, one can even envision a therapeutic strategy for HIV PR inhibitors that could apply specific, structure-guided selection pressure to influence clinical outcome (51,56). Such strategies would represent a fundamental departure from currently held concepts of infectious disease treatment, and will require a great deal of experimental virology, assuming that the appropriate inhibitors can be designed in the first place.

An interesting approach to solving resistance and pharmacokinetics simultaneously has been taken by the Abbott group. In appropriate combination dosages, ritonavir has been demonstrated to dramatically improve the bioavailability and plasma levels of saquinavir, the bioavailability of which is particularly limited by first-pass P450 metabolism (61). This strategy is currently being explored in clinical experiments designed to boost plasma levels of other antiretroviral drugs. This type of therapy poses risks from drug interactions and other side effects of using P450 inhibitors. However, a major motivating factor in these studies is the goal to maintain steady-state drug concentrations that are well above the antiviral IC_{90} levels 24 h/day, 7 days a week. The idea here is to prevent low-level HIV replication that can occur immediately on the fall of drug levels below some maximal inhibitory threshold. Such transient replication, even for less than 1 day, in the presence of sufficient drug selection pressure gives rise to the emergence of drug-resistant strains of virus.

IX. VIRAL PROTEASES AS TARGETS FOR ANTIVIRAL DISCOVERY: LESSONS LEARNED

Do HIV-1 protease inhibitors really represent a successful new paradigm that will be generally applicable to antiviral drug discovery? A common strategy in the life cycle of positive-strand RNA viruses is the utilization of polycistronic mRNAs that can be translated into precursor polyproteins that subsequently are processed enzymatically into mature, functional proteins during virus assembly (62,63). Processing enzymes may be either cellular or viral-encoded, and offer novel targets for intervention. Virus-encoded proteases afford particularly attractive therapeutic targets for the design of antiviral agents that are highly specific and nontoxic to their host cells. However, before the discovery of HIV protease inhibitors, viral proteases were largely ignored as potential therapeutic targets. The success achieved with HIV PR changed all this, and many drug discovery teams set out to discover and develop protease inhibitors for herpesviruses, adenovirus, plant viruses, picornaviruses, and hepatitis C virus (for review, see Ref. 4). However, unlike the case with HIV PR, the design of potent and selective inhibitors for some of these other viral proteases has proved to be unexpectedly difficult. The ability to successfully design inhibitors for HIV PR was, to a certain extent, foreshadowed from the literature on aspartic proteases and the track record at designing potent and selective inhibitors for renin. Unfortunately, retroviruses are the only family of human disease-causing viruses the proteases of which have adopted an aspartic proteinase mechanism and fold. All the other viral proteases are either serine or cysteine proteases, often with more shallow and less extensive substrate-binding pockets than HIV PR. The lesson here is that all targets are not created equal, and that judicious evaluation of all aspects of a target should be exercised before setting off on a major discovery expedition.

X. SUMMARY AND CONCLUSIONS

The introduction of PR inhibitors to the armamentarium of AIDS drugs has had a profoundly successful influence on the AIDS community. These drugs are prolonging life and providing new hope for long-term health and well-being until such time as a more effective cure or vaccine becomes available. HIV PR has also become a valuable instructional target for the design of mechanism- and structure-based drugs, and for providing novel strategies for designing antiviral therapeutics. However, the success of PR inhibitor therapy is a qualified one. A growing number of individuals are developing multidrug-resistant strains of HIV and there is emerging evidence that these strains can be readily transmitted. At present, there are no effective treatment options available for many of these patients. This brings us to the brink of a new forefront in antiviral research that requires novel approaches to deal with the problem of drug resistance. This problem poses different challenges from those we faced in the design of the first-line drugs, as it forces us to think about selection pressure mechanisms in addition to the usual issues of potency, pharmacology, safety, and mechanism of drug action. New drugs and new treatment strategies need to be found that will suppress drug-resistant virus, and that will prevent the evolution of drug-resistant mutants. These challenges will push us to the limits of our knowledge in structure-based design, in pharmacology and drug metabolism, and in the underlying bases of pharmacokinetic phenomena. With the growing number of resistance mutations, improved approaches to structure-based inhibitor design and the development of quantitative methods of modeling and binding affinity prediction take on added significance. Meanwhile, it is important to continue to search for combinations of drugs with complementary resistance profiles and to explore new treatment modalities that can be used to exert specific selection pressure on HIV in the hopes that at least a stalemate, if not an end game, strategy can be found to control the progression of AIDS. Lessons learned with HIV PR may provide valuable paradigms that can be used to improve the probability of success for the design of effective drugs and treatment strategies for other diseases.

XI. PROSPECTS FOR GLOBAL CONTROL OF AIDS

That there are today more than 15 drugs that can be used to treat HIV infection is an overwhelming testimonial to progress in the field of antiviral drug discovery, especially considering that there was only one approved antiviral AIDS drug less than a decade ago. In contrast, that less than 10% of all HIV-infected individuals have ready access to these drugs is a tragic testimonial to the lack of global economic parity. The death rate of AIDS has recently overtaken that of malaria on the African continent, where the human population is actually decreasing. Where drugs become available, resistant strains emerge. The global pandemic is becoming worse, despite our sense that AIDS is under control at home. Indeed, faced with headlines that AIDS deaths are decreasing locally how can we be too worried? However, the growing incidence of drug-naïve HIV infections that are resistant to most current antiretroviral agents portends a new epidemic of drug-resistance that may be more difficult to control than the initial epidemic. Control is relative to where you stand: In most Third World countries, the epidemic is already out of control, and has been for many years. In these high incidence regions, HIV discriminates neither sex nor age.

How can we begin to deal with AIDS at a global level? Obviously, economic aid is required, as is access to effective therapy. But is it reasonable to suggest that with sufficient economic aid and drug supplies, the problem can be effectively addressed? With the current drug treatment strategies, multiple pills must be taken several times daily to have any effect at all. The smallest problem with compliance can result in the drugs becoming ineffective, and worse, can lead to the generation of resistant virus, which can then be transmitted. There are obvious problems with the distribution of drugs to the general population, not the least of which is the vanishingly small number of primary care physicians relative to the infected population in endemic areas. Clearly, what is not needed are more drugs that are inaccessible to the Third World. What is needed is a strategy to deal with the difficult problems of compliance and costs—not just financial costs, but costs in terms of human and material resources required to manufacture and maintain sufficient quantities of drug. Depot formulations come to mind. Specifically, a formulation that will deliver a constant inhibitory amount of a potent antiretroviral agent over a period of weeks to months would be highly desirable. The central problem with depot formulations for HIV drugs is that these formulations simply do not exist. Why? Because depot formulations have not been a major R&D focus of drug developers in the First World. Are depot formulations easy to develop? No. Are they impossible to develop? No. Would we use a depot at home? Yes. Could a depot prevent resistance? Quite likely, if used wisely.

So, let us get started.

ACKNOWLEDGMENTS

I wish to thank Dr. Jack Collins for assistance with preparation of the graphics figures and Ms. Christine Ray for preparation of the manuscript.

REFERENCES

1. Hammer SM, Yeni P. Antiretroviral therapy: where are we? AIDS 1998; 12:S181–S188.
2. Boden D, Markowitz M. Resistance to HIV-1 protease inhibitors. Antimicrob Agents Chemother 1998; 42:2775–2783.
3. Shafer R, Winters M, Palmer S, Merigan T. Multiple concurrent reverse transcriptase and protease mutations and multidrug resistance of HIV-1 isolates from heavily treated patients. Ann Intern Med 1998; 128:906–911.
4. Erickson JW, Eissenstat ME. HIV protease as a target for the design of antiviral agents for AIDS. In: Dunn B, ed. Proteases of Infectious Agents. San Diego: Academic Press, 1999:1–60.
5. Coffin J, Hughes S, Varmus H. Retroviruses. New York: Cold Spring Harbor Laboratory Press, 1997.
6. Mitsuya H, Broder S. Strategies for antiviral therapy in AIDS. Nature 1987; 325: 773–778.
7. De Clercq E. Toward improved anti-HIV chemotherapy: therapeutic strategies for intervention with HIV infections. J Med Chem 1995; 38:2491–2517.
8. Swanstrom R, Wills JW. Synthesis, assembly, and processing of viral proteins. In: Coffin JM, Hughes SH, Varmus HE, eds. Retroviruses. New York: Cold Spring Harbor Laboratory Press, 1997:263–334.
9. Debouck C. The HIV-1 protease as a therapeutic target for AIDS. AIDS Res Hum Retroviruses 1992; 8:153–164.
10. Poorman RA, Tomasselli AG, Heinrikson RL, Kézdy FJ. A cumulative specificity model for proteases from human immunodeficiency virus types 1 and 2, inferred from statistical analysis of an extended substrate data base. J Biol Chem 1991; 266: 14554–14561.
11. Chou K–C, Zhang C–T. Studies on the specificity of HIV protease: an application of Markov chain theory. J Protein Chem 1993; 12:709–724.
12. Dunn BM, Gustchina A, Wlodawer A, Kay J. Subsite preferences of retroviral proteinases. In: Kuro LC, Shafer JA, eds. Methods Enzymology. San Diego: Academic Press, 1994:254–278.
13. Katz RA, Skalka AM. The retroviral enzymes. Annu Rev Biochem 1994; 63:133–173.
14. Hellen CUT. Assay methods for retroviral proteases. In: Kuo LC, Shafer JA, eds. Methods in Enzymology. San Diego: Academic Press, 1994:46–58.
15. Krafft GA, Wang GT. Synthetic approaches to continuous assays of retroviral proteases. In: Kuo LC, Shafer JA, eds. Methods Enzymology. San Diego: Academic Press, 1994:70–86.
16. Shoeman RL, Sachse B, Honer E, Mothes E, Kaufmann M, Traub P. Cleavage of human and mouse cytoskeletal and sarcomeric proteins by human immunodeficiency virus type 1 protease: actin, desmin, myosin, and tropomyosin. Am J Pathol 1993; 142:221–230.
17. Roberts M, Oroszlan S. The preparation and biochemical characterization of intact capsids of equine infectious anemia virus. Biochem Biophys Res Commun 1989; 160:486–494.

18. Baboonian C, Dalgleish A, Bountiff L, et al. HIV-1 proteinase is required for synthesis of pro-viral DNA. Biochim Biophys Acta 1991; 179:17–24.
19. Nagy K, Young M, Baboonian C, Merson J, Whittle P, Oroszlan S. Antiviral activity of human immunodeficiency virus type 1 protease inhibitors in a single cycle of infection: evidence for a role of protease in the early phase. J Virol 1994; 68:757–765.
20. Jacobsen H, Ahlborn L, Gugel R, Mous J. Progression of early steps of human immunodeficiency virus type 1 replication in the presence of an inhibitor of viral protease. J Virol 1992; 66:5087–5091.
21. Uchida H, Maeda Y, Mitsuya H. HIV-1 protease does not play a critical role in the early stages of HIV-1 infection. Antiviral Res 1997; 36:107–113.
22. Kaplan AH, Manchester M, Smith T, Yang YL, Swanstrom R. Conditional human immunodeficiency virus type 1 protease mutants show no role for the viral protease early in virus replication. J Virol 1996; 70:5840–5844.
23. Kohl NE, Emini EA, Schleif WA, et al. Active human immunodeficiency virus protease is required for viral infectivity. Proc Natl Acad Sci USA 1988; 85:4686–4690.
24. Seelmeier S, Schmidt H, Turk V, von der Helm K. Human immunodeficiency virus has an aspartic-type protease that can be inhibited by pepstatin A. Proc Natl Acad Sci USA 1988; 85:6612–6616.
25. McQuade TJ, Tomasselli AG, Liu L, et al. A synthetic HIV-1 protease inhibitor with antiviral activity arrests HIV-like particle maturation. Science 1990; 247:454–456.
26. Crawford S, Goff SP. A deletion mutation in the 5′ part of the *pol* gene of Moloney murine leukemia virus blocks proteolytic processing of the *gag* and *pol* polyproteins. J Virol 1985; 53:899–907.
27. Katoh I, Yoshinaka Y, Rein A, Shibuya M, Odaka T, Oroszlan S. Murine leukemia virus maturation: protease region required for conversion from ''immature'' to ''mature'' core form and for virus infectivity. Virology 1985; 145:280–292.
28. Toh H, Ono M, Saigo K, Miyata T. Retroviral protease-like sequence in the yeast transposon *Ty1*. Nature 1985; 315:691–692.
29. Katoh I, Yasunaga T, Ikawa Y, Yoshinaka Y. Inhibition of retroviral protease activity by an aspartyl proteinase inhibitor. Nature 1987; 329:654–656.
30. Davies DR. The structure and function of the aspartic proteinases. Annu Rev Biophys Bioeng 1990; 19:189–215.
31. Tang J, James MNG, Hsu IN, Jenkins JA, Blundell TL. Structural evidence for gene duplication in the evolution of the acid proteases. Nature 1978; 271:618–621.
32. Pearl LH, Taylor WR. A structural model for the retroviral proteases. Nature 1987; 329:351–354.
33. Wlodawer A, Erickson J. Structure-based inhibitors of HIV-1 protease. Annu Rev Biochem 1993; 62:543–585.
34. Collins JR, Burt SK, Erickson JW. Flap opening in HIV-1 protease simulated by ''activated'' molecular dynamics. Nat Struct Biol 1995; 2:334–338.
35. Appelt K. Crystal structuers of HIV-1 protease-inhibitor complexes. Perspect Drug Dispos Design 1993; 1:23–48.
36. Erickson JW. Design and structure of symmetry-based inhibitors of HIV-1 protease. Perspect Drug Dispos Design 1993; 1:109–128.

37. Wlodawer A, Vondrasek J. Inhibitors of HIV-1 protease: a major success of structure-assisted drug design. Annu Rev Biophys Biomol Struct 1998; 27:249–284.
38. Szelke M. Chemistry of renin inhibitors. In: Kostka V, ed. Aspartic Proteinases and Their Inhibitors. New York: de Gruyter, 1985; 421–441.
39. Roberts NA, Martin JA, Kinchington D, et al. Rational design of peptide-based HIV proteinase inhibitors. Science 1990; 248:358–361.
40. Prasad JV, Boyer FE, Domagala JM, et al. Nonpeptidic HIV protease inhibitors possessing excellent antiviral activities and therapeutic indices. PD 178390: a lead HIV protease inhibitor. Bioorg Med Chem 1999; 7:2775–2800.
41. Thaisrivongs S, Romero DL, Tommasi RA, et al. Structure-based design of HIV protease inhibitors: 5,6-dihydro-4-hydroxy-2-pyrones as effective, nonpeptidic inhibitors. J Med Chem 1996; 39:4630–4642.
42. Erickson J, Neidhart DJ, VanDrie J, et al. Design, activity,and 2.8 Å crystal structure of a C_2 symmetric inhibitor complexed to HIV-1 protease. Science 1990; 249:527–533.
43. Kempf DJ, Codacovi L, Wang XC, et al. Structure-based, C2 symmetric inhibitors of HIV protease. J Med Chem 1990; 33:2687–2689.
44. Rao JKM, Erickson JW, Wlodawer A. Structural and evolutionary relationships between retroviral and eucaryotic aspartic proteinases. Biochemistry 1991; 30:4663–4671.
45. Kempf DJ. Progress in the discovery of orally bioavailable inhibitors of HIV protease. Perspect Drug Dispos Design 1994; 2:427–436.
46. Sommadossi JP. Pharmacological considerations in antiretroviral therapy. Antiviral Ther 1998; 3:9–12.
47. Hoetelmans RM, Meenhorst PL, Mulder JW, Burger DM, Koks CH, Beijnen JH. Clinical pharmacology of HIV protease inhibitors: focus on saquinavir, indinavir, and ritonavir. Pharm World Sci 1997; 19:159–175.
48. Deeks SG, Smith M, Holodniy M, Kahn JO. HIV-1 protease inhibitors. A review for clinicians. JAMA 1997; 277:145–153.
49. Fischl MA, Richman DD, Flexner C, et al. Phase I/II study of the toxicity, pharmacokinetics, and activity of the HIV protease inhibitor SC-52151. J Acquir Immune Defic Syndr Hum Retrovirol 1997; 15:28–34.
50. Humphrey RW, Wyvill KM, Nguyen BY, et al. A phase I trial of the pharmacokinetics, toxicity, and activity of KNI-272, an inhibitor of HIV-1 protease, in patients with AIDS or symptomatic HIV infection. Antiviral Res 1999; 41:21–33.
51. Erickson J, Gulnik S, Markowitz M. Protease inhibitors: resistance, cross-resistance, fitness and the choice of initial and salvage therapies. AIDS 1999; 13:S189–S204.
52. Zhang XQ, Schooley RT, Gerber JG. The effect of increasing $alpha_1$-acid glycoprotein concentration on the antiviral efficacy of human immunodeficiency virus protease inhibitors. J Infect Dis 1999; 180:1833–1837.
53. Ho DD, Neumann AU, Perelson AS, Chen W, Leonard JM, Markowitz M. Rapid turnover of plasma virions and CD4 lymphocytes in HIV-1 infection. Nature 1995; 373:123–126.
54. Wei X, Ghosh SK, Taylor ME, et al. Viral dynamics in human immunodeficiency virus type 1 infection. Nature 1995; 373:117–122.

55. Coffin JM. HIV population dynamics in vivo: implications for genetic variation, pathogenesis and therapy. Science 1995; 267:483–489.
56. Erickson JW. The not-so-great escape. Nat Struc Biol 1995; 2:523–529.
57. Richman DD. Antiretroviral drug resistance: mechanisms, pathogenesis, clinical significance. Adv Exp Med Biol 1996; 394:383–395.
58. Hecht FM, Grant RM, Petropoulos CJ, et al. Sexual transmission of an HIV-1 variant resistant to multiple reverse-transcriptase and protease inhibitors. N Engl J Med 1998; 339:307–311.
59. Kozal M, Shah N, Shen N, et al. Extensive polymorphisms observed in HIV-1 clade B protease gene using high-density oligonucleotide arrays. Nat Med 1996; 2:753–759.
60. Carr A, Samaras K, Burton S, Law M, Freund J, Chisholm D, Cooper D. A syndrome of peripheral lipodystrophy, hyperlipidaemia, and insulin resistance in patients receiving HIV protease inhibitors. AIDS 1998; 12:F51–F58.
61. Kempf DJ, Marsh KC, Kumar G, et al. Pharmacokinetic enhancement of inhibitors of the human immunodeficiency virus protease by coadministration with ritonavir. Antimicrob Agents Chemother 1997; 41:654–660.
62. Kräusslich H–G, Schneider H, Zybarth G, Carter CA, Wimmer E. Processing of in vitro-synthesized *gag* precursor proteins of human immunodeficiency virus (HIV) type 1 by HIV proteinase generated in *Escherichia coli*. J Virol 1988; 62:4393–4397.
63. Kay J, Dunn BM. Viral proteinases: weakness in strength. Biochim Biophys Acta 1990; 1048:1–18.
64. Tomasselli AG, Sarcich JL, Barrett LJ, et al. Human immunodeficiency virus type-1 reverse transcriptase and ribonuclease H as substrates of the viral protease. Protein Sci 1993; 2:2167–2176.
65. Schechter I, Berger A. On the size of the active site in proteases. I. Papain. Biochem Biophys Res Commun 1967; 27:157–162.
66. Baldwin ET, Bhat TN, Liu B, Pattabiraman N, Erickson J. Structural basis of drug resistance for the V82A mutant of HIV-1 protease: backbone flexibility and subsite repacking. Nat Struc Biol 1995; 2:244–249.
67. Erickson JW, Burt SK. Structural mechanisms of HIV drug resistance. Annu Rev Pharm 1996; 36:545–571.
68. Erickson JW, Kempf D. Structure-based design of symmetric inhibitors of HIV-1 protease. Arch Virol 1994; 9:19–29.

2

Discovery and Early Development of Saquinavir

Ian B. Duncan and Sally Redshaw
Roche Discovery Welwyn, Welwyn Garden City, Hertfordshire, England

Saquinavir, the first human immunodeficiency virus (HIV) protease inhibitor to reach the market, introduced an important new option for therapy against HIV. Moreover, as the product of a rational drug design program, it was recognized as heralding a new era for antiviral therapy in general.

I. BACKGROUND

First-generation treatments for HIV-related disease, the reverse transcriptase inhibitors, had only limited efficacy, largely because of toxicities caused by interference with human cell metabolism. This created a need for novel antiretrovirals that could be administered for long enough to allow recovery of patients' immune function. Reduced viral drug sensitivity (''resistance'') should also be minimized, or at least diverted down a different genetic route. Both of these factors called for an examination of alternative antiviral targets.

In the mid 1980s, by analogy with other retroviruses, it was suggested (1) that HIV encoded a protease. Many groups, including our own, quickly recognized the opportunity of an exciting new therapeutic target, and a large effort was devoted to characterizing the enzyme. The Molecular Genetics Department at Hoffmann–La Roche in Nutley, New Jersey, was the first center to confirm that active protease was essential for processing the *gag* and *gag–pol* gene products (2). By 1985, a mechanistic relation with the known aspartic proteases had been suggested (3), and some of the substrate specificities of HIV protease had also been predicted (4).

Early in 1986, strategic meetings were held within Roche to review progress in an ongoing reverse transcriptase inhibitor program and to discuss new approaches for the treatment of AIDS. The viral protease emerged as the most attractive alternative target, and an international program was coordinated to design inhibitors. It was decided that the Roche centers in the United States (Nutley, New Jersey) and Switzerland (Basel) would undertake the necessary molecular biology, biochemistry, and X-ray crystallography, whereas inhibitor design, biochemistry, chemistry, and in vitro virology would be carried out in the United Kingdom (Welwyn and associates).

II. SCREENING ASSAY

A suitably robust and sensitive assay for enzyme activity and its inhibition is essential to any program of inhibitor design. Recombinant DNA technology clearly offered the best source of HIV protease in sufficient quantities. Roche molecular biologists in Nutley (5) and Basel (6) set out to clone, express, and purify the enzyme, and its protein substrates, to allow testing of potential inhibitors and further mechanistic studies (7). We considered a discontinuous enzyme assay would offer better sensitivity, because the conditions for detection could be adjusted independently from those for enzyme activity.

Although the ability of HIV protease to cleave amide bonds between aromatic amino acids and proline was indicated at this time only by peptide sequences isolated from infected cells (4), our plan was nevertheless to devise an assay based on the cleavage of a small synthetic substrate containing a phenylalanine–proline (Phe–Pro) linkage. The heptapeptide (1) based on the P_5–P_2' sequence of the tyrosine–proline (Tyr–Pro) cleavage site in the *gag* polyprotein, but in which Tyr is replaced by Phe, was chosen as a potential substrate. The N- and C-termini were protected to prevent cleavage of the substrate by extraneous exopeptidases, and an N-terminal succinyl group was introduced to enhance solubility.

We strongly favored a colorimetric assay for routine testing, and aimed to detect the N-terminal proline of the dipeptide cleavage product **2** by its reaction with isatin. The reaction of isatin with free proline to give a deep blue color had been known since the end of the last century, but reaction with peptides needed harsh conditions (8) and had not been used quantitatively. We succeeded in identifying much milder conditions for the reaction (9), to provide the basis of a new assay. Because N-terminal proline is very rare in normal peptides and proteins, our assay also allowed us to use unpurified enzyme preparations without interference by contaminants, and so to proceed more quickly with our program.

Succ.Val.Ser.Gln.Asn.Phe.Pro.Ile.NH^{i}Bu **1**

H.Pro.Ile.NH^{i}Bu **2**

We were delighted to find that the heptapeptide **1** was a substrate for HIV protease with a K_m of 0.79 mM, and that reaction of the cleavage product **2** with isatin occurred quantitatively in the presence of an acid catalyst, allowing spectrophotometric measurement of the blue product at a wavelength of 599 nm. This simple, specific, and high-throughput system allowed us to test potential inhibitors very rapidly, giving us an early advantage over groups using more usual technologies, such as high-pressure liquid chromatography (HPLC).

III. DESIGNING INHIBITORS

A. Strategy

The important problem of how to design selective inhibitors remained. Inhibition of the human aspartic proteases—renin, pepsin, gastricsin, and cathepsins D and E—would probably lead to undesirable side effects. Additionally, pepsin, gastricsin, and cathepsin E are present in high concentration in the gut, and binding to these enzymes could also reduce oral absorption.

When our synthetic chemistry program began in late 1986, only very preliminary details about the structure and biochemistry of the target protease were available. The viral enzyme had been provisionally classified as an aspartic protease, and two putative cleavage sites (Met–Met and Tyr–Pro) in the *gag* protein had been identified (4). Having already decided to base our assay on scissile Phe–Pro, we were also very optimistic that inhibitors based on this motif would be highly selective for the target enzyme.

Despite the lack of detailed information about HIV-1 protease, it seemed likely that the catalytic mechanism would be essentially similar to the general acid–general base mechanism proposed for the monomeric aspartic proteases (10,11). Catalysis in these cases was thought not to entail an acyl enzyme intermediate but, instead, a high-energy transition state formed by addition of water to the scissile amide bond of the substrate. We were confident, therefore, that we could design inhibitors of the new viral enzyme using the "transition-state mimetic concept" which had already been used to design inhibitors of other aspartic proteases, especially renin.

B. Our First Lead Compounds

Many different transition-state mimetics could potentially be used to provide inhibitors of an aspartic protease, each of these imitating some, but not all, of the features of the presumed transition state **3**. Of the possible transition-state mimet-

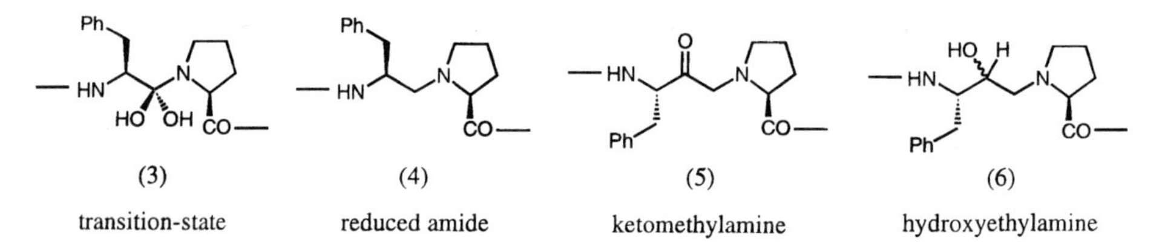

Figure 1 Transition-state mimetics compatible with a Phe–Pro cleavage site: The unusual ability of the HIV protease to cut at Phe–Pro allowed the rational design of a very selective as well as potent inhibitor.

ics, we felt that three (Fig. 1) were especially suited to mimic scissile Tyr/Phe–Pro; namely, the reduced amide **4**, the ketomethylamine **5**, and the diastereomeric hydroxyethylamines **6**.

Our first compounds (Table 1) were based on the supposed Asn.Phe–Pro cleavage sequence in the *pol* polyprotein. The reduced amide derivative **7** proved to be a rather poor inhibitor of HIV protease; similar findings have also been reported by other laboratories (12–14). The ketomethylamine derivative **8** was somewhat more active, but the best activity was seen with the diastereomeric hydroxyethylamines, **9** and **10**.

We next undertook to define the smallest peptide mimetic with which we could achieve acceptable levels of inhibition by preparing a series of hydroxy-

Table 1 Effect of Different Transition-State Mimetics on Potency[a]

Compound	Structure		IC_{50} (μM)
7	Ph; Cbz.Asn.NH; N; $CO_2{}^tBu$	Reduced amide	50
8	Ph; Cbz.Asn.NH; O; N; $CO_2{}^tBu$	Ketomethylamine	0.87
9, 10	Ph; Cbz.Asn.NH; OH; N; $CO_2{}^tBu$	(*R*)- and (*S*)-Hydroxyethylamine	0.14, 0.3

[a] The hydroxyethylamine transition-state mimetic was clearly the best of those considered.

Table 2 Effect of Peptide Size on Potency[a]

Phe-Ψ[CH(OH)CH_2N]—Pro

Compound	Structure	IC_{50} (nM)
11, **12**	Cbz.PheΨ[CH(OH)CH_2N]Pro.O^{t}Bu	6,500; 30,000
9, **10**	Cbz.Asn.PheΨ[CH(OH)CH_2N]Pro.O^{t}Bu	140; 300
13, **14**	Cbz.Leu.Asn.PheΨ[CH(OH)CH_2N]Pro.O^{t}Bu	600; 1,100
15, **16**	Cbz.Asn.PheΨ[CH(OH)CH_2N]Pro.Ile.NH^{i}Bu	130; 2,400
17, **18**	Cbz.Leu.Asn.PheΨ[CH(OH)CH_2N]Pro.Ile.NH^{i}Bu	750; 10,000

[a] A tripeptide mimetic afforded the best activity and was selected for further development.

ethylamines of differing lengths (Table 2). The protected dipeptide derivatives, **11** and **12**, showed detectable inhibitory activity although this was some 50- to 100-fold weaker than we had observed for the tripeptide mimetics **9** and **10**. Further extension towards the N-terminus **13** and **14**, or toward the C-terminus, **15** and **16**, or indeed, in both directions, **17** and **18**, gave no further improvement in potency (15).

The more active of the tripeptide mimetics **9** and **10** had the *R*-configuration at the secondary alcohol function, and we chose this compound for optimization. At the time, no X-ray crystallographic data were available for either the native enzyme or enzyme-inhibitor complexes. (The first data for the native enzyme were to appear 2 years later in 1989; 16). In the absence of structural data that might help design more potent inhibitors, we set out to explore our lead compound systematically, modifying each amino acid residue in turn.

C. DEVELOPING THE LEAD

Replacement of the N-terminal benzyl carbamate (Cbz) group by nonaromatic groups, such as acetyl or ^{t}butyl carbamate (BOC) reduced activity significantly. However, introduction of bicyclic aromatic groups such as β-naphthoyl or, especially, quinoline carbonyl (**20**; Table 3) improved potency compared with the parent compound. At the P_2 subsite, conservative changes were tolerated, but no improvement over asparagine was identified. At the S_1 position, we found that aromatic groups were strongly preferred, but again identified no significant im-

Table 3 Activities of Improved Analogues[a]

Compound	Structure	HIV-1 protease IC_{50} (nM)	HIV-2 protease IC_{50} (nM)	Antiviral activity (HIV-1) EC_{50} (nM)
19	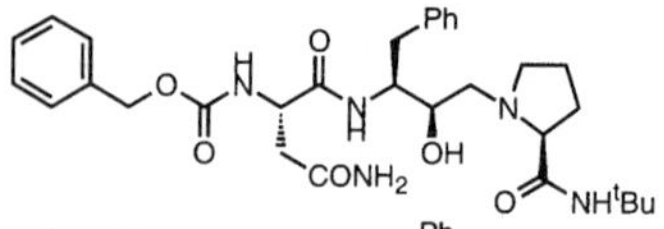	210	330	400
20	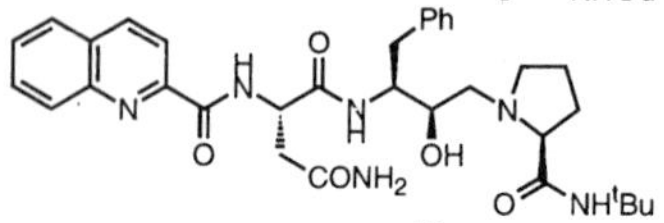	52	50	130
21	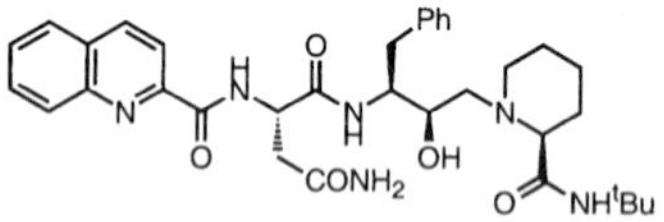	2	10	17
22	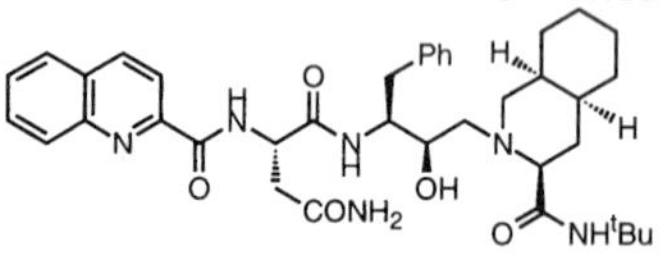	<0.4	<0.8	2

[a] Optimization of the lead structure gave >500-fold improvement in potency.

provement over the parent compound. The most dramatic changes in potency were achieved by modifying the propyl residue that occupies the P_1' subsite. Within a series of imino acids, ring size was very important for activity. Replacing the proline five-membered ring by a four-membered azetidine ring almost abolished activity; whereas incorporation of a six-membered ring improved potency approximately 12-fold (**21**). Replacement of proline by fused bicyclic imino acids led to the greatest enhancement of activity (**22**), and *S, S, S*-decahydroisoquinoline carboxylic acid was the best replacement for proline that we identified. At the C-terminus, medium-sized lipophilic residues appeared to be preferred, with little difference between esters and amides (**9** and **19**). On the basis of chemical, and possibly metabolic, stability, the ^{t}butylamide group was chosen as the C-terminal residue.

Having located two regions of the molecule in which changes substantially altered binding affinities, we determined by further syntheses that the combined effects of individual changes were at least additive. The compounds synthesized in the process showed the same order of potencies against HIV-2 protease, although they were somewhat more active against the HIV-1 enzyme. Within this series of compounds, the order of potencies in a preliminary antiviral assay also paralleled the enzyme inhibitory potency very closely, indicating good penetration into cells. One of the most potent of our hydroxyethylamine derivatives was compound **22** (Ro 31-8959; later saquinavir), which we selected for further evaluation.

So far, we had prepared only 10 g of saquinavir, and most of the critical in vitro studies had been carried out using only 10 mg of the compound. The original synthesis was a rather tedious 26-step process, with an overall yield of only about 10%. Despite this, we were able to advance quickly through hundred-gram to kilogram batches, making improvements to the synthesis as we did so. Within 2½ years, we had prepared almost 300 kg of material. The chemistry has since been improved dramatically by our process research group in Basel who devised a convergent 11-step synthesis with an overall yield of 50% (17), providing a secure basis for clinical trials and commercial-scale manufacture.

IV. PROTEASE INHIBITION

A. Viral Proteases

In our colorimetric assay, saquinavir showed 50% inhibitory concentration (IC_{50}) values of 0.4 nM, or less, against HIV-1 protease and 0.8 nM, or less, against HIV-2 protease. We recognized that the true potency of the compound was likely somewhat greater, but at the time, we were unable to obtain more accurate K_i values because of "mutual depletion" (i.e., the limitation of an assay when the enzyme concentration is of the same order as, or higher than, the K_i value of the inhibitor). Our collaborators, however, calculated K_i values of 0.12 nM against HIV-1 protease and 0.1 nM against HIV-2 protease [J. Kay, personal communication]. Other laboratories have subsequently reported measured K_i values for saquinavir as low as 0.011 nM against HIV-1 protease (18), and 0.3 nM against HIV-2 protease (19).

Saquinavir also inhibited the processing by HIV protease of its natural viral protein substrates. At 1 μM, the inhibitor completely abolished processing of HIV-1 *gag* polyprotein by both HIV-1 and HIV-2 proteases in mixed lysates of *Escherichia coli* expressing the individual protein components of the assay (15). It also inhibited baculovirus-expressed *gag–pol* polyprotein autoprocessing in a dose-dependent fashion (20).

B. Selectivity Relative to Other Proteases

Saquinavir inhibited the human aspartic proteases renin, pepsin, gastricsin, cathepsin D, and cathepsin E by less than 50% at 10 μM (i.e., a concentration at least 10^5 higher than the K_i against the HIV enzymes) (15) (Table 4). When tested against representatives of the three other mechanistic classes of mammalian proteases (serine, cysteine, and metallo), the compound again gave less than 50% inhibition at 10 μM. In other words, there was no significant effect on mechanisti-

Table 4 Selectivity of Saquinavir Relative to Other Proteases[a]

Protease	Mechanistic class	IC_{50} (nM)
HIV-1	Aspartic	<0.37[b] (K_i = 0.12)[c]
HIV-2	Aspartic	<0.80[b] (K_i = 0.10)[c]
Human renin	Aspartic	>10,000
Human pepsin	Aspartic	>10,000
Human gastricsin	Aspartic	>10,000
Human cathepsin D	Aspartic	>10,000
Human cathepsin E	Aspartic	>10,000
Human leukocyte elastase	Serine	>10,000
Bovine cathepsin B	Cysteine	>10,000
Human synovial fibroblast collagenase	Metallo	>10,000
Prolidase	Dipeptidase	>10,000

[a] Saquinavir proved to be an extremely potent inhibitor of HIV-1 and 2 proteases, without conflicting activity against mammalian enzymes.
[b] Results limited by mutual depletion.
[c] Calculated values.

cally related human proteases of the immune or gastrointestinal systems. These results confirmed that, as intended, saquinavir was a highly selective inhibitor of the HIV proteases, and promised limited side effects in clinical use.

V. BIOLOGICAL ASSESSMENTS

A. Antiviral Activities

From the very outset of the program we were acutely aware of the lack of a suitable animal model, and this situation had not improved by 1989 when we were attempting to select a development candidate from among a number of highly potent options. Because of this, we felt that it was important to demonstrate consistency of antiviral activity in a wide range of assay systems. A collaboration was established with St. Mary's Hospital Medical School in London, and a high-containment laboratory for HIV work was later commissioned in the Roche U.K. facility. In addition to the assays performed in these centers, saquinavir was included in a Medical Research Council multicenter, blinded testing program (21). Typical antiviral IC_{50} values from all of these laboratories were in the range of 1–10 nM against HIV-1, with similar potencies against HIV-2 (21) and SIV (22). The results showed that saquinavir retained its potent antiviral activity in both lymphoblastoid and macrophage cell lines, as well as in primary cells, and against both laboratory viral strains and clinical isolates, including zidovudine (ZDV)-resistant strains (23,24). The consistently high degree of efficacy provided reassurance that the compound passed readily across cell membranes, which might have presented a barrier. No subsequent protease inhibitor has been shown to have better efficacy or breadth of activity.

Changes in host cell metabolism of zidovudine had been reported to result in loss of antiviral activity (25–27). Saquinavir, however, which requires no metabolic activation, retained full potency in the JM lymphoblastoid cell line, in which an apparent failure to undertake the necessary metabolic activation (P. Wong-Kai-In, personal communication) rendered ZDV essentially inactive. This result indicated a likely superiority of saquinavir over nucleosides in quiescent cells, in which ZDV probably underwent insufficient metabolic activation (28).

The antiviral effect seen after delayed addition of saquinavir to infected cells (23,29), confirmed its late point of action in the infectious cycle, as did its effectiveness against chronic infection (29,30). This contrasted with the reverse transcriptase inhibitors, which act at an early point in viral replication and are ineffective if added more than 2 or 3 h postinfection. Electron microscopy of virions produced in chronically infected cells confirmed that saquinavir prevents maturation of new viral particles (Fig. 2) (29), and this was accompanied by a lack of infectivity (31).

In a range of antiviral tests, therefore, saquinavir proved to be of at least equal potency with the then standard therapy, zidovudine. To enable phase 1 clinical studies, we needed to demonstrate a good safety profile for saquinavir. No indication of cytotoxicity appeared with saquinavir treatment until micromolar concentrations were reached. Comparison of these values with the antiviral activities demonstrated that, although there was some variation in host cell sensitivity, there was an overall differential of at least three orders of magnitude between the antiviral and cytotoxic effects of saquinavir. We were becoming increasingly confident in our original premise that a highly selective inhibitor of the viral protease would show little toxicity, and be suitable for long-term administration.

This principle was supported by a series of studies (32) in which no evidence of infectious virus or viral components (including intracellular DNA) was found after saquinavir treatment of HIV-infected cell cultures for 87 days at 100 nM, or during a further 35 days after drug removal. Control cultures, by comparison, died by 32 days postinfection. This result was interpreted as showing the regrowth of healthy cells following the progressive death of the infected population, and provided a graphic, if idealized, demonstration of the potential of this new, selective antiviral strategy.

B. Reduced Sensitivity ("Drug Resistance")

It was of considerable interest and importance to determine in advance of clinical studies whether viral variants with reduced sensitivity to saquinavir could be selected. HIV-1 strain GB8 was serially passaged in CEM cells in the presence of increasing concentrations of saquinavir, in parallel with selection by ZDV and

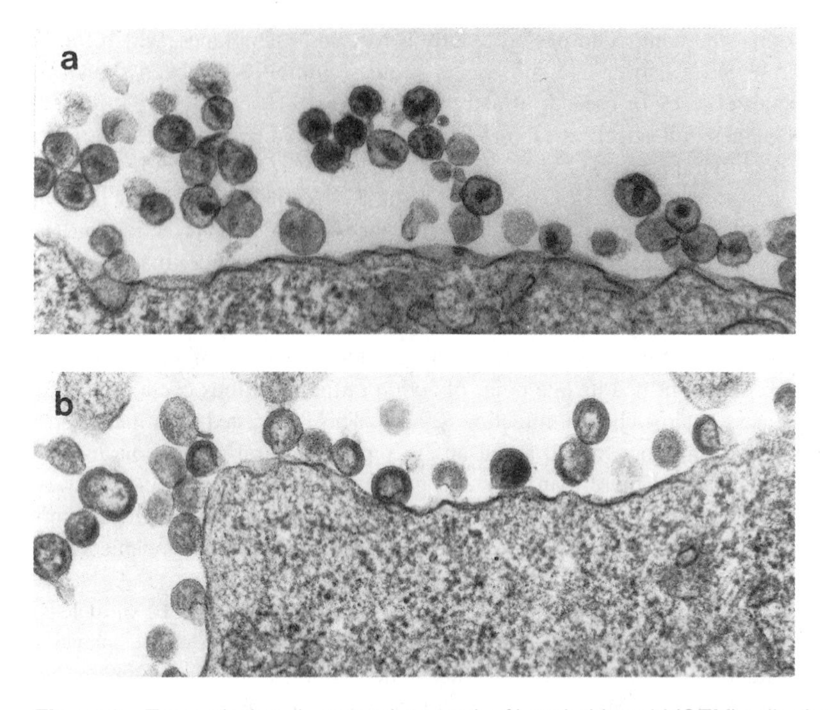

Figure 2 Transmission electron micrograph of lymphoblastoid (CEM) cells chronically infected with HIV-1$_{IIIB}$: (a) Untreated culture: released virus particles show the condensed cores that characterize mature virus. (b) Culture treated for 24 h with 100 nM of saquinavir: particles show a clear center and dark periphery that characterize the immature form. (From Ref. 29.)

a non-nucleoside reverse transcriptase inhibitor, the TIBO analogue, R82150 (Fig. 3). It proved possible to reduce drug sensitivity in all three cases, although reduced sensitivity to saquinavir appeared at a rather later time point than with either of the reverse transcriptase inhibitors (33). We were much encouraged by these findings, which we felt augured very well for the clinical performance of our compound.

After having generated virus with reduced sensitivity to saquinavir, we were obviously intrigued to identify the underlying changes in the protease. Cloning and sequencing of the protease from resistant virus showed that G48V tended to be the first or only amino acid substitution to occur and was associated with a relatively low coefficient of resistance. The L90M substitution, where present, emerged later, and the double substitution was associated with a markedly greater loss of sensitivity than either single change.

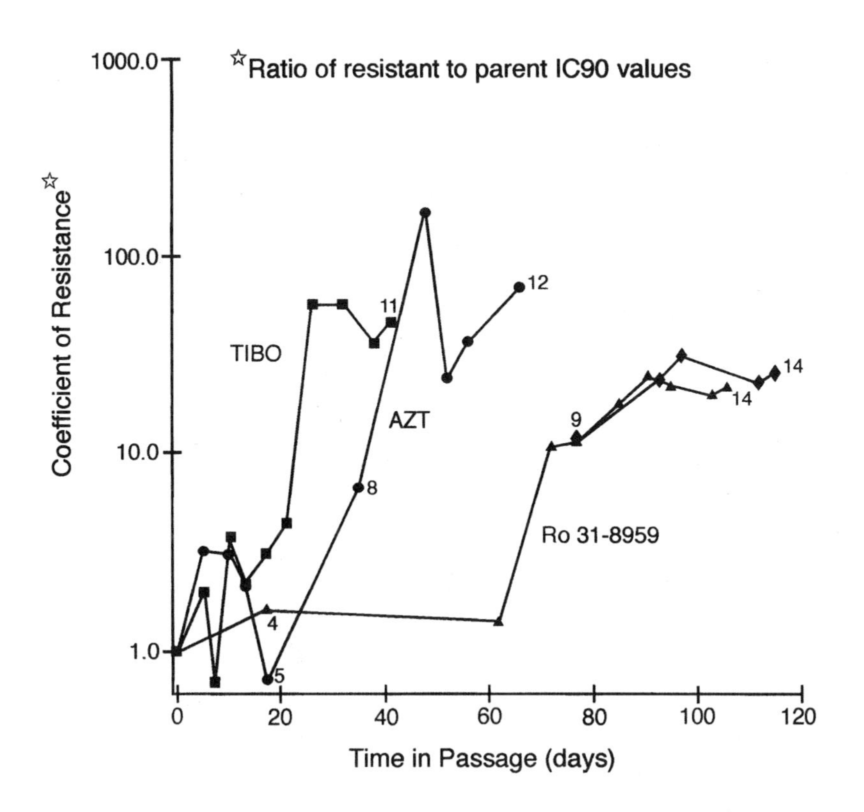

Figure 3 Selection of HIV-1 with reduced sensitivity to saquinavir in vitro: repeated passage of HIV-1 in vitro selected for virus with reduced sensitivity to saquinavir, but apparently more slowly than using representative reverse transcriptase inhibitors. Passage numbers are indicated. (From Ref. 33.)

Mutagenesis of the standard strain HIV-1_{HXB2} protease at positions 48 and 90 also suggested that each mutation alone contributed only moderately to decreased drug sensitivity (34), whereas a variant containing both mutations showed a more substantial (20-fold) change in IC_{50} value.

A model of the double-mutant protease based on X-ray analysis is shown in Figure 4. A valine residue at position 48 in the flap region of the protease clearly would have an unfavorable steric interaction with saquinavir, but the effect of a methione residue at position 90 is more indirect. We were encouraged by the discovery that the intracellular cleavage of HIV-1_{HXB2} polyprotein by proteases carrying G48V, L90M, or both substitutions was abnormal (34), indicating possible detrimental effects on viral growth and pathogenicity that might be beneficial in the clinic. Although these in vitro studies gave us an insight into the

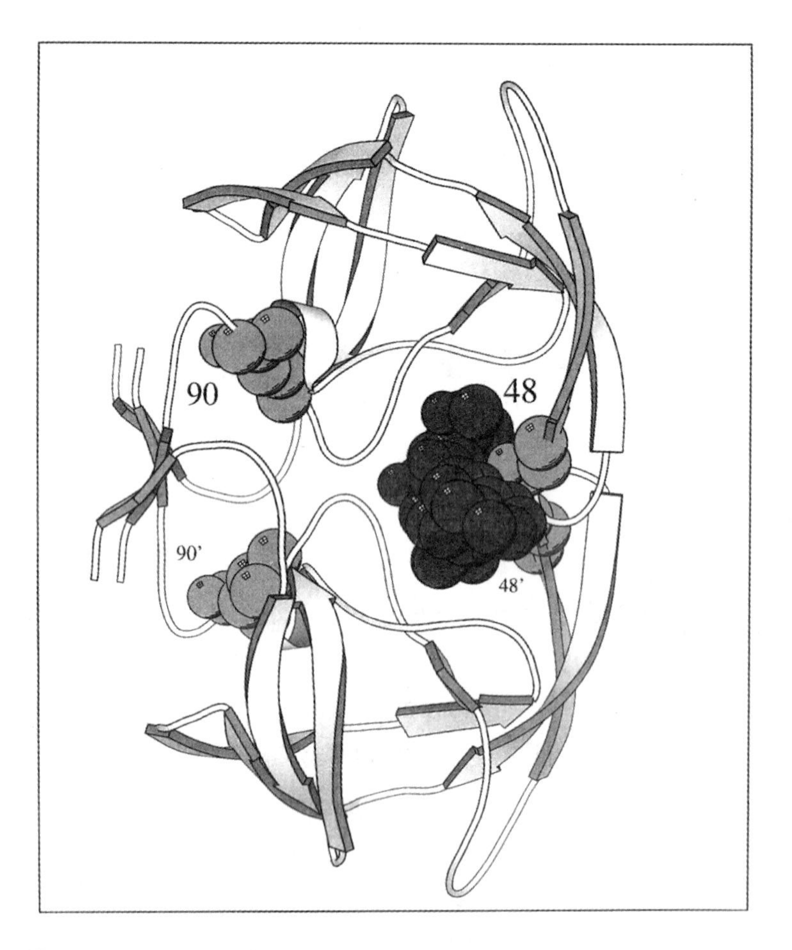

Figure 4 Location of key mutations that reduce sensitivity to saquinavir: of the two substitutions that have been linked consistently with reduced sensitivity to saquinavir, G48V is located in the flaps next to the active site, whereas L90M acts remotely.

genetic changes likely to be seen in the clinic, nevertheless, we undertook to monitor the full protease sequence and the accompanying sensitivity of viral isolates in our clinical program to ensure the best possible understanding of the basis for reduced drug sensitivity.

It became clear that the G48V and L90M substitutions, which have consistently associated with reduced sensitivity to saquinavir, differ from the genetic "signatures" selected by other HIV protease inhibitors (35). This distinctive profile might be expected to offer saquinavir special advantages in clinical use, for it would not anticipate a general cross-resistance among saquinavir and any other protease inhibitors in development.

C. Combination with Other Antiretrovirals

From first principles (36), combining drugs with different antiviral profiles should enhance suppression of viral replication and allow dose reduction of the individual agents. An additive to synergistic antiviral effect was obtained (24,37–40) by two- or three-way combinations of saquinavir with nucleoside analogues or other antiretrovirals, irrespective of the mathematical method applied. This would offer the possibility of polypharmacy, with advantages for improved efficacy and control over drug resistance by reducing viral turnover. An example of a three-way combination of saquinavir, ZDV, and zalcitabine (ddC) is shown in Figure 5.

D. Saquinavir and Immune Function

In preliminary in vitro experiments (41), exposure of dendritic cells to saquinavir during incubation with HIV-1 opposed the viral suppression of allogeneic stimulation and enhanced the autologous immune response. These cells are key to the process of antigen presentation and are numerically, as well as functionally, depleted in early HIV-infected patients (42; later reviewed in 43), even before $CD4^+$ T-cell depletion is apparent. These preliminary data might illustrate a potential to restore protective immune responses against the virus as well as against other antigens, including opportunistic infections and, therefore, be of benefit in saquinavir treatment beyond its immediately visible effects on virus load.

VI. TOWARD AN IND

In 1991, we reached the point of clinical trials with saquinavir. Although we had shown very potent antiviral activity for the compound as well as minimal cytotoxicity, the lack of a suitable animal model now meant that we had to progress into development without any true efficacy data. This meant that we had to extrapolate from in vitro antiviral concentrations to the dosage needed for a clinical effect. Representative IC_{50} and IC_{90} values of 2 nM (1.5 ng/mL) and 16 nM (12 ng/mL), respectively, were adopted for projecting to clinical dosage; at the same time, by varying the protein content of the culture medium a twofold increase in concentration was calculated to allow for plasma protein binding in vivo. Although laboratory data (unpublished) had indicated that the antiviral activity of saquinavir was not rapidly reversible, we felt that plasma concentrations

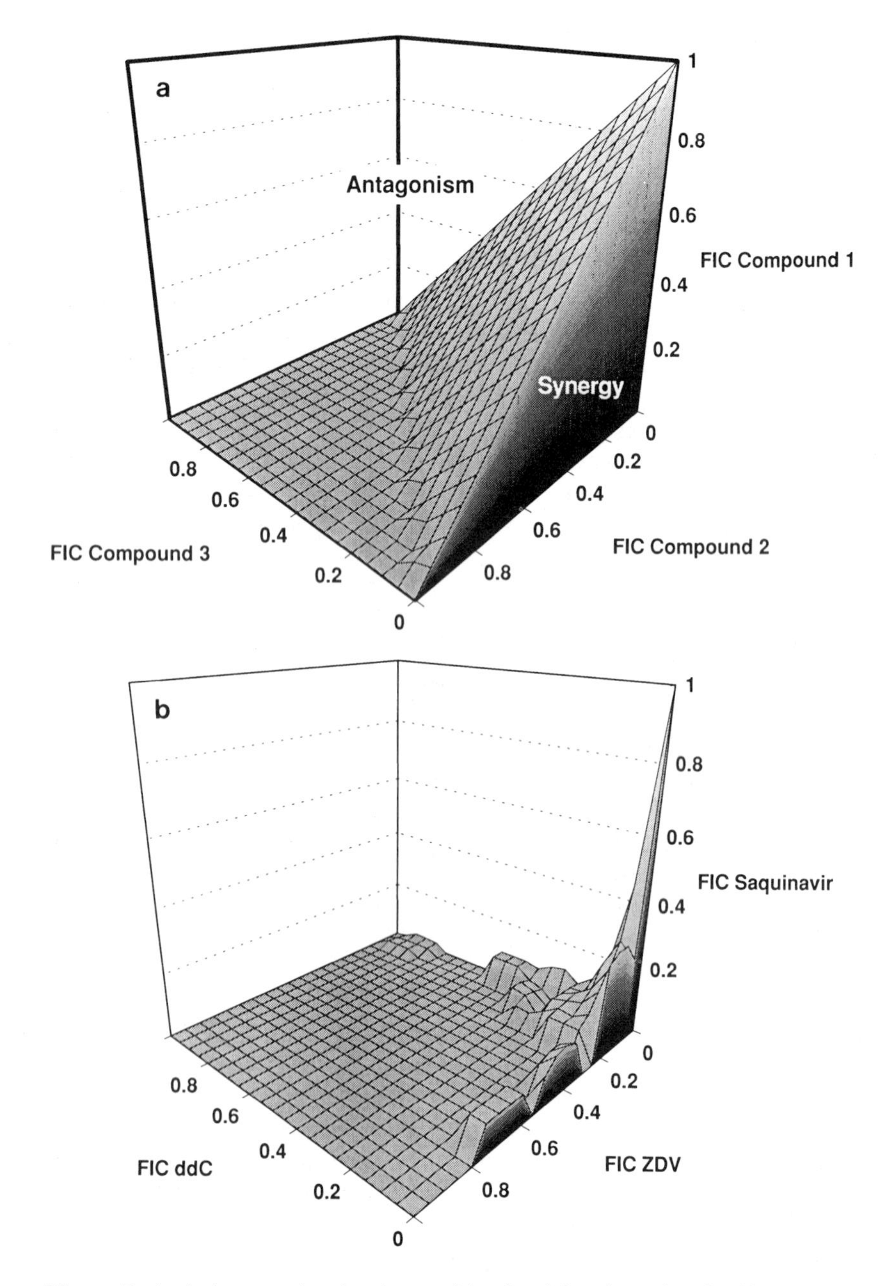

Figure 5 Isobologram showing the combined activity of saquinavir with other antiretrovirals: saquinavir has a potential for synergy in combination (here, triple) with other antiretrovirals. (a) Additive results would fall on the inclined plane of antiviral effect; (b) the actual result shown for saquinavir + ZDV + ddC (and pairwise on the "walls" of this 3D plot) shows potential synergy of action. (From Ref. 39.)

should be maintained as high as possible to minimize any risk of losing control over virus replication. Our ultimate goal had always been an orally active agent; therefore, we ensured that a pharmacokinetic evaluation formed part of the early assessment of selected compounds. These preliminary studies indicated that it should be possible to achieve clinically relevant doses of saquinavir by dosing orally.

The in vivo toxicity of saquinavir proved to be minimal. When administered orally to various animal species, saquinavir produced only minor effects in general pharmacology, with little or no effect on the gastrointestinal, immune, or central nervous systems. Oral toxicity and toxicokinetic studies of up to 12 months duration in the mouse, rat, and marmoset demonstrated excellent tolerance to saquinavir at high plasma exposure levels.

Functional evidence of hepatotoxicity (raised liver enzyme functions) was seen in male rats, but the effects were slight and inconsistent, and not dose-related. A slight but dose-related decrease in total leukocyte count, with an associated reduction in lymphocytes and neutrophils, was observed in a 6-month rat study; similar changes were not found, however, at higher plasma exposures to saquinavir in the rat or in other species. Intravenous administration of saquinavir to rats and marmosets for up to 4 weeks induced some tissue reaction at the sites of injection, but only minor drug-related systemic findings. More severe local intolerance was seen in rats when the drug was given intravenously over 2 weeks by continuous infusion at high-dose levels, but no histological evidence of disseminated toxicity was seen in this study. Because the route of administration differed from that proposed for clinical development, these observations were, in any event, of little relevance.

No reproductive, teratogenic, or developmental defects were seen in statutory Segment I, II, or III reproductive toxicity studies and no activity was seen in repeated mutagenicity and genotoxicity assays. The data from these studies confirmed the lack of toxicity of saquinavir and laid the foundation for safe entry into humans.

VII. FIRST CLINICAL STUDIES

A. Pharmacokinetics and Metabolism

These topics are dealt with more fully elsewhere (44). The mean absolute bioavailability of a single (600-mg) oral dose of saquinavir in its initial hard-gelatin formulation, Invirase, was governed by its limited absorption, and more especially by extensive first-pass metabolism in the liver (45,46). Importantly, the absorption level of the drug was increased (from 24 ng h^{-1} mL^{-1} to 161 ng h^{-1} mL^{-1}) when the same single dose was administered to volunteers after food, leading to the stipulation that this medication be taken within 2 h of a meal. Only a twofold difference was found in favor of a high-calorie, high-fat meal versus a low-calorie, low-fat one. Steady-state concentrations of saquinavir after multiple dosing also appeared to be higher in HIV-infected individuals than in healthy volunteers (46) for unidentifiable reasons. Saquinavir is rapidly metabolized to various mono- and dihydroxylated derivatives by the cytochrome $P450_{3A4}$ (CYP3A4) isoenzyme (46,47), and subsequent elimination is mostly in the feces (8%), rather than in urine (1%). Drugs in clinical or recreational use that induce or block CYP3A4 would thus be expected to reduce or raise the level of saquinavir correspondingly. For example, levels of saquinavir (600 mg tid) were increased threefold when the antifungal drug ketoconazole (a CYP3A4 inhibitor) was administered concomitantly, even in healthy volunteers (48). Such factors would obviously influence the exposure level, and the antiviral response, achieved by an infected individual in the clinic.

B. Phase I/II Clinical Trials

Preliminary safety and efficacy trials on Invirase began in August 1991, forming an integrated program of monotherapy and double- or triple-combination therapy with reverse transcriptase inhibitors, in naive and pretreated patients at different disease stages.

For expediency, these early clinical studies followed surrogate markers of disease—viral load (as measured by quantitative amplification of plasma viral RNA) and $CD4^+$ cell count—to provide a general measure of immune status. This avoided the large numbers of patients and lengthy studies that would be needed to demonstrate statistically significant changes using clinical endpoints.

1. Monotherapy

Invirase doses ranging from 25 to 600 mg thrice daily (tid) as monotherapy were assessed initially in randomized, double-blind studies in Europe. The results of these studies indicated a positive dose–response relationship for saquinavir (44,49, 50), and advocated the 600-mg–tid regimen. Subsequently, in an independently managed study at Stanford University in 1994–1995 (51), the efficacy of Invirase monotherapy was studied at elevated doses of 3600 or 7200 mg/day. Increased activity was observed at higher doses and decreases in viral load of near 1.1 and 1.3 $\log_{10}$ units, respectively, were maintained to 24 weeks of assessment, while $CD4^+$ cell counts remained elevated above baseline.

2. Combination Therapy

Early in vitro studies had supported the principle of combining saquinavir with one or more reverse transcriptase inhibitors. In advanced, previously untreated Italian patients, the peak reduction in circulating viral load (as measured by RNA-

PCR) and the median average increase in $CD4^+$ cell count over 16 weeks were greater in patients receiving 600 mg tid of saquinavir in combination with ZDV than in those receiving lower saquinavir doses or monotherapy (44,50).

A further study [NV14255 = ACTG229] advanced this concept by making a comparison of saquinavir plus zalcitabine with zalcitabine plus zidovudine, or a combination of all three agents in advanced, heavily zidovudine-pretreated patients. The median increase in $CD4^+$ cell count over 24 or 48 weeks was higher in the triple-therapy group than with either of the double-therapies (44). The same was true for the mean normalized areas under the plots of $CD4^+$ cell count versus time, and for reduction of viral load.

In these early clinical studies, Invirase showed a beneficial effect on surrogate markers (49) and a more profound effect when used at higher dose or in combination with reverse transcriptase inhibitors (44).

3. Tolerability

Invirase was exceptionally well-tolerated in the clinic, which was reflected in extremely good compliance. Side effects occurred infrequently; gastrointestinal events (mainly diarrhea) were most common, and seen in 3.8% of patients (44). Other adverse events (headache, paresthesia, asthenia, skin rash, or musculoskeletal pain) were reported at 1% or lower incidence. The most common marked laboratory abnormalities were increased creatine phosphokinase (4%) and low plasma glucose levels (5%). All adverse events were fully reversible.

VIII. APPROVAL AND RECOGNITION

These studies formed the basis for the approval of Invirase in December 1995 in the United States, and later in Europe and elsewhere, as an antiretroviral agent for human use. This was the first HIV protease inhibitor to reach the market, and heralded a new era of designing antiviral drugs from first principles as selective inhibitors of specific viral functions.

Encouraged by the increased antiviral effect achieved in the Stanford University high-dose study with Invirase (51), Roche proceeded to develop a new, soft-gelatin capsule, formulation of saquinavir to improve the clinical efficacy of this intrinsically very potent compound. This later formulation, which increased exposure levels and antiviral potency in the clinic (details outside the remit of this chapter), was approved (United States, November 1997) under the tradename Fortovase.

In 1996 and 1997, the team in Roche Discovery Welwyn was delighted by the honor of pharmaceutical industry and other awards, including Prix Galiens in the United Kingdom, Spain, and Portugal, and the U.K. Society for Medicines

Research Award for Drug Discovery, for the innovation of saquinavir and the new clinical benefit offered by Invirase.

ACKNOWLEDGMENTS

The successful and timely outcome of this project was the result of a substantial collaboration not only within the company, but also with external academic groups and national AIDS programs in the United States, United Kingdom, France, and Italy. Many individuals provided important contributions that made the entire discovery and development process possible, and the inspiration, enthusiasm, and sheer hard work of all of those involved is gratefully acknowledged.

REFERENCES

1. Ratner L, Haseltine W, Patarca R, Livak KJ, Starcich B, Josephs SF, Doran ER, Rafalski JA, Whitehorn EA, Baumeister K, Ivanoff L, Petteway SR, Jr, Pearson ML, Lautenberger JA, Papas TS, Ghrayeb J, Chang NT, Gallo RC, Wong–Staal F. Complete nucleotide sequence of the AIDS virus, HTLV-III. Nature 1985; 313:277–284.
2. Kramer RA, Schaber MD, Skalka AM, Ganguly K, Wong–Staal F, Reddy EP. HTLV-III *gag* protein is processed in yeast cells by the virus *pol*-protease. Science 1986; 231:1580–1584.
3. Toh H, Ono M, Saigo K, Miyata T. Retroviral protease-like sequence in the yeast transposon. Nature 1985; 315:691–692.
4. Sanchez–Pescador R, Power MD, Barr PJ, Steimer KS, Stempien MM, Brown Shimer SL, Gee WW, Renard A, Randolph A, Levy JA, Dina D, Luciw PA. Nucleotide sequence and expression of an AIDS-associated retrovirus (ARV-2). Science 1985; 227:484–492.
5. Graves MC, Lim JJ, Heimer EP, Kramer R. An 11-kD form of human immunodeficiency virus protease expressed in *Escherichia coli* is sufficient for enzymatic activity. Proc Natl Acad Sci USA 1988; 85:2449–2453.
6. Mous J, Heimer EP, Le Grice SFJ. Processing protease and reverse transcriptase from human immunodeficiency virus type 1 polyprotein in *Escherichia coli*. J Virol 1988; 62:1433–1436.
7. Le Grice SFJ, Mills J, Mous J. Active site mutagenesis of the AIDS virus protease and its alleviation by trans complementation. EMBO J 1988; 7:2547–2553.
8. Grassman W, von Arnim K. Über neue Farbreaktionen des Pyrrolidins und Prolins. Justus Liebigs Ann Chem 1935; 519:192–208.
9. Broadhurst AV, Roberts NA, Ritchie AJ, Handa BK, Kay C. Assay of HIV-1 proteinase: a colorimetric method using small peptide substrates. Anal Biochem 1991; 193:280–286.

10. James MNG, Hsu I–N, Delbaere LJJ. Mechanism of acid protease catalysis based on the crystal structure of penicillopepsin. Nature 1977; 267:808–813.
11. James MNG, Sielecki AR. Stereochemical analysis of peptide bond hydrolysis catalysed by the aspartic proteinase penicillopepsin. Biochemistry 1985; 24:3701–3713.
12. Moore ML, Bryan WM, Fakhoury SA, Maagard VW, Huffman WF, Dayton BD, Meek TD, Hyland L, Dreyer GB, Metcalf BM, Strickler JE, Gorniak JG, Debouck C. Peptide substrates and inhibitors of the HIV-1 protease. Biochem Biophys Res Commun 1989; 159:420–425.
13. Dreyer GB, Metcalf BW, Tomaszek TA Jr, Carr TJ, Chandler AC III, Hyland L, Fakhoury SA, Magaard VW, Moore ML, Strickler JE, Debouk C, Meek TD. Inhibition of human immunodeficiency virus 1 protease in vitro: rational design of substrate analog inhibitors. Proc Natl Acad Sci USA 1989; 86:9752–9756.
14. Tomasselli AG, Olsen MK, Hui JO, Staples DJ, Sawyer TK, Henrikson RL, Tomich CSC. Substrate analogue inhibition and active site titration of purified recombinant HIV-1 protease. Biochemistry 1990; 29:264–269.
15. Roberts NA, Martin JA, Kinchington D, Broadhurst AV, Craig JC, Duncan IB, Galpin SA, Handa BK, Kay J, Kröhn A, Lambert RW, Merrett JM, Mills JS, Parkes KEB, Redshaw S, Ritchie AJ, Taylor DL, Thomas GJ, Machin PJ. Rational design of peptide-based HIV proteinase inhibitors. Science 1990; 248:358–361.
16. Navia MA, Fitzgerald PMD, McKeever BM, Leu C–T, Heimbach JC, Herber WK, Sigal IS, Darke PL, Springer JP. Three-dimensional structure of aspartyl protease from human immunodeficiency virus HIV-1. Nature 1989; 337:615–620.
17. Gohring W, Gokhale S, Hilpert H, Roessler F, Schlageter M, Vogt P. Synthesis of the HIV-proteinase inhibitor saquinavir: a challenge for process research. Chimia 1996; 50:532–537.
18. Daenke S, Schramm HJ, Bangham CRM. Analysis of substrate cleavage by recombinant protease of human T cell leukaemia virus type 1 reveals preferences and specificity of binding. J Gen Virol 1994; 75:2233–2239.
19. Griffiths JT, Tomchak LA, Mills JS, Graves MC, Cook ND, Dunn BM, Kay J. Interactions of substrates and inhibitors with a family of tethered HIV-1 and HIV-2 homo- and heterodimeric proteinases. J Biol Chem 1994; 269:4787–4793.
20. Overton H, McMillan D, Gridley SJ, Brenner J, Redshaw S, Mills J. Effect of two novel inhibitors of human immunodeficiency virus protease on the maturation of the HIV *gag* and *gag*-polypolyproteins. Virology 1990; 179:508–511.
21. Holmes HC, Mahmood N, Karpas A, Petrik J, Kinchington D, O'Connor T, Jeffries DJ, Desmyter J, De Clerq E, Pauwels R, Hay A. Screening of compounds for activity against HIV: a collaborative study. Antiviral Chem Chemother 1991; 2:287–293.
22. Martin JA, Mobberley MA, Redshaw S, Burke A, Tyms AS, Ryder TA. The inhibitory activity of a peptide derivative against the growth of simian immunodeficiency virus in C8166 cells. Biochem Biophys Res Commun 1991; 176:180–188.
23. Galpin S, Roberts NA, O'Connor T, Jeffries DJ, Kinchington D. Antiviral properties of the HIV-proteinase inhibitor, Ro 31-8959. Antiviral Chem Chemother 1994; 5: 43–45.
24. Johnson VA, Merrill DP, Chou T–C, Hirsch MS. Human immunodeficiency virus type-1 (HIV-1) inhibitory interactions between protease inhibitor Ro 31-8959 and

zidovudine, 2,3-dideoxycytidine or recombinant interferon αA against zidovudine-sensitive or -resistant HIV-1 in vitro. J Infect Dis 1992; 166:1143–1146.
25. Antonelli G, Turriziani O, Cianfriglia M, Riva E, Dong G, Fattorossi A, Dianzani F. Resistance of HIV-1 to AZT might also involve the cellular expression of multidrug resistance P-glycoprotein. AIDS Res Hum Retroviruses 1992; 8:1839–1843.
26. Avramis VI, Kwock R, Solorzano MM, Gomperts E. Evidence of in vitro development of drug resistance to azidothymidine in T-lymphocytic leukemia cell lines (Jurkat E6-1/AZT-100) and in pediatric patients with HIV infection. J Acquir Immune Defic Syndr 1993; 6:1287–1296.
27. Dianzani F, Antonelli G, Turriziani O, Riva E, Simeoni E, Signoretti C, Strosselli S, Cianfriglia M. Zidovudine induces the expression of cellular resistance affecting its antiviral activity. AIDS Res Hum Retroviruses 1994; 10:1471–1478.
28. Gao W–Y, Shirasaka T, Johns DG, Broder S, Mitsuya H. Differential phosphorylation of azidothymidine, dideoxycytidine, and dideoxyinosine in resting and activated peripheral blood mononuclear cells. J Clin Invest 1993; 91:2326–2333.
29. Craig JC, Duncan IB, Hockley D, Grief C, Roberts NA, Mills JS. Antiviral properties of Ro 31-8959, an inhibitor of human immunodeficiency virus (HIV) proteinase. Antiviral Res 1991; 16:295–305.
30. Craig JC, Grief C, Mills JS, Hockley D, Duncan IB, Roberts NA. Effects of a specific inhibitor of HIV proteinase (Ro 31-8959) on virus maturation in a chronically infected promonocytic cell line (U1). Antiviral Chem Chemother 1991; 2:181–186.
31. Roberts NA, Craig JC, Duncan IB. HIV proteinase inhibitors. Biochem Soc Trans 1992; 20:513–516.
32. Nitschko H, Lindhofer H, Schaetzl H, Eberle J, Deby G, Kranz B, von der Helm K. Long-term treatment of HIV-infected MT-4 cells in culture with HIV proteinase inhibitor Ro31-8959 leads to complete cure of infection. Antiviral Chem Chemother 1994; 5:236–242.
33. Craig JC, Whittaker L, Duncan IB, Roberts NA. In vitro resistance to an inhibitor of HIV proteinase (Ro31-8959) relative to inhibitors of reverse transcriptase (AZT and TIBO). Antiviral Chem Chemother 1993; 4:335–339.
34. Jacobsen H, Yasargil K, Winslow DL, Craig JC, Krohn A, Duncan IB, Mous J. Characterisation of human immunodeficiency virus type 1 mutants with decreased sensitivity to proteinase inhibitor Ro 31-8959. Virology 1995; 206:527–534.
35. Mellors JW, Larder BA, Schinazi RF. Mutations in HIV-1 reverse transcriptase and protease associated with drug resistance. Int Antiviral News 1995; 3:8–13.
36. Hall MJ, Duncan IB. Antiviral drug and interferon combinations In: Field HJ, ed. Antiviral Agents: The Development and Assessment of Antiviral Chemotherapy. Boca Raton, FL: CRC Press, 1988; 2:29–84.
37. Craig JC, Duncan IB, Whittaker LN, Roberts NA. Antiviral synergy between inhibitors of HIV proteinase and reverse transcriptase. Antiviral Chem Chemother 1993; 4:161–166.
38. Taylor DL, Brennan TM, Bridges CG, Kang MS, Tyms AS. Synergistic inhibition of human immunodeficiency virus type 1 in vitro by 6-*O*-butanoylcastanospermine (MDL 28574), in combination with inhibitors of the virus encoded reverse transcriptase and proteinase. Antiviral Chem Chemother 1995; 6:143–152.
39. Craig, JC, Whittaker LN, Duncan IB, Roberts NA. In vitro anti-HIV and cytotoxico-

logical evaluation of the triple combination: AZT and ddC with HIV proteinase inhibitor saquinavir (Ro 31-8959). Antiviral Chem Chemother 1994; 5:380–386.

40. Connell EV, Hsu M–C, Richman DD. Combinative interactions of a human immunodeficiency virus (HIV) *tat* antagonist with HIV reverse transcriptase inhibitors and an HIV proteinase inhibitor. Antimicrob Agents Chemother 1994; 38:348–352.
41. Duncan IB, Macatonia SE, Knight SC. Improved function of dendritic cells exposed in vitro to human immunodeficiency virus type 1 (HIV-1) on treatment with viral protease inhibitor (Ro 31-8959). Abstracts of the 9th International Conference on AIDS, Berlin, June 6–11, 1993. abstr PO-A25-0601.
42. Macatonia SE, Lau R, Patterson S, Pinching AJ, Knight SC. Dendritic cell infection, depletion and dysfunction in HIV-infected individuals. Immunology 1990; 71:38–45.
43. Knight SC, Patterson S. Bone marrow-derived dendritic cells, infection with human immunodeficiency virus, and immunopathology. Annu Rev Immunol 1997; 15:593–615.
44. Noble S, Faulds D. Saquinavir: a review of its pharmacology and clinical potential in the management of HIV infection. Drugs 1996; 52:93–112.
45. Williams PEO, Muihead GJ, Madigan MJ, Mitchell AM, Shaw T. Disposition and bioavailability of the HIV proteinase inhibitor, Ro 31-8959, after single doses in healthy volunteers. Br J Clin Pharmacol 1992; 34:155P–156P.
46. Hoffmann-La Roche Ro 31-8959 Saquinavir Invirase Investigational Drug Brochure, 1996.
47. Farrar G, Mitchell AM, Hooper H, Stewart F, Malcolm SL. Prediction of potential drug interactions of saquinavir (Ro 31-8959) from in vitro data. Br J Clin Pharmacol 1994; 38:162P.
48. Hoffmann-La Roche Inc. Invirase (saquinavir mesylate) capsules prescribing information (package insert). Nutley, NJ, 1996.
49. Kitchen VS, Skinner C, Ariyoshi K, Lane EA, Duncan IB, Burckhardt J, Burger HU, Bragman K, Pinching AJ, Weber JN. Safety and activity of saquinavir in HIV infection. Lancet 1995; 345:952–955.
50. Vella S, Galluzzo C, Giannini G, Pirillo MF, Duncan I, Jacobsen H, Andreoni M, Sarmati L, Ercoli L. Saquinavir/zidovudine combination in patients with advanced HIV infection and no prior antiretroviral therapy: $CD4^+$ lymphocyte/plasma RNA changes, and emergence of strains with reduced phenotypic sensitivity. Antiviral Res 1996; 29:91–93.
51. Schapiro JM, Winters MA, Stewart F, Efron B, Norris J, Kozal MJ, Merigan TC. The effect of high-dose saquinavir on viral load and $CD4^+$ T-cell counts in HIV-infected patients. Ann Intern Med 1996; 124:1039–1050.

3
Discovery and Early Development of Ritonavir and ABT-378

Dale J. Kempf
Abbott Laboratories, Abbott Park, Illinois

I. EARLY STUDIES

The discovery and development of the HIV protease inhibitor ritonavir (Norvir, initially designated ABT-538) and the second-generation inhibitor ABT-378 (lopinavir) illustrate an interdisciplinary approach to drug development that yielded potent therapies for AIDS with unprecedented speed. Ritonavir and ABT-378 are the third and fourth of a series of symmetry-based protease inhibitors from Abbott Laboratories to undergo clinical investigation. Given the nature of HIV infection and the need for prolonged therapy, the goal of this and other HIV protease inhibitor programs has been the development of orally effective agents that can be safely and conveniently administered. Because aspartic proteases, including HIV protease, recognize their natural polypeptide substrates in an extended, β-conformation, inhibitors that potently bind into the active site must mimic this conformation and assume a size sufficient for high affinity. The most difficult challenge for such peptidomimetic agents has been the identification of compounds with the proper physicochemical and pharmacokinetic properties for oral efficacy. Before the advent of HIV protease inhibitors, the engineering of peptidomimetics for other diseases had, in general, failed to produce compounds with acceptable oral bioavailability. The design of ritonavir and ABT-378, as detailed in the following, represents the integration of structure-based drug design and traditional medicinal chemistry, with an emphasis on understanding the relation of chemical structure to pharmacokinetic profile. The successful solution of this problem not only produced antiviral agents with high potency and efficacy,

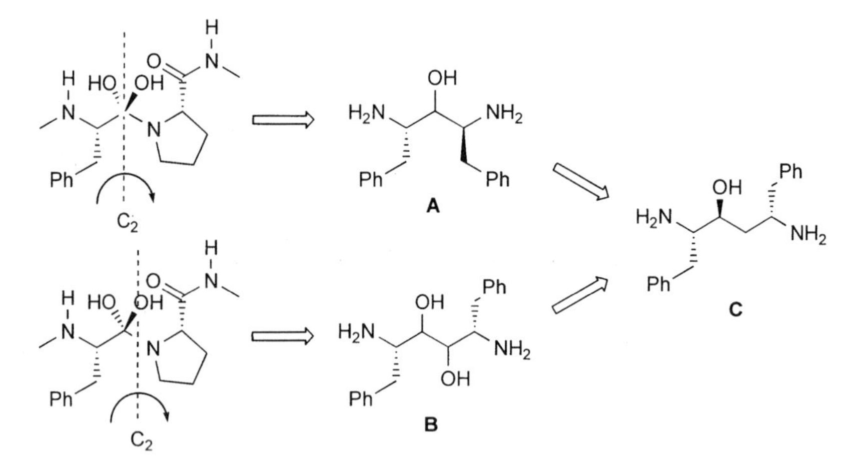

Figure 1 Design of symmetry-based HIV protease inhibitor core units. (From Ref. 4.)

but also allowed studies that brought a new understanding of the pathogenesis of HIV and the development of AIDS.

The discovery of both ritonavir and ABT-378 arose ultimately out of systematic studies of symmetry-based inhibitors of HIV protease, originally designed to match the C_2-symmetric nature of HIV protease as well as mimic its substrate sequence. Previous studies had demonstrated that dipeptide analogues containing a secondary hydroxyl group produced potent inhibitors when incorporated into the substrate sequence of the human aspartic protease renin (1). This hydroxyl group both accepted and donated a hydrogen bond to the two catalytic aspartates within the active site, and mimicked the transition state for cleavage of the scissile peptide bond. Similar approaches produced asymmetric substrate-based inhibitors of HIV protease (2,3). For the design of symmetric inhibitors, an additional three-step conceptual process (4) was used (Fig. 1). First, a putative axis of symmetry was defined either (1) through the tetrahedral carbon, or (2) through the middle of the carbon–nitrogen scissile bond of the tetrahedral intermediate for substrate cleavage. In the second step, the C-terminus of the substrate was deleted. Finally, duplication of the remaining N-terminus using a C_2 operation around the symmetry axis, retaining a secondary hydroxyl group(s), produced two unique symmetric or pseudosymmetric core units (see Fig. 1A and B). The same conceptual process beginning with a reduced peptide inhibitor of the fungal aspartic protease rhizopuspepsin docked into the structure of Rous sarcoma virus protease led to a similar design (5). Attachment of protected amino acids to the terminal amino groups of either the mono-ol **A** or diastereomeric diols **B** produced inhibitors of HIV protease (e.g., A-74704 and A-75925; Fig. 2) with lower nanomolar or subnanomolar potency against recombinant HIV protease in a biochemical assay and submicromolar activity against HIV in tissue culture (4). Both series of inhibitors were shown through X-ray crystallographic studies to bind in an overall symmetric fashion (5,6). Notably, each of the diastereomeric diols **B** adopted one of two unique conformations within the enzyme active site. In the first conformation (observed with *S,S*-diols), the axis of symmetry of the inhibitor and enzyme, were superimposed. In the second, a deviation from a perfectly symmetric binding mode by up to one-half bond length was observed (6). This latter, asymmetric binding mode (observed with *R,R*- and *R,S*-diols) as inhibitors derived from core **C** [vide infra] projected one hydroxyl group directly between the catalytic aspartate residues for optimal hydrogen-bonding. These structural studies not only provided insight into the detailed interactions between the inhibitors and the symmetric enzyme active site, but additionally served to validate the original design paradigm that led to the novel inhibitor structures.

The evolution of structural changes leading from initial lead compounds to ritonavir and ultimately to ABT-378 is summarized in Figure 2. The hydrophobic nature of the subsites of HIV protease dictated that initial inhibitors based on core diamines **A** and **B** were highly hydrophobic and insoluble in aqueous buffer. Examination of the crystal structure of inhibitor A-74704, derived from **A** (5), suggested that placement of polar heterocyclic groups at the symmetry-related S3 and S3′ subsites might be tolerated because those groups were solvent-accessible (Fig. 3). Accordingly, corresponding pyridyl-substituted inhibitors were sufficiently soluble for pharmacokinetic evaluation in animals (7). Interestingly, certain inhibitors based on core **A** exhibited significant (approximately 20%) oral bioavailability in rats, whereas analogous inhibitors based on **B** were uniformly poorly bioavailable. However, the potency of inhibitors derived from **B** were about tenfold more potent than those from **A**. Being unable at this point to optimize both antiviral activity and oral bioavailability simultaneously in the same molecule, a representative of series **B**, A-77003 (see Fig. 3), with sufficient (>100 μg/mL) aqueous solubility for intravenous formulation, was selected for clinical evaluation of ''proof of principle'' in spite of a limited pharmacokinetic profile (7,8).

A-77003 exhibited potent (EC_{50} 0.03–0.30 μM) activity against a variety of HIV strains in various lymphocyte cultures (7,9) and blocked proteolytic *gag* polyprotein processing and mature virion production from chronically HIV-infected cells (10,11). In general, the concentration of the inhibitors in series **A** or **B** required for in vitro antiviral activity was two to three orders of magnitude higher than that required to inhibit purified HIV protease in a fluorogenic (12) biochemical assay (K_i = 0.2 nM for A-77003). This difference may be due to the combination of several factors, including the optimized conditions used in

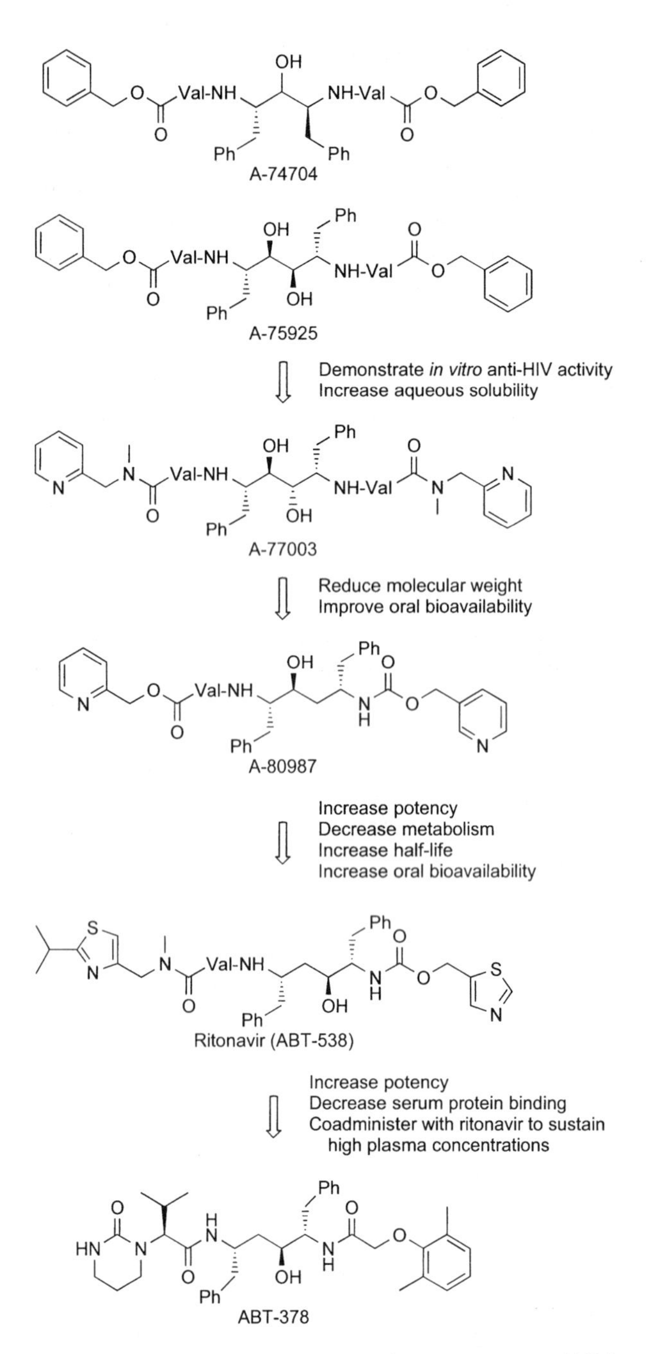

Figure 2 Evolution of structures of symmetry-based HIV protease inhibitors leading to the identification of ritonavir and ABT-378. Obstacles to development that were addressed by subsequent lead compounds are shown between respective structures.

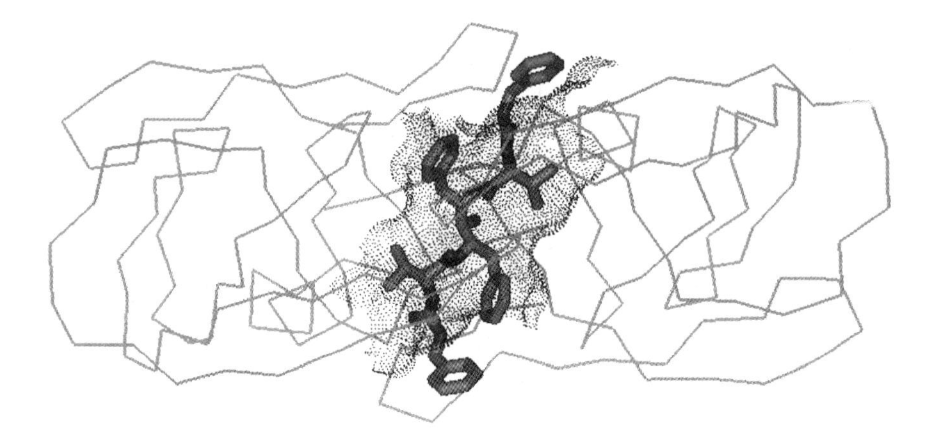

Figure 3 Crystal structure of the symmetry-based HIV protease inhibitor A-74704 in the active site of HIV-1 protease, looking down the axis of symmetry. For simplicity, only the α-trace of the enzyme is shown (gray lines). The shape of the active site is represented by the dotted surface. The solvent accessibility of the terminal phenyl groups of the inhibitor permitted substitution with polar, solubilizing functionality without loss of affinity.

the biochemical assay (pH 4.5 and 1 M NaCl) that are not representative of the environment within nascent virions, limited cellular penetration by the inhibitors, and binding to the 10% fetal bovine serum present in the tissue culture medium. Because of its poor oral bioavailability, A-77003 was administered to HIV-infected individuals by continuous intravenous infusion. Unfortunately, the biliary clearance of A-77003 in humans was extremely high (62 L/h), and plasma levels in excess of the 50% effective concentration EC_{50} were achieved only at infusion rates that produced severe phlebitis (13).

II. PHARMACODYNAMICS

Systematic structure–activity studies on the more potent series of inhibitors derived from diamine core unit **B** continued with the goal of identifying analogues with improved oral bioavailability. The incorporation of functionality that significantly increased aqueous solubility often led to inhibitors with lower antiviral potency in vitro, possibly owing to decreased cellular penetration. Furthermore, analogues with aqueous solubility less than 1 μg/mL rarely produced detectable plasma levels in rats after oral administration, even on formulation with organic cosolvents. Two structural changes within the series based on core diamine **B** significantly affected the pharmacokinetic properties of the resulting inhibitors, producing several compounds with more than 20% oral bioavailability in rats (14). First, removal of one of the symmetry-related P2 amino acid groups, with resulting attachment of the solubilizing heterocyclic group directly to one of the amino groups of **B**, reduced the molecular weight of the inhibitors by about 100 mass units. Second, removal of one of the two hydroxyl groups of **B** produced inhibitors based on core **C** (see Fig. 1), which incorporated the more potent carbon framework of core **B** with the single hydroxyl group (and corresponding superior pharmacokinetics) of core **A**. Inhibitors containing core **C** displayed significantly higher oral bioavailability (median 18%, $n = 14$) than otherwise analogous inhibitors containing core **B** (median 2.1%, $n = 7$). The inverse association of the presence of a hydrogen-bonding functional group with bioavailability observed in this series is consistent with the results of previous studies of the absorption of *N*-methylated peptides through epithelial cells (15,16). Removal of the hydroxyl group improved the affinity of the inhibitors toward HIV protease and, following optimization, overcame the loss of potency resulting from the reduction in molecular weight. Accordingly, the optimized inhibitor A-80987 displayed antiviral potency similar to that of A-77003 and more than 20% oral bioavailability in rats and dogs (14). Initial clinical studies with A-80987 produced plasma concentrations in humans that were higher than those in any animal. However, similar to A-77003, the hepatic clearance of A-80987 was rapid, and even with frequent dosing, plasma concentrations estimated to be highly suppressive of HIV replication could not be maintained.

The rapid clearance of both A-77003 and A-80987 prompted further structural changes designed to increase the circulating half-life of the inhibitors. In vitro and in vivo studies on the metabolism of A-77003 and A-80987 identified *N*-oxidation of the terminal pyridyl residues of each compound by hepatic cytochrome P450 as the major metabolic pathway (17,18). Consequently, efforts aimed at stabilizing those groups against oxidation were initiated. Substituents on the rings that either sterically encumbered or electronically modified the pyridyl nitrogen were unsuccessful in improving oral bioavailability. However, replacement of the pyridyl rings with heterocyclic groups (e.g., thiazole or oxazole) that were less electron-rich, but still sufficiently basic for adequate solubility in a low-pH formulation, produced a significant enhancement of pharmacokinetic properties (18,19). Substitution of one pyridyl ring with a thiazolyl group slowed metabolism in vitro by fivefold. Similar substitution of the second pyridyl group further slowed metabolism by an additional fourfold. Consequently, the oral bioavailability in rats increased from 26 (A-80987) to 78% (ritonavir). More importantly, plasma concentrations of ritonavir were sustained over its respective EC_{50} for inhibition of HIV for more than 8–12 h in three species. The foregoing structural changes also allowed the simultaneous optimization of potency by attachment of an isopropyl group to the P3-thiazolyl ring (18,19). This group, which improved

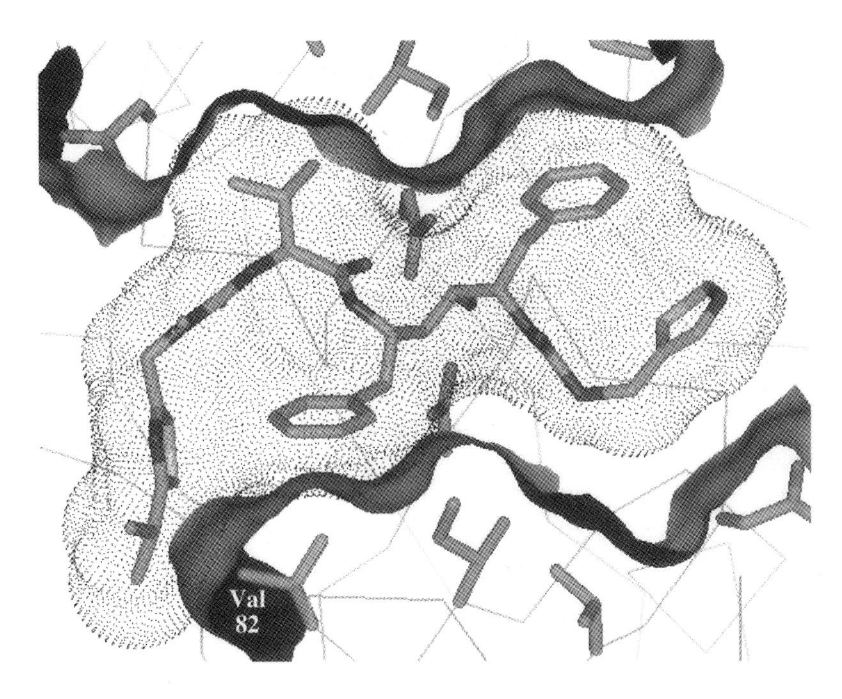

Figure 4 Crystal structure of ritonavir in the active site of HIV-1 protease. The inhibitor surface is shown in dots, and the active site surface is shown in solid gray. The hydrophobic interaction between the (isopropyl)thiazolyl group of ritonavir and the side chain of valine-82 of the protease both enhanced inhibitor potency and determined the initial mutation pattern leading to viral resistance.

the antiviral activity of ritonavir by tenfold over A-80987, was shown by crystallographic studies to undergo a hydrophobic interaction with the side chain of the valine-82 residue within the HIV protease active site (Fig. 4). Importantly, this optimization also ultimately determined the pattern of mutations leading to resistance to ritonavir (vide infra).

III. ANTIVIRAL POTENCY

The antiviral potency of ritonavir against a battery of laboratory strains and clinical isolates ranged from 0.022–0.13 μM. However, in a manner similar to many protease inhibitors, ritonavir was determined to be highly (>98%) bound to human serum proteins, potentially reducing the fraction available for antiviral activity. To more accurately predict the in vivo potency of ritonavir, the in vitro anti-

HIV assay was adapted to tolerate the addition of 50% human serum (20). Under these conditions, a 20-fold attenuation in the anti-HIV activity of ritonavir was observed, thus dictating that the plasma concentrations of ritonavir required for efficacy in vivo be substantially higher than had been previously assumed. Fortunately, 500 or 600-mg–bid doses of ritonavir in humans produced steady-state plasma concentrations consistently in excess of the EC_{50} in the presence of human serum (21). Initial treatment of HIV-infected individuals with ritonavir in two dose-ranging phase 2 studies produced a rapid and profound decline in plasma viral RNA (22,23). The rapid rate of decline in this and subsequent studies first allowed calculation of the rapid dynamics of HIV turnover in plasma (24–26). In a phase 3 study, ritonavir, added to existing antiretroviral therapy, produced a highly significant reduction in mortality and disease progression in patients with advanced HIV infection (27). In healthier patients, triple therapy with ritonavir and two reverse transcriptase inhibitors resulted in a sustained reduction of plasma viral RNA to unquantifiable levels both in the plasma and lymph (28,29), and in the substantial reconstitution of immune function (30–32).

IV. VIRAL RESISTANCE

The inherently high-mutation rate of HIV during reverse transcription results in the rapid emergence of resistance to antiretroviral agents if replication is not completely suppressed. To initially define the development of resistance to ritonavir, HIV was serially passaged with increasing concentrations of drug to produce—after 22 passages—viral strains that primarily contained mutations at amino acids 46, 63, 71, 82, and 84 in the HIV protease gene (33). The passage 22 virus exhibited a 10- to 25-fold decrease in drug susceptibility compared with the parent virus. In vivo, resistance to ritonavir was characterized by systematic analysis of plasma viral HIV protease sequences following rebound in viral load during ritonavir monotherapy (34). Mutations at a total of nine residues (positions 82, 54, 36, 71, 46, 20, 84, 33, and 90) were documented as being selected by drug therapy. In general, a consistent pattern was observed, with a primary (active site) mutation at Val82 emerging first, followed by secondary (nonactive site) mutations at positions 54, 71, or 36. Infectious HIV molecular clones containing single mutations remained highly susceptible to ritonavir, and multiple mutations were required for high level resistance. The in vivo selection rate (number of mutations/time on therapy) for 13 patients receiving ritonavir monotherapy was inversely correlated with the trough concentrations of the drug maintained during therapy (Fig. 5). These results supported the conclusion that drug resistance arises from residual low-level viral replication that is incompletely suppressed by drug therapy (34). This conclusion was further substantiated by studies showing that the duration of viral suppression before the emergence of resistance correlated

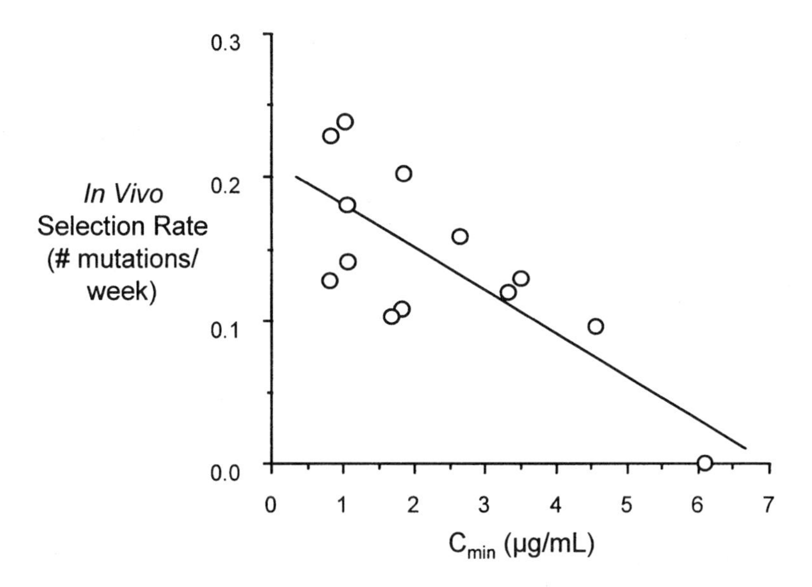

Figure 5 Relation of the in vivo selection rate (number of resistance mutations per week on therapy) and steady-state trough plasma drug concentrations for patients receiving monotherapy with ritonavir. (From Ref. 34.)

with the nadir in the RNA curve, following the initial decline from baseline, in patients receiving monotherapy or combination therapy with ritonavir (35). Taken together, these findings indicate that resistance is best avoided by initial treatment with combination regimens that maximally suppress viral replication, both in the plasma and in other compartments where replication occurs.

Further in vitro studies determined that the favorable pharmacokinetic profile of ritonavir was a consequence of its potent inhibition of the cytochrome P450 3A4 (C3A4) hepatic isozyme, thereby slowing its own metabolism (36). Because the metabolism of other peptidomimetic HIV protease inhibitors is also mediated primarily by C3A4, the in vitro and in vivo interaction of ritonavir and other inhibitors was examined to determine whether the plasma levels of those inhibitors could be enhanced by coadministration with ritonavir. In rat and human liver microsomes, the presence of ritonavir potently inhibited the metabolism of saquinavir, indinavir, nelfinavir, and amprenavir (37). Furthermore, in rats, codosing with ritonavir enhanced the 0- to 8-h–plasma exposure of each of the other protease inhibitors by 8- to 36-fold (Fig. 6). Importantly, the reduction in hepatic clearance resulted in the maintenance of plasma levels of each inhibitor in excess of its EC_{50} in the presence of human serum for more than 8 h. The remarkable pharmacokinetic enhancement by ritonavir coadministration was con-

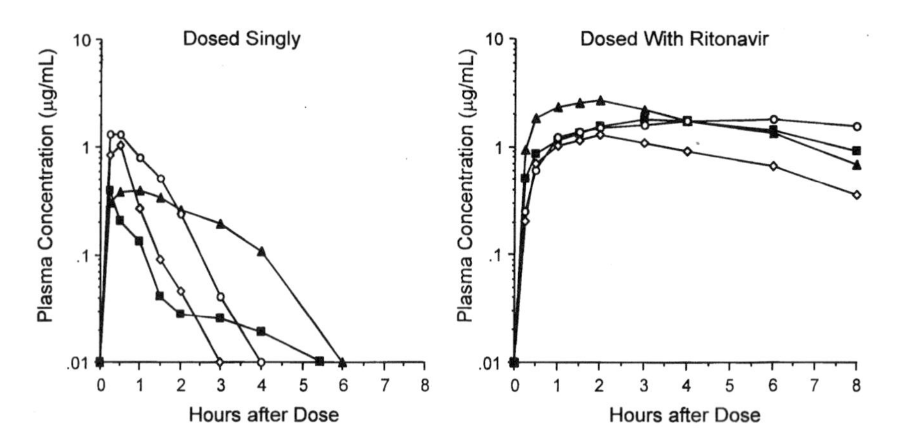

Figure 6 Enhancement of the plasma levels of protease inhibitors in rats by coadministration with ritonavir. HIV protease inhibitors were dosed orally in rats either singly (10 mg/kg) or in combination with 10 mg/kg of ritonavir: closed squares, saquinavir; open diamonds, indinavir; closed triangles, nelfinavir; open circles, amprenavir. (From Ref. 37.)

firmed in human studies (38,39), and forms the basis for powerful therapeutic regimens using dual protease inhibitors (40–42).

V. ABT-378 STUDIES

Following the clinical development of ritonavir, the search for a second-generation protease inhibitor with improved properties continued, with the ultimate identification of ABT-378 (43). Several key insights derived from the ritonavir development program enabled the discovery of ABT-378 (Fig. 2). The first was the recognition of the significant attenuation of the in vitro antiviral potency of many protease inhibitors, including ritonavir, as a result of high binding to human serum proteins in vivo. Consequently, in the search for ABT-378, the in vitro screening anti-HIV assay was modified to routinely assess the effects of 50% human serum on the EC_{50} of new protease inhibitors. The attenuation of the in vitro activity of ABT-378 by human serum (about sixfold) was substantially less than that observed for ritonavir (20-fold or greater) (20). This factor, combined with the greater intrinsic affinity of ABT-378 for HIV protease, resulted in a tenfold increase in the antiviral potency of ABT-378, in the presence of human serum, against wild type HIV compared with ritonavir (43).

A second key insight leading to the discovery of ABT-378 was the recognition of the importance of maintaining high trough plasma levels between doses to impede or even completely block the emergence of resistance (34) and the ability of ritonavir coadministration to increase those levels (37). The correlation of trough plasma levels of protease inhibitors with virological response (44,45) support a pharmacodynamic model wherein the in vivo potency of protease inhibitors is estimated by the ratio of their mean trough concentrations in patients to their respective EC_{50} in the presence of 50% human serum (46). The relevance of the relation between plasma drug concentrations and required inhibitory concentrations is recognized for other anti-infective agents, and is referred to as the inhibitory quotient (47). This ratio is relatively modest (fourfold or less) for ritonavir and other protease inhibitors (46), and it was recognized that to display more profound suppression of viral replication in vivo, an optimal second-generation protease inhibitor should achieve a trough/EC_{50} ratio significantly higher than observed for existing agents. To circumvent the difficulty in identifying new protease inhibitors with sufficient pharmacokinetic properties, as outlined earlier, new analogues were dosed both alone in animal models, as well as with ritonavir. Ultimately, ABT-378 was selected for development because of the combination of high antiviral potency in the presence of human serum and exquisite sensitivity to pharmacokinetic enhancement by ritonavir (43), and is being developed as a coformulated product (ABT-378–ritonavir) at dose of 400:100 mg twice daily. The plasma concentrations of ABT-378 in HIV-infected subjects produced by this combination are shown in Figure 7. Mean trough levels of ABT-378 were 75-fold or more above its serum-adjusted EC_{50} for wild-type HIV (48). Conse-

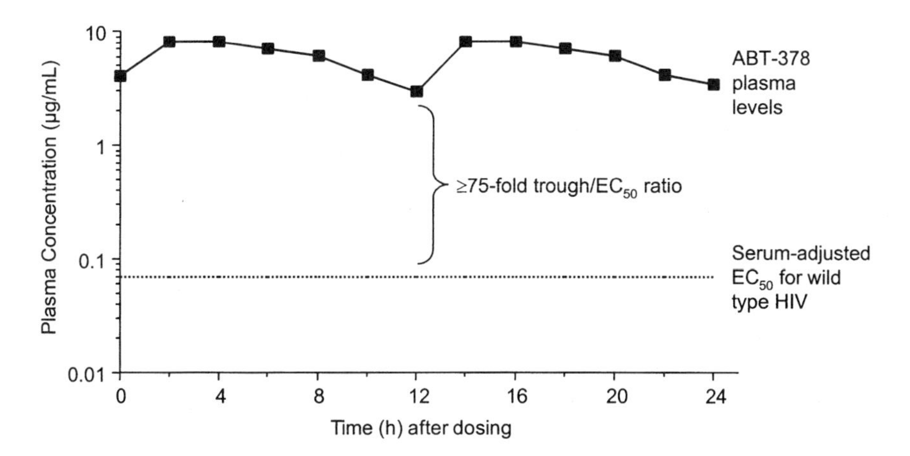

Figure 7 Plasma levels of ABT-378 in HIV-infected subjects after dosing ABT-378/r (400/100 mg) twice daily. (From Ref. 48.)

quently, in phase 2 studies, combination regimens containing ABT-378–ritonavir plus reverse transcriptase inhibitors have produced a decline in plasma HIV RNA to unquantifiable levels in a vast majority of treatment-naïve and single protease inhibitor-experienced subjects (49,50).

VI. FUTURE DRUG DESIGN CHALLENGES

The discovery and development of ritonavir and ABT-378, as described in the foregoing, illustrates the unique interdisciplinary approach required for modern drug discovery. The time period between the initial synthesis of symmetric inhibitors to the discovery of ritonavir covered 3 years. The subsequent development of ritonavir to licensure by the U.S. Food and Drug Administration (FDA) was accomplished in 4 years. ABT-378 was synthesized in 1995 and a New Drug Application (NDA) filed in June 2000. The studies described here leading to and utilizing these protease inhibitors illustrate the diverse scientific contribution of this developmental program to disciplines such as peptidomimetic chemistry, structure-based drug design, drug metabolism, HIV dynamics and resistance, and clinical pharmacology. More importantly, since their introduction, the widespread use of protease inhibitors has been associated with a significant decline in the death toll from AIDS and an improvement in the quality of life of people living with HIV infection.

ACKNOWLEDGMENTS

The assistance of Drs. Kent Stewart and Rick Bertz (Abbott Laboratories) in the preparation of Figures 3 and 4 and Figure 7, respectively, is gratefully acknowledged. The assistance of Dr. Eugene Sun in reviewing this manuscript is also acknowledged.

REFERENCES

1. Greenlee WJ. Renin inhibitors. Med Res Rev 1990; 10:173.
2. Dreyer GB, Metcalf BW, Tomaszek TA Jr, Carr TJ, Chandler AC III, Hyland L, Fakhoury SA, Magaard VW, Moore ML, Strickler JE, Debouck C, Meek TD. Inhibition of human immunodeficiency virus 1 protease in vitro: rational design of substrate analogue inhibitors. Proc Natl Acad Sci USA 1989; 86:9752–9756.
3. Vacca JP. Design of tight-binding human immunodeficiency virus type 1 protease inhibitors. Methods Enzymol 1994; 241:311–334.
4. Kempf DJ, Norbeck DW, Codacovi L, Wang XC, Kohlbrenner WE, Wideburg NE,

Paul DA, Knigge MF, Vasavanonda S, Craig–Kennard A, Saldivar A, Rosenbrook W Jr, Clement JJ, Plattner JJ, Erickson J. Structure-based, C_2 symmetric inhibitors of HIV protease. J Med Chem 1990; 33:2687–2689.

5. Erickson J, Neidhart DJ, Van Drie J, Kempf DJ, Wang XC, Norbeck DW, Plattner JJ, Rittenhouse JW, Turon M, Wideburg N, Kohlbrenner WE, Simmer R, Helfrich R, Paul DA, Knigge M. Design, activity, and 2.8 Å crystal structure of a C_2 symmetric inhibitor complexed to HIV-1 protease. Science 1990; 249:527–533.
6. Hosur MV, Bhat TN, Kempf DJ, Baldwin ET, Liu B, Gulnik S, Wideburg NE, Norbeck DW, Appelt K, Erickson JW. Influence of stereochemistry on activity and binding modes for C_2 symmetry-based inhibitors of HIV-1 protease. J Am Chem Soc 1994; 116:847–855.
7. Kempf DJ, Marsh KC, Paul DA, Knigge MF, Norbeck DW, Kohlbrenner WE, Codacovi L, Vasavanonda S, Bryant P, Wang XC, Wideburg NE, Clement JJ, Plattner JJ, Erickson J. Antiviral and pharmacokinetic properties of C_2-symmetric inhibitors of the human immunodeficiency virus type 1 protease. Antimicrob Agents Chemother 1991; 35:2209–2214.
8. Kempf DJ, Codacovi L, Wang XC, Kohlbrenner WE, Wideburg NE, Saldivar A, Vasavanonda S, Marsh KC, Bryant P, Sham HL, Green BE, Betebenner DA, Erickson J, Norbeck DW. Symmetry-based inhibitors of HIV protease. Structure–activity studies of acylated 2,4-diamino-1,5-diphenyl-3-hydroxypentane and 2,5-diamino-1,6-diphenylhexane-3,4-diol. J Med Chem 1993; 36:320–330.
9. Kageyama S, Weinstein J, Shirasaka T, Kempf DJ, Norbeck DW, Plattner JJ, Erickson J, Mitsuya H. In vitro inhibition of human immunodeficiency virus (HIV) type 1 replication by C_2 symmetry-based HIV protease inhibitors as single agents or in combinations. Antimicrob Agents Chemother 1992; 36:926–933.
10. Kaplan AH, Zack JA, Knigge M, Paul DA, Kempf DJ, Norbeck DW, Swanstrom R. Partial inhibition of the human immunodeficiency virus, type 1 protease results in aberrant virus assembly and formation of noninfectious particles. J Virol 1993; 67:4050–4055.
11. Kageyama S, Hoekzema DT, Murakawa Y, Kojima E, Shirasaka T, Kempf DJ, Norbeck DW, Erickson J, Mitsuya H. A C_2–symmetry-based HIV protease inhibitor, A77003, irreversibly inhibits infectivity of HIV-1 in vitro. AIDS Res Hum Retroviruses 1994; 10:735–743.
12. Matayoshi ED, Wang GT, Krafft GA, Erickson J. Novel fluorogenic substrates for assaying retroviral proteases by resonance energy transfer. Science 1990; 247:954–958.
13. Reedijk M, Boucher CAB, Vanbommel T, Ho DD, Tzeng TB, Sereni D, Veyssier P, Jurriaans S, Granneman R, Hsu A, Leonard JM, Danner SA. Safety, pharmacokinetics, and antiviral activity of A77003, a C-2 symmetry-based human immunodeficiency virus protease inhibitor. Antimicrob Agents Chemother 1995; 39:1559–1564.
14. Kempf DJ, Marsh KC, Fino LC, Bryant P, Craig–Kennard A, Sham HL, Zhao C, Vasavanonda S, Kohlbrenner WE, Wideburg NE, Saldivar A, Green BE, Herrin T, Norbeck DW. Design of orally bioavailable, symmetry-based inhibitors of HIV protease. Bioorg Med Chem 1994; 2:847–858.
15. Conradi RA, Hilgers AR, Burton PS, Hester JB. Epithelial cell permeability of a

series of peptidic HIV protease inhibitors: aminoterminal substituent effects. J Drug Target 1994; 2:167–171.
16. Karls MS, Rush BD, Wilkinson KF, Vidmar TJ, Burton PS, Ruwart MJ. Desolvation energy: a major determination of absorption, but not clearance, of peptides in rats. Pharm Res 1991; 8:1477–1481.
17. Denissen J, Marsh K, Grabowski B, Johnson M. Metabolism and disposition of the HIV protease inhibitor A-77003. Pharm Res 1993; 10:S-376.
18. Kempf D, Marsh KC, Denissen JF, McDonald E, Vasavanonda S, Flentge CA, Green BE, Fino L, Park CH, Kong X–P, Wideburg NE, Saldivar A, Ruiz L, Kati WM, Sham HL, Robins T, Stewart KD, Hsu A, Plattner JJ, Leonard JM, Norbeck DW. ABT-538 is a potent inhibitor of human immunodeficiency virus protease and has high oral bioavailability in humans. Proc Natl Acad Sci USA 1995; 92:2484–2488.
19. Kempf DJ, Sham HL, Marsh KC, Flentge CA, Betebenner D, Green BE, McDonald E, Vasavanonda S, Saldivar A, Wideburg NE, Kati WM, Ruiz L, Zhao C, Fino L, Patterson J, Molla A, Plattner JJ, Norbeck DW. Discovery of ritonavir, a potent inhibitor of HIV protease with high oral bioavailability and clinical efficacy. J Med Chem 1998; 41:602–617.
20. Molla A, Vasavanonda S, Kumar G, Sham H, Johnson M, Grabowski B, Denissen JF, Kohlbrenner W, Plattner JJ, Norbeck DW, Leonard JM, Kempf D. Human serum attenuates the activity of protease inhibitors toward wild-type and mutant human immunodeficiency virus. Virology 1998; 250:255–262.
21. Hsu A, Granneman GR, Witt G, Locke C, Denissen J, Molla A, Valdes J, Smith J, Erdman K, Lyons N, Niu P, Decourt J–P, Fourtillan J–B, Leonard JM. Multiple-dose pharmacokinetics of ritonavir in human immunodeficiency virus-infected subjects. Antimicrob Agents Chemother 1997; 41:898–905.
22. Danner SA, Carr A, Leonard JM, Lehman LM, Gudiol F, Gonzales J, Raventos A, Rubio R, Bouza E, Pintado V, Aguado AG, Delomas JG, Delgado R, Borleffs JCC, Hsu A, Valdes JM, Boucher CAB, Cooper DA, Gimeno C, Clotet B, Tor J, Ferrer E, Martinez PL, Moreno S, Zancada G, Alcami J, Noriega AR, Pulido F, Glassman HN. A short-term study of the safety, pharmacokinetics, and efficacy of ritonavir, an inhibitor of HIV-1 protease. N Engl J Med 1995; 333:1528–1533.
23. Markowitz M, Saag M, Powderly WG, Hurley AM, Hsu A, Valdes JM, Henry D, Sattler F, Lamarca A, Leonard JM, Ho DD. A preliminary study of ritonavir, an inhibitor of HIV-1 protease, to treat HIV-1 infection. N Eng J Med 1995; 333:1534–1539.
24. Ho DD, Neumann AU, Perelson AS, Chen W, Leonard JM, Markowitz M. Rapid turnover of plasma virions and CD4 lymphocytes in HIV-1 infection. Nature 1995; 373:123–126.
25. Wei X, Ghosh SK, Taylor ME, Johnson VA, Emini EA, Deutsch P, Lifson JD, Bonhoeffer S, Nowak MA, Hahn BH, Saag MS, Shaw GM. Viral dynamics in human immunodeficiency virus type 1 infection. Nature 1995; 373:117–122.
26. Perelson AS, Neumann AU, Markowitz M, Leonard JM, Ho DD. HIV-1 dynamics in vivo: virion clearance rate, infected cell lifespan, and viral generation time. Science 1996; 271:1582–1586.
27. Cameron DW, Heath–Chiozzi M, Danner S, Cohen C, Krabcik S, Maurath C, Sun E, Henry D, Rode R, Potoff A, Leonard J. Randomised placebo-controlled trial of ritonavir in advanced HIV-1 disease. Lancet 1998; 351:543–549.

28. Cavert W, Notermans DW, Staskus K, Wietgrefe SW, Zupancic M, Gebhard K, Henry K, Zhang Z–Q, Mills R, McDade H, Goudsmit J, Danner SA, Haase AT. Kinetics of response in lymphoid tissues to antiretroviral therapy of HIV-1 infection. Science 1997; 276:960–964.
29. Notermans DW, Jurriaans S, deWolf F, Foudraine NA, deJong JJ, Cavert W, Schuwirth CM, Kauffmann RH, Meenhorst PL, McDade H, Goodwin C, Leonard JM, Goudsmit J, Danner DA. Decrease of HIV-1 RNA levels in lymphoid tissue and peripheral blood during treatment with ritonavir, lamivudine and zidovudine. AIDS 1998; 12:167–173.
30. Autran B, Carcelain G, Li TS, Blanc C, Mathez D, Tubiana R, Katlama C, Debre P, Leibowitch J. Positive effects of combined antiretrovial therapy on $CD4^+$ T cell homeostasis and function in advanced HIV disease. Science 1997; 277:112–116.
31. Pakker NG, Notermans DW, de Boer RJ, Roos MTL, de Wolf F, Hill A, Leonard JM, Danner SA, Miedema F, Schellekens PTA. Biphasic kinetics of peripheral blood T cells after triple combination therapy in HIV-1 infection: a composite of redistribution and proliferation. Nat Med 1998; 4:208–214.
32. Gorochov G, Neumann AU, Kereveur A, Parizot C, Li T, Katlama C, Karmochkine M, Raguin G, Autran B, Debre P. Perturbation of $CD4^+$ and $CD8^+$ T-cell repertoires during progression to AIDS and regulation of the $CD4^+$ repertoire during antiviral therapy. Nat Med 1998; 4:215–221.
33. Markowitz M, Mo H, Kempf DJ, Norbeck DW, Bhat TN, Erickson JW, Ho DD. Selection and analysis of human immunodeficiency virus type 1 variants with increased resistance to ABT-538, a novel protease inhibitor. J Virol 1995; 69:701.
34. Molla A, Korneyeva M, Gao Q, Vasavanonda S, Schipper PJ, Mo H–M, Markowitz M, Chernyavskiy T, Niu P, Lyons N, Hsu A, Granneman GR, Ho D, Boucher CAB, Leonard JM, Norbeck DW, Kempf DJ. Ordered accumulation of mutations in HIV protease confers resistance to ritonavir. Nat Med 1996; 2:760–766.
35. Kempf DJ, Rode RA, Xu Y, Sun E, Heath-Chiozzi ME, Valdes J, Japour AJ, Danner S, Boucher C, Molla A, Leonard JM. The duration of viral suppression during protease inhibitor therapy for HIV-1 infection is predicted by plasma HIV-1 RNA at the nadir. AIDS 1998; 12:F9–F14.
36. Kumar GN, Rodrigues AD, Buko AM, Denissen JF. Cytochrome P450-mediated metabolism of the HIV-1 protease inhibitor ritonavir (ABT-538) in human liver microsomes. J Pharmacol Exp Ther 1996; 277:423–431.
37. Kempf DJ, Marsh KC, Kumar G, Rodrigues AD, Denissen JF, McDonald E, Kukulka MJ, Hsu A, Granneman GR, Baroldi PA, Sun E, Pizzuti D, Plattner JJ, Norbeck DW, Leonard JM. Pharmacokinetic enhancement of inhibitors of the human immunodeficiency virus protease by coadministration with ritonavir. Antimicrob Agents Chemother 1997; 41:654–660.
38. Hsu A, Granneman GR, Cao G, Carothers L, El-Shourbagy T, Baroldi P, Erdman K, Brown F, Sun E, Leonard JM. Pharmacokinetic interactions between two HIV-protease inhibitors, ritonavir and saquinavir. Clin Pharmacol Ther 1998; 63:453–464.
39. Hsu A, Granneman GR, Cao G, Carothers L, Japour A, El–Shourbagy T, Dennis S, Berg J, Erdman K, Leonard JM, Sun E. Pharmacokinetic interaction between ritonavir and indinavir in healthy volunteers. Antimicrob Agents Chemother 1998; 42:2784–2791.

40. Cameron DW, Japour AJ, Xu Y, Hsu A, Cohen C, Farthing C, Follansbee S, Markowitz M, Mellors J, Poretz D, Angel J, Ho D, McMahon D, Devanarayan V, Rode R, Salgo M, Kempf DJ, Granneman R, Leonard JM, Sun E. Ritonavir and saquinavir combination therapy for the treatment of HIV infection. AIDS 1998; 13:213–224.
41. Kirk O, Katzenstein T, Gerstoft J, Mathiesen L, Nielsen H, Pedersen C, Lundgren JD. Combination therapy containing ritonavir plus saquinavir has superior short-term antiretroviral efficacy: a randomized trial. AIDS 1999; 13:F9–F16.
42. Rockstroh JK, Bergmann F, Wiesel W, Rieke A, Nadler M, Knechten H, the German Ritonavir/Indinavir Study Group. Efficacy and safety of BID firstline ritonavir/indinavir plus double nucleoside combination therapy in HIV-infected individuals. 6th Conference on Retroviruses and Opportunistic Infections, Chicago, IL, Jan 31–Feb 4, 1999, abstr 631.
43. Sham HL, Kempf DJ, Molla A, Marsh KC, Kumar GN, Chen C–M, Kati W, Stewart K, Lal R, Hsu A, Betebenner D, Korneyeva M, Vasavanonda S, McDonald E, Saldivar A, Wideburg N, Chen X, Niu P, Park C, Jayanti V, Grabowski B, Granneman GR, Sun E, Japour AJ, Leonard JM, Plattner JJ, Norbeck DW. ABT-378, a highly potent inhibitor of the human immunodeficiency virus protease. Antimicrob Agents Chemother 1998; 42:3218–3224.
44. Acosta EP, Havlir DV, Richman DD, Zhou XJ, Hirsch M, Collier AC, Tebas P, Sommadossi J–P. Pharmacodynamics (PD) of indinavir (IDV) in protease-naïve HIV-infected patients receiving ZDV and 3TC. 7th Conference on Retroviruses and Opportunistic Infections, San Francisco, CA, Jan 30–Feb 2, 2000, abstr 455.
45. Garraffo R, Durant J, Clevenbergh P, Icard S, Shapiro JM, Dellamonica P. Relevance of protease inhibitor plasma levels in patients treated with genotypic adapted therapy: pharmacological data from the Viradapt study. 3rd International Workshop on HIV Drug Resistance and Treatment Strategies, San Diego, CA, June 23–26, 1999, abstr 109.
46. Kempf DJ, Molla A, Hsu A. Protease inhibitors as anti-HIV agents. In: DeClerq E, ed. Antiretroviral Therapy. American Society for Microbiology Press, in press.
47. Ellner PD, Neu HC. JAMA 1981; 246:1575.
48. Bertz R, Lam W, Brun S, Kumar G, Fields C, Orth K, Jennings J, Hsu A, Granneman GR, Japour A, Sun E. Multiple-dose pharmacokinetics (PK) of ABT-378/ritonavir (ABT-378/r) in HIV+ subjects. 39th Interscience Conference on Antimicrobial Agents and Chemotherapy, San Francisco, CA, Sept 26–29, 1999; abstr 327.
49. Gulick R, King M, Brun S, Real K, Murphy R, Hicks C, Eron J, Thommes J, Thompson M, White C, Benson C, Albrecht M, Kessler H, Hsu A, Bertz R, Kempf D, Sun E, Japour A. ABT-378/ritonavir (ABT-378/r) in antiretroviral naive HIV+ patients: 72 weeks. 7th Conference on Retroviruses and Opportunistic Infections, San Francisco, CA, Jan 30–Feb 2, 2000; abstr 515.
50. Deeks S, Brun, S, Xu Y, Real K, Benson C, Kessler H, Murphy R, Wheeler D, Hicks C, Eron J, Feinberg J, Gulick R, Sax P, Stryker R, Riddler S, Thompson M, King M, Potthoff A, Hsu A, Bertz R, Molla A, Mo H, Kempf D, Japour A, Sun E. ABT-378/ritonavir (ABT-378/r) suppresses HIV RNA to <400 copies/mL in 84% of PI-experienced patients at 48 weeks. 7th Conference on Retroviruses and Opportunistic Infections, San Francisco, CA, Jan 30–Feb 2, 2000; abstr 532.

4

Discovery and Early Development of Indinavir

Bruce D. Dorsey and Joseph P. Vacca
Merck Research Laboratories, West Point, Pennsylvania

On December 21, 1988, over the skies of Lockerbie, Scotland, a terrorist bomb blew apart the airliner Pan Am 103, killing everyone on board. The tragic death of 259 passengers and crew affected the lives of families, friends, and colleagues from across the world. One of those who died was a scientist from the Merck Research Laboratories, Dr. Irving Sigal. Sigal, a senior director in the Department of Molecular Biology, was the leading member of the Merck team of HIV researchers. This chapter is dedicated to Irving Sigal's vision and drive to discover and develop something of value to society.

I. INTRODUCTION

The importance of the HIV protease as a target for antiviral therapy has been known for the past 13 years. Currently, there are five HIV protease inhibitors approved in the United States for the treatment of acquired immunodeficiency syndrome (AIDS). All of these compounds, as detailed in the chapters of this book, have resulted from intense research efforts and represent an important step forward in the understanding and treatment of AIDS. The following is a detailed description of the Merck research efforts that resulted in indinavir (Crixivan), a potent and orally bioavailable HIV protease inhibitor.

A. HIV Protease

The human immunodeficiency virus (HIV), the etiologic agent of AIDS, has initiated an urgent pursuit to comprehend and control this disease. The virally encoded homodimeric aspartyl protease (1), which is responsible for processing the *gag* and *gag–pol* gene products that allow for the organization of core structural proteins and release of viral enzymes, appeared to be one such target. Kramer (2) first proposed that an inhibitor of HIV protease represented a viable strategy for the development of antiviral agents. To substantiate Kramer's hypothesis, Sigal and colleagues, designed several experiments to define the role that HIV-1 protease played in viral replication (3). Site-directed mutagenesis of the critical aspartic acid residue 25 in wild-type HIV-1 protease to an asparagine residue resulted in an inactive enzyme that was unable to cleave its natural substrates. Also, HIV-1 proviral DNA, which incorporated the single-point mutation Asp35 to Asn25, was transfected into nonlymphoid SW480 cells. Viral particles produced as a result of this transfection were unable to infect MT-4 lymphoid cells. These results confirm that HIV-1 protease is an essential enzyme and suggest that inactivation of the protease by a small molecule inhibitor should render the virus noninfectious. Later, this hypothesis was confirmed with inhibitors of HIV-1 protease in cell culture. Much later, the supposition that it would be beneficial to treat humans infected with the HIV virus with a HIV-1 protease inhibitor would be established in clinical trials. However, in the mid-1980s, this ''proof-of-concept'' experiment provided the necessary momentum for Merck to commit the resources needed to pursue HIV-1 protease as a viable target for drug discovery.

B. Crystallography

Concurrent with Merck's efforts to establish the critical role of HIV protease in the viral life cycle, the Department of Biophysical Chemistry set out to determine the crystal structure of native HIV-1 protease. This success resulted in conformation of the homodimeric architecture of the enzyme and its mechanistic class (4). Subsequently, several crystal structures of enzyme–inhibitor complexes were reported allowing, for the first time, a glimpse at the important lipophilic and hydrophilic interactions of the enzyme–inhibitor complex (5). The ability to determine the structures of enzyme–inhibitor complexes would prove invaluable in the design of potent inhibitors. This would allow the drug discovery effort the opportunity to design inhibitors, evaluate them in assays, determine how they bind to the enzyme with atom level analysis, and then optimize the binding interactions of the inhibitors.

II. INHIBITOR IDENTIFICATION AND DESIGN OF INDINAVIR

A. Lead Generation

There are several general strategies to identify a lead structure for modification and evaluation in vitro and in vivo assays. The classic approaches are (1) random

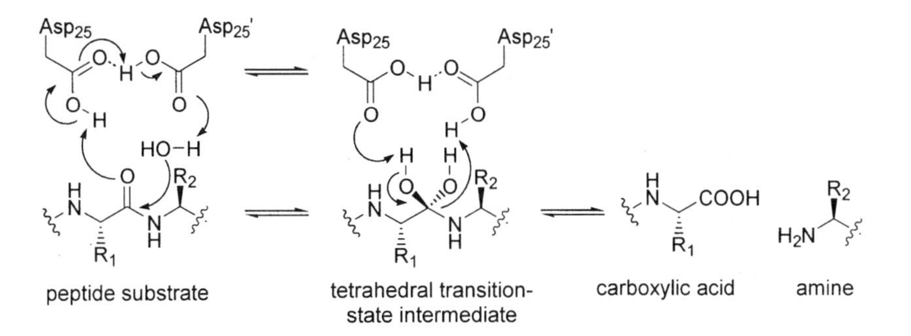

Figure 1 Cleavage mechanism of an aspartyl acid protease.

screening of sample collections or combinatorial libraries, (2) rational screening of inhibitors of a related enzyme target, (3) modifying inhibitors disclosed in the literature, (4) substrate-based designed inhibitors, and (5) de novo design, utilizing structural knowledge of the target derived from crystal structure or nuclear magnetic resonance (NMR) analysis. All of these strategies have been successfully exploited in the design of HIV protease inhibitors.

When HIV protease was first identified as a potential target for antiviral therapy, there were no known inhibitors disclosed in the literature. However, the HIV protease enzyme is closely related to a mammalian aspartic acid protease, renin, involved with the regulation of blood pressure. Incorporation of a transition-state mimic into substrate analogues has proved highly successful in producing potent renin inhibitors (6).

Figure 1 depicts a simplified illustration of how an aspartic acid protease cleaves a protein substrate. At the core of the protease enzyme reside two aspartic acid residues that are critical in stabilizing the addition of water across an amide bond and generating a tetrahedral transition-state intermediate. Once this intermediate is formed, it can break down to generate the C-terminal carboxylic acid and N-terminal amine, resulting in cleavage of the polypeptide substrate. Inhibitors that can mimic the tetrahedral transition-state intermediate should bind tightly to the enzyme and prevent substrate hydrolysis. Numerous transition-state peptidomimetics of the aspartic acid protease renin have been designed based on this concept. Several hydrolytically stable, dipeptide isostere replacements for the scissile bond in aspartyl protease substrates are illustrated in Figure 2. Although this concept has provided potent renin inhibitors, there are no renin inhibitors marketed as drugs for the treatment of high blood pressure. In general, these peptidomimetic compounds retain a substantial amount of peptide character, and as a result, they possess poor aqueous solubility and inadequate oral bioavailabil-

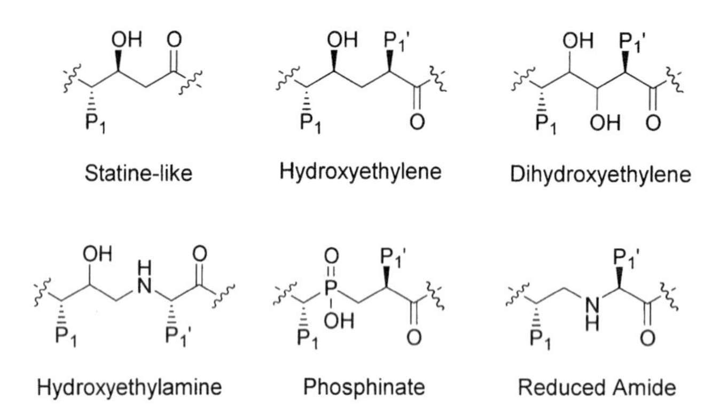

Figure 2 Transition-state mimics employed in aspartyl protease inhibition.

ity in animal models. These difficulties have limited their usefulness as therapeutic agents, and these issues would become critical in the pursuit of HIV protease inhibitors.

The initial approach that Merck researchers used to identify a lead structure involved the screening of compounds that had previously been synthesized as renin inhibitors (7,8). First, to evaluate the HIV protease potency of the compounds in our renin library, a peptide cleavage assay was developed. The assay determines the concentration of compound needed to inhibit by 50% the cleavage of substrate (Val-Ser-Gln-Asn-β-naphthyl-alanine*Pro-Ile-Val) by HIV-1 protease. The identification of the hexapeptide analogue **1** (IC_{50} [HIVP] = 1 nM), a potent lead structure, resulted from this effort. As illustrated in Figure 3, this compound incorporates a hydroxyethylene isostere replacement of the scissile amide bond in Phe–Phe. The absolute configuration of this transition-state isostere was critical in maintaining intrinsic potency, and inversion of any of the chiral centers resulted in greater than a 1000-fold loss in potency. Although the compound was potent, it was essentially inactive in our cell-based assay. Our cell-based assay, developed in-house, determines the concentration of compound needed to completely inhibit the spread of HIV-1 infection in MT-4 cells with a 5% error range (CIC_{95}).

The initial objective of our medicinal chemistry effort was to eliminate the renin activity of our lead structure and improve the ability of the compound to

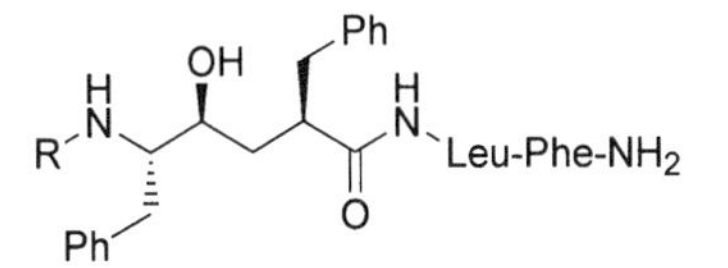

1 R = Boc-Phe-Phe- (IC_{50} HIV-1PR = 1.0 nM, Renin = 73 nM; CIC_{95} = >50 μM)
2 R = Boc- (IC_{50} HIV-1PR = 0.6 nM, Renin = >10,000 nM; CIC_{95} = 6 μM)

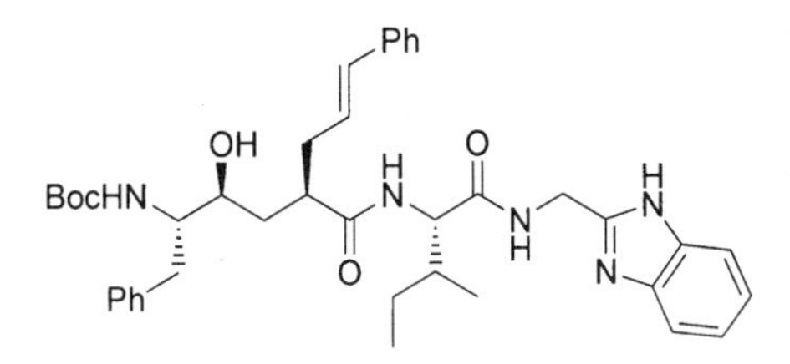

3 (IC_{50} HIV-1PR = <0.3 nM, Renin = >10,000 nM; CIC_{95} = 12 nM)

Figure 3 Identification of initial renin screening lead.

inhibit the spread of the virus in cell culture. Removal of the N-terminal phenylalanine moieties provided the tetrapeptide mimic **2** (IC_{50} = 0.6 nM; CIC_{95} = 6 μM). The compound maintained potency against HIV protease, was devoid of renin activity ($IC_{50} > 10$ μM), and now stopped viral replication at a concentration of 6 μM. Several analogues were prepared based on this structure, and this culminated in the discovery of isostere **3** (IC_{50} = <0.03 nM; CIC_{95} = 12 nM), a compound that met our requirement for both cell-based potency and selectivity (7). However, when inhibitor **3** was dosed orally in animal models, very little intact drug was observed in the plasma. This low oral bioavailability in animals was attributed to the compound's peptidic nature and lack of aqueous solubility.

As depicted in the foregoing, inhibitor **3** possesses an isoleucine amino acid in the P2′ position and a 2-benzimidazole aminomethyl moiety in the P3′ position of the inhibitor. To further reduce molecular weight and decrease peptidyl character of this molecule, we next optimized binding interactions by constraining the C-terminus (P2′–P3′) into a bioactive conformer. This was achieved using the tools of medicinal chemistry; intuition, molecular modeling, and rational design. Figure 4 illustrates how the P2′–P3′ moieties were constrained into a phenylglycinol mimic. Incorporation of a simple C-terminus benzyl amine provided an inhibitor with low molecular weight (MW 502 g/mol) and weak intrinsic potency (**4**; IC_{50} = 110 nM). The P3′ phenyl ring is in position to interact with S2′ pocket of the enzyme, but the flexibility of the side chain limits its effective interaction with the enzyme. The addition of a hydroxymethyl group to the benzyl group of **4**, which reintroduces oxygen to serve as a potential hydrogen bond

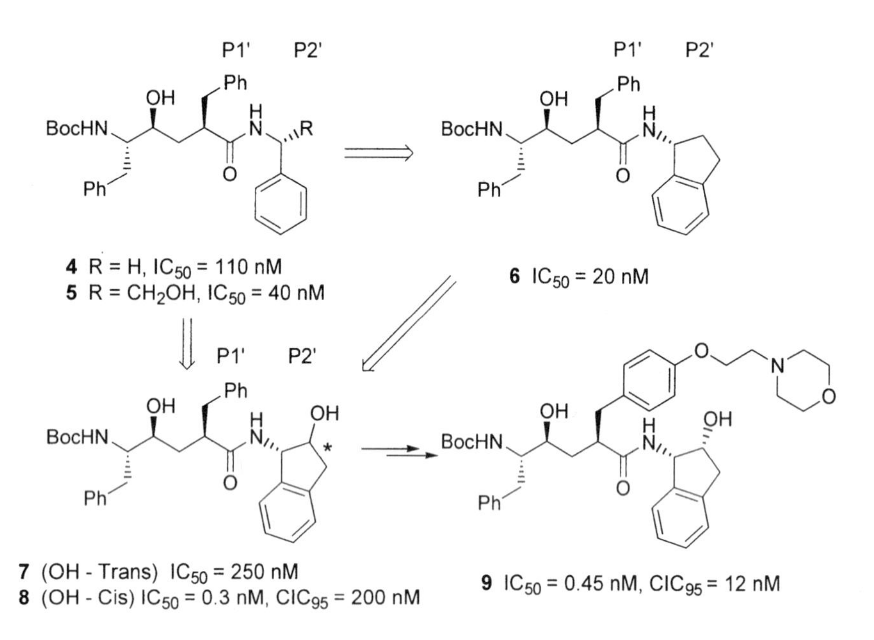

Figure 4 Optimization of a ring-constrained phenylglycine P2′ ligand.

acceptor and restricts conformational freedom of the side chain, resulted in 2.5-fold enhancement in potency (**5**;R = CH_2OH, IC_{50} = 40 nM). Constraining the benzyl moiety of **4** as an indanamide, to enforce a bioactive conformation, led to a fivefold increase in potency (**6**; IC_{50} = 20 nM). The combination of the cyclic constraint, together with the placement of the hydroxyl group in position to act as a hydrogen bond acceptor then afforded **8** (IC_{50} = 0.3 nM; CIC_{95} = 200 nM). Subsequent modeling studies of **8** docked into the active site of HIV-1 protease reinforced the notion that phenyl ring fits tightly into the S2′ pocket and hydroxyl moiety can interact with Asp39 of the enzyme. Later, these results were confirmed by X-ray crystal structure analysis of compound **9** bound in the active site of the enzyme (9).

B. First Safety Assessment Candidate

Although inhibitor **8** no longer contained any natural amino acids, had a lower molecular weight, and less peptide character then **3**, it was not orally bioavailable in animals. Again, we speculated that poor solubility in aqueous solvents limited the absorption of the drug. Incorporation of solubilizing groups, weakly basic amines, into the series of hydroxyethylene isosteres was used to improve aqueous solubility. To help direct our efforts, results from molecular modeling studies of **8** suggested that substituents added to the 4 position of the P1 and P1′ phenyl rings would occupy space in the aqueous medium outside the cavity of the enzyme. If weakly basic amines were appended onto these positions, they should improve water solubility and not interact with the lipophilic active site of the enzyme; therefore, potency should not be affected. Conformation of this hypothesis was achieved with isostere **9** (9) (IC_{50} = 0.45 nM; CIC_{95} = 12 nM, 5% oral bioavailability in dogs). The oral bioavailability of **9**, albeit low, was the most promising seen for any compound in the hydroxyethylene isostere series and was entered into safety assessment studies. Unfortunately, **9** failed in safety studies because of liver hepatotoxicity observed in dogs (10). The toxicity was compound-specific and not a general phenomenon of the hydroxyethylene transition-state series. Despite this setback, the compound did confirm our hypothesis that increasing aqueous solubility in the hydroxyethylene isostere series would increase the oral bioavailability in dogs.

C. HAPA Isostere Design

Shortly after inhibitor **9** failed in safety assessment, researchers at Hoffmann–La Roche published a series of potent hydroxyethylamine HIV-PR inhibitors, exemplified by Ro31-8959 (saquinavir) (11). This compound is potent against the HIV-1 protease (IC_{50} = 0.3 nM), it incorporates a weakly basic amine into the backbone of the inhibitor, and is somewhat orally bioavailable in rats (4%). At Merck, we hypothesized that incorporation of a basic amine into the backbone of inhibitors such as **8** might also improve the bioavailability of this series of compounds. To understand if this hypothesis was possible, a series of modeling studies were undertaken. The results suggested that replacement of the P2/P1 ligands, the *tert*-butyl carbamate and Phe moieties, of compound **8** with the P2′/P1′ ligands, the decahydroisoquinoline *tert*-butyl amide, of **10** would generate a novel class of hydroxylamine pentaneamide (HAPA) isosteres. This unique class of transition-state inhibitors is represented by compound **11**, as illustrated in Figure 5 (12).

Despite the relatively high intrinsic potency of compound **11** (IC_{50} = 7.8 nM), the inhibition of the spread of viral infection in MT-4 human T-lymphoid cells infected with the IIIb isolate was weak (CIC_{95} = 200 nM). However, because **11** did possess a favorable pharmacokinetic profile (J. Lin, personal communication, 1991), when compared with **9**, other analogues were pursued. Initially, L-proline, *cis*-4-substituted L-proline, and L-pipecolinic *tert*-butylamides were explored as replacements for the decahydroisoquinoline *tert*-butylamide (Table 1). These N-terminal analogues provided no improvement in potency. Next, 2-*tert*-butylcarboxamide-4-substituted-piperazines were examined. The piperazine analogues would provide two potential advantages over the decahydroisoquinoline *tert*-butylamide or the substituted L-proline *tert*-butylamides. First,

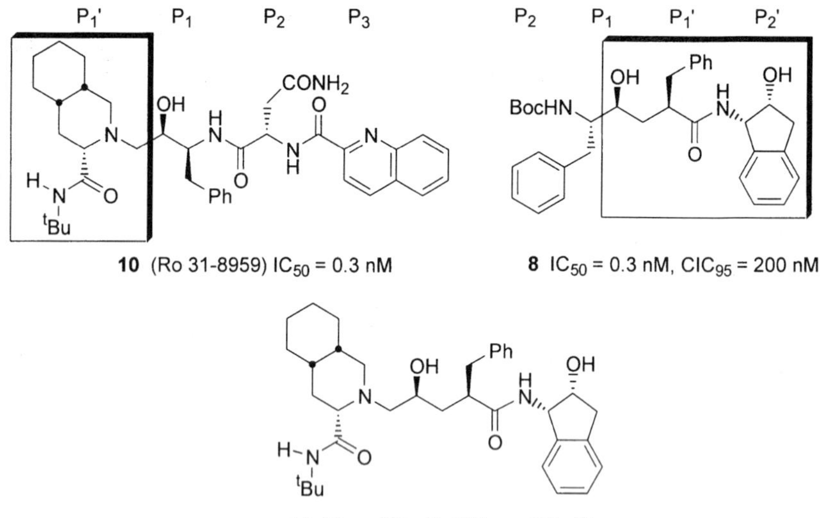

Figure 5 Design of the hydroxylamine pentaneamide isosteres.

the nitrogen in the 4 position of the piperazine ring could be easily functionalized. This would permit the introduction and optimization of a P3 ligand that could balance both the hydrophobic and hydrophilic requirements of our target molecules. Second, the additional amine in the piperazine ring should provide improved aqueous solubility, which might improve oral bioavailability.

One of the first compounds prepared in the piperazine series possessed a benzyloxycarbonyl moiety attached to the N4 position of the piperazine ring. This compound showed an improvement in both intrinsic potency and in the ability to inhibit viral spread in infected cells. To better understand the significant increase in potency of compound **15**, a cocrystallization with HIV-1 protease was undertaken (Fig. 6). Consistent with the modeling observations, the ligands from P3 to P3′ tightly bound into the S2 to S2′ region of the HIV-1 PR. Also observed is the critical water molecule bridging the two carbonyl moieties of the P1 and P1′ ligands. The most important observation was that the benzyloxycarbonyl moiety fills the lipophilic S3 domain of HIV-1 PR. This interaction is the primary reason inhibitor **15** is 22-fold more potent then **11**, our original hybrid molecule, which lacks a P3-binding ligand. These interactions combined to generate the first subnanomolar compound in the HAPA isostere series (13).

A small subset of the hundreds of compounds prepared in the piperazine series is illustrated in Table 2, which exemplify the guiding principles to the discovery of indinavir. First, replacement of the benzyloxycarbonyl moiety with

Table 1 Exploration of HIV-1 Protease S_1 and S_3 Domains

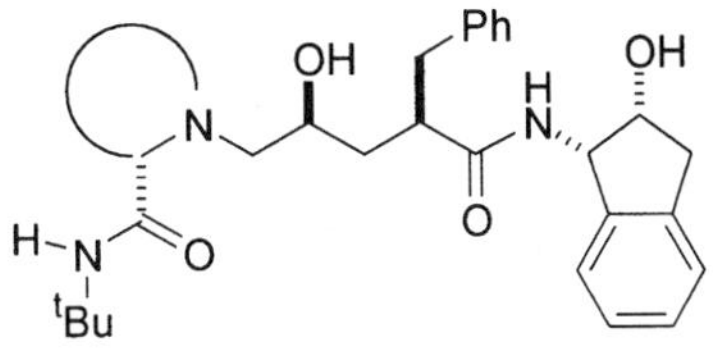

Compd	R	IC_{50} (nM)	CIC_{95} (nM)
11	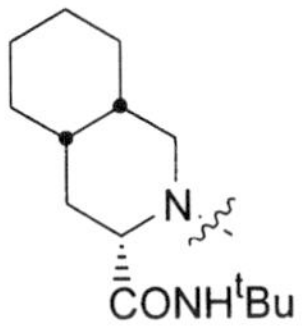 CONH^{t}Bu	7.8	200
12	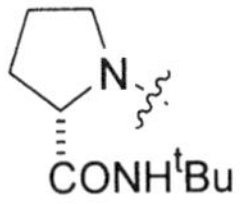 CONH^{t}Bu	347	1500
13	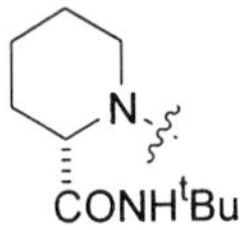 CONH^{t}Bu	80	not determined
14	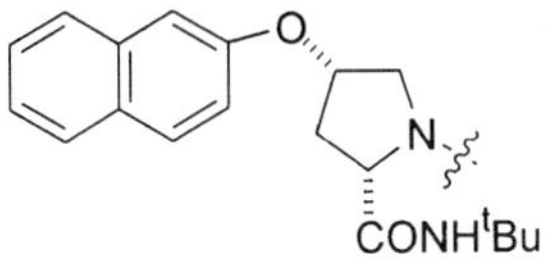 O; CONH^{t}Bu	15	>400
15	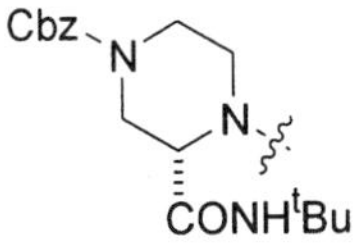 Cbz; CONH^{t}Bu	0.35	100

an acyl or sulfonyl moiety resulted in, for most examples, compounds with subnanomolar potency. However, this did not always translate into increased potency in the cell-based assay. One critical factor in cell-based potency is the ability of the inhibitors to cross a cell membrane. In many examples, if the piperazine ring is acylated or sulfonylated, the ability to penetrate the cell membrane appears to have been severely restricted.

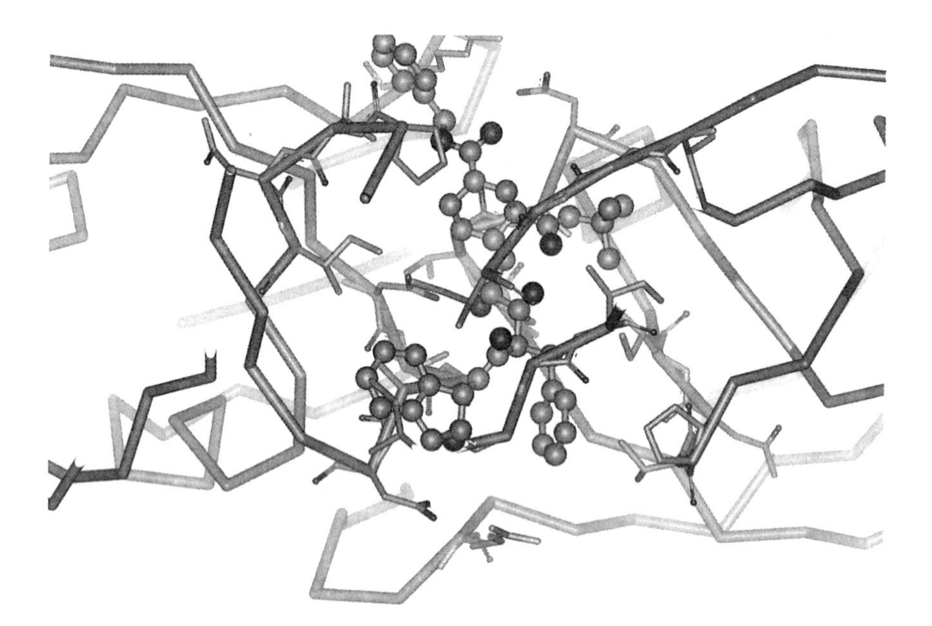

Figure 6 X-ray crystal structure of the first potent HAPA isostere.

Concurrent with the foregoing series, alkylated piperazine analogues were also pursued. Structure–activity data from this set of examples revealed that a variety of arylmethyl substitutions increased potency in the cell-based assay by two to threefold over the previously presented acylated or sulfonylated piperazines. The ability to modify the P3 ligands without adversely affecting potency proved crucial in our search for an orally bioavailable inhibitor. Large and highly lipophilic P3 ligands (e.g., phenylmethyl and 2-(benzyloxy)ethyl, were effective at increasing potency, but also significantly decreased aqueous solubility. Smaller P3 ligands improved solubility, but lost potency. The 3-pyridylmethyl group found in compound **18** provided the best balance of lipophilicity for binding to the protease and a weakly basic nitrogen that increased aqueous solubility. This combination of the 3-pyridylmethyl and the piperazine basic amines proved very successful in improving oral bioavailability.

D. Pharmacokinetics in Animals

The connection between oral bioavailability in animals and aqueous solubility is highlighted in Table 2. Compound **15**, which was very insoluble at pH 7.4, showed no appreciable plasma levels when administered to dogs as a citric acid

Table 2 Potency, Oral Bioavailability, and Solubility of Selected HAPA Isosteres

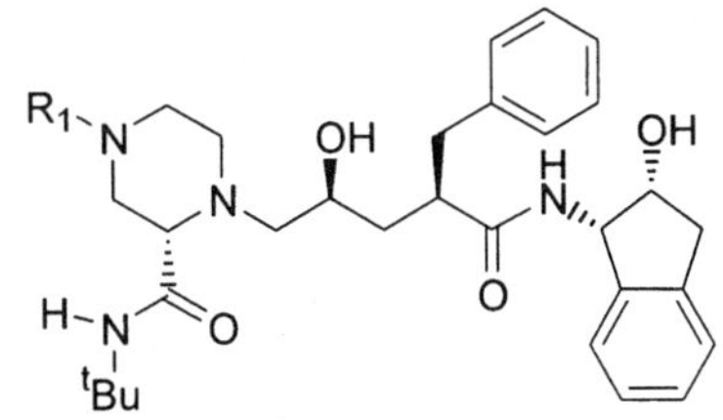

Compd	R_1	IC_{50} (nM)	CIC_{95} (nM)	C_{max} (nM)[a]	Sol@pH 7.4[b] (5.2) (mg/μL)
15	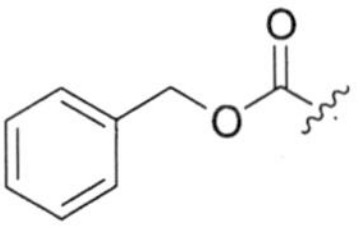	0.35	100	<100	<1
16	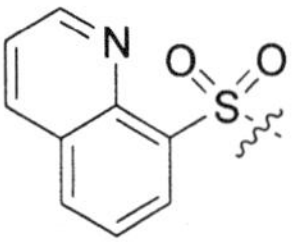	0.013	100	<100	<1
17	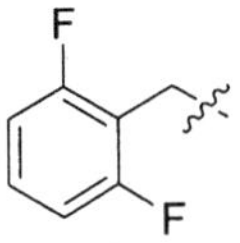	0.31	25	900	1.2 (3)
18	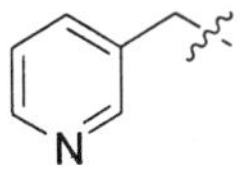	0.41	50	10,500	70 (690)
indinavir					

[a] C_{max}, concentration maximum in plasma after oral dosing at 10 mg/kg in dogs.
[b] Solubility measured at pH 7.4 abd 5.2

solution (10 mg/kg). The same result was obtained for sulfonamide **16**. Acylation or sulfonylation completely removes the basic character of the N4 nitrogen and also decreases the basicity of the N1 nitrogen of the piperazine ring. This lack of basicity translated into a lack of aqueous solubility that was detrimental to bioavailability. Adequate plasma levels were obtained with the slightly soluble difluorophenylmethyl **17**. Still further improvement occurred when the lipophilic aromatic moiety was replaced by a more soluble pyridine derivative. Of the variety of ligands explored, the most exciting results were obtained with **18**, later given the generic name indinavir. The maximum plasma concentration levels achieved with an oral dose of 10 mg/kg in 0.05-M citric acid solution for dogs ($n = 4$) was 10.5 ± 2.3 μM. The C_{max} levels for rat ($n = 4$; 20 mg/kg), and monkey ($n = 4$; 10 mg/kg) after oral dosing as a citric acid solution was found to be 2.80 ± 1.05 μM and 0.71 ± 0.24 μM, respectively (not shown). The oral bioavailability for this compound in the three animal species was 70, 22, and 13%, respectively, when compared with intravenous studies. In all animal models examined, the plasma concentrations after 6 h were twice the levels needed to completely stop viral growth in the cell-based assay. Solid dosage formulations with the crystalline free-base, although more variable, gave levels comparable with the solution formulations. An improvement in formulation was found with the sulfate salt of indinavir. A crystalline sulfate salt was prepared with both improved aqueous solubility (>450 mg/mL) and consistency of bioavailability in the solid-dosage formulation studies.

E. Protein Binding and Antiviral Activity

The ability of plasma proteins to adversely affect the efficacy of HIV-1 protease inhibitors has been well documented (14). The binding of indinavir to human, dog, and rat plasma was determined by an ultrafiltration method. The unbound fraction of the drug in plasma was approximately 38% for rats, 15% for dogs, and 40% for humans. In the cell-based antiviral assay the addition of a variety of proteins had minimal effects on potency. The addition of 10% fetal bovine serum, standard conditions for the assay, resulted in an CIC_{95} determination of 50 nM for indinavir. The addition of either 50% normal human serum, 500 μM human serum albumin, or 2.0 mg/mL of α_1-acid glycoprotein resulted in, at most, a twofold reduction in potency for indinavir. This is in contrast with the results observed with most other HIV protease inhibitors, which lose between 10- and 100-fold in potency depending on cell type (14). These results demonstrated that indinavir was effective in suppressing the spread of acute HIV-1 infection in cells in the presence of biologically relevant proteins.

F. Synthetic Route to Indinavir

The first milligrams of indinavir were synthesized in the laboratory in January 1992. Tens of grams of material would be needed immediately to support the compound's rapid development timeline. This was accomplished with the synthesis illustrated in Scheme 1. The known (*S*)-2-piperazine carboxylic acid bis-(*S*)-(+)-camphorsulfonic acid salt **19** (15) was converted into the differentially protected piperazine, following the procedure of Bigge and Hays (16). The resulting acid was then coupled with *tert*-butylamine following standard peptide-coupling procedures. Hydrogenolysis then removed the N1 benzyloxycarbonyl protecting

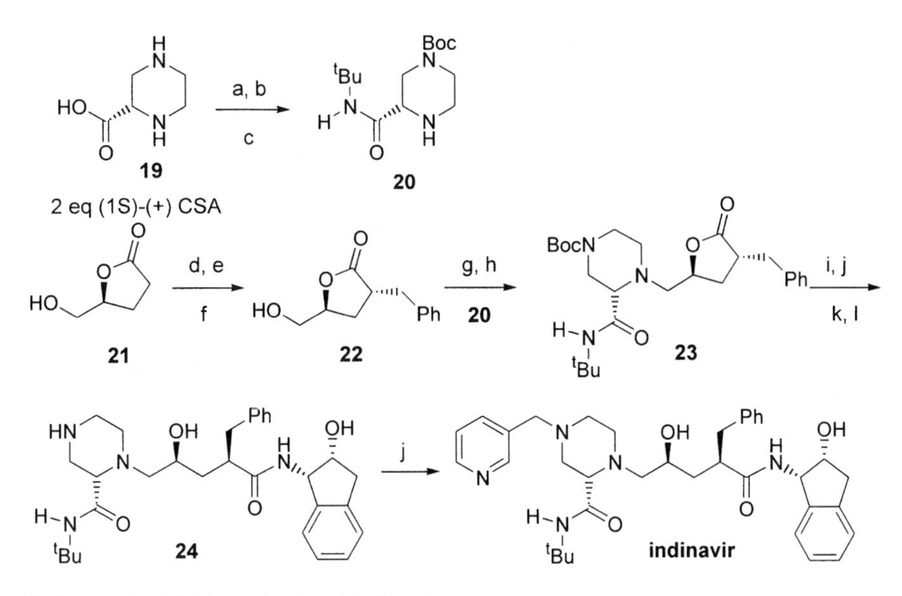

Scheme 1 Initial synthesis of indinavir.

group to provide amine **20** in 78% yield for the three steps. Commercially available (*S*)-(+)-dihydro-5-(hydroxymethyl)-2(3*H*)-furanone, **21**, was converted into the *tert*-butyldimethylsilyl ether under standard conditions. The protected lactone was deprotonated with LDA and alkylated with benzyl bromide to afford a 6:1 mixture of diastereomers that were separated by flash column chromatography. The major diastereomer was then treated with HF and resulting alcohol **22** was treated with triflic anhydride and 2,6-lutidine to provide the activated lactone in 96% yield. Displacement of the triflate with amine **20** at room temperature in isopropanol with *N,N*-diisopropylethylamine provided lactone **23** in 83% yield. Hydrolysis, protection, amine coupling, and final deprotection then provided penultimate amine **24**. This was converted into indinavir through a reaction with 3-picolyl chloride in DMF with triethylamine. This sequence provided the desired target molecule in 12 steps, from (*S*)-(+)-dihydro-5-(hydroxymethyl)-2(3*H*)-furanone, generally in 35% overall yield. Most importantly, 10-g batches could be conveniently processed in our laboratories, which supported the preclinical evaluation of indinavir.

Concurrent with the efforts of the medicinal chemists, process chemists in Rahway, New Jersey, quickly stepped in to supply material needed to support safety assessment studies and initial human clinical trials. Herculean efforts from our colleagues in the Departments of Process Chemistry and Chemical Engineering resulted in the preparation of kilogram quantities of indinavir. Subsequent campaigns in the pilot plant provided hundreds of kilograms of drug and later, after positive results from human clinical trials, factory production would provide several hundred thousand kilograms of indinavir on a yearly basis (17,18).

III. CLINICAL RESULTS WITH INDINAVIR

A. Phase 1/2

Indinavir successfully completed initial safety assessment studies. In early 1993, Clinical Pharmacology and Drug Metabolism started phase 1 human clinical trials to establish tolerability, safety, bioavailability, and antiviral activity of indinavir in monotherapy. The results from phase 1 trials were encouraging; they established that indinavir was generally well tolerated, had an absolute oral bioavailability of 60%, and demonstrated an antiviral effect. However, in early phase 2 trials the results did not appear to be as encouraging. After being suppressed for a couple of months, serum viral RNA levels increased in several patients. Unfortunately, the Merck Clinical Group had just identified the first signs of viral resistance to indinavir. Even though the viral RNA levels were increasing, the CD4 cell counts remained relatively high, which indicated that the drug was having a desirable effect. One patient in that early clinical group had a serum viral RNA level that decreased to below the level of detection and remained there. The dose of indinavir used in the study was increased, and patients who initiated therapy with the higher doses generally achieved sustained viral suppression. These results suggested that the higher doses of indinavir might prevent or delay the appearance of viral resistance. Subsequent studies suggested the combination of reverse transcriptase inhibitors and protease inhibitors, attacking two sites of viral replication, resulted in better viral suppression than monotherapy of either drug (19,20).

B. Phase 3 and Product Launch

In the spring of 1995, Clinical Research proceeded with very large phase 3 efficacy studies in Brazil, Canada, Europe, and the United States. These trials would provide critical data on clinical endpoints and combination therapy. This included the pilot study (035) which was investigating the use of indinavir in combination of AZT and 3TC. Encouraging results from this study together with results from phase 1 and 2 studies, would support the filing of an accelerated New Drug Application (NDA) with the Food and Drug Administration (FDA) on January 31, 1996. The FDA advisory committee meeting occurred on March 1, 1996, and heard data presented by the Clinical Group that established the scientific basis for indinavir's safety, low level of toxicity, and effectiveness in the treatment of HIV/AIDS. Besides scientific results, the public was also permitted to voice opinions at the meeting. Linda Grinberg, an AIDS patient and advocate, urged the committee and agency to approve Merck's product. She stated, "Last year, I sincerely doubted whether I would live to see Christmas, let alone the new year. Today, I come before you with renewed health and energy to implore you to approve this drug. The Merck protease inhibitor has given me back my life.''

The FDA accelerated its decision-making process and approved indinavir on March 13, 1996; a record 42 days after the submission of the New Drug Application. Indinavir was first synthesized in the laboratory in January 1992, and in just 4 years and 2 months was placed into the hands of patients in need.

Clinical trials with indinavir have established its daily recommended dose, which is 800 mg every 8 h on an empty stomach or with a light, low-fat meal. Nephrolithiasis has been reported in approximately 9.3% of the patients receiving indinavir in clinical trials. In general, these events were not associated with renal dysfunction and resolved with temporary stoppage of treatment. Patients are advised to maintain adequate hydration to minimize the occurrence of nephrolithiasis. Indirect hyperbilirubinemia has occurred in approximately 10% of the patients taking indinavir; however, there has been no clinical significance or evidence of hepatic involvement or failure associated with the incidence of hyperbilirubinemia. Indinavir is a good substrate for the CYP3A4 isozyme; it has the ability to affect plasma levels of other drugs metabolized by this system, but has few clinically significant drug–drug interactions.

C. Study 035, 100-Week Follow-Up

Clinically, indinavir is considered to be potent and effective HIV protease inhibitor. It has been approved for use in combination with reverse transcriptase inhibitors. Protocol 035 studied the combination of zidovudine–lamivudine versus indinavir–zidovudine–lamivudine and indinavir monotherapy in lamivudine- and indinavir-naïve patients who had extensive zidovudine experience. After at least 24 weeks of blinded therapy, all patients received open-label therapy. In 1998, Merck researchers and their colleagues disclosed the 100-week follow-up data from these studies (21). These results demonstrated that triple combination therapy using indinavir, zidovudine, and lamivudine reduced the levels of HIV RNA to fewer than 500 copies per milliliter for as long as 100 weeks, in 78% of the contributing HIV-infected patients. When the drugs were initiated sequentially the treatment was much less effective. Only 30–45% of the patients had a significant reduction of serum viral RNA to fewer than 500 copies per milliliter. These results established the durability of triple combination regimen; however, some of the patients did not have viral RNA levels below the levels of detection, suggesting the emergence of viral resistance.

D. Resistant Virus

The emergence of drug-resistant mutants has been, and will continue to be, a critical issue in the treatment of HIV infection (22). In vivo viral resistance has been demonstrated for all classes of HIV inhibitors; the nucleoside and nonnucleoside reverse transcriptase inhibitors, as well as protease inhibitors (23). For protease inhibitors, the appearance of virus with reduced susceptibility has occurred both in cell culture experiments (24) and in human clinical trials (25,26). The use of suboptimal doses of indinavir (monotherapy) in initial clinical trials resulted in a lowering of viral load. However, over time, mutant variants that exhibited resistance to indinavir began to emerge, several of these have been characterized both phenotypically and genotypically (26a). Not only was resistance to indinavir observed, but also cross-resistance to a structurally diverse group of protease inhibitors. Also, researchers (27) have reported that, in patients treated with other protease inhibitors, cross-resistant variants have developed.

IV. SUMMARY

In summary, a novel series of HIV-PR transition-state isosteres have been developed. Starting from an initial peptide renin screening lead, we developed the highly potent and selective hydroxyethylene isostere **8**. Based on this achievement, we were able to incorporate a basic amine into the backbone of this series to provide a novel HAPA isostere series of HIV-PR inhibitors. By modifying the physical properties of this series of inhibitors (i.e., solubility and lipophilicity) and concurrently maintaining potency, we were able to design indinavir. From the efforts of clinical researchers throughout the world, indinavir has developed as an appropriate foundation for combination therapy in the treatment of AIDS, and it continues to enhance the lives of hundreds of thousands of patients infected with the human immunodeficiency virus.

ACKNOWLEDGMENTS

The authors would like to thank the hundreds of Merck employees for going beyond the call of duty by devoting most, if not all, of their time and efforts to the success of indinavir. The hard work and self-sacrifice of everyone on the indinavir team was very inspiring. Finally, we thank the patients themselves, for their willingness to participate in clinical trials and undergo that hardship to forward our knowledge of medical science.

REFERENCES

1. (a) Ratner L, Haseltine W, Patarca R, Livak KJ, Starcich B, Josephs SF, Doran ER, Rafalski JA, Whitehorn EA, Baumeister K, Ivanoff L, Petteway SR Jr, Pearson ML,

Lautenberger JA, Papas TS, Ghrayeb J, Chang NT, Gallo RC, Wong–Staal F. Complete nucleotide sequence of the AIDS virus, HTLV-III. Nature 1985; 313:277–284. (b) Toh H, Ono M, Saigo K, Miyata T. Retroviral protease-like sequence in the yeast transposon *Tyl*. Nature 1985; 313:691. (c) Pearl LH, Taylor WR. A structural model for the retroviral protease. Nature 1987; 329:351–356.

2. Kramer RA, Schaber MD, Skalka AM, Ganuyl K, Wong–Staal F, Reddy EP. Science 1986; 231:1580–1584.
3. (a) Kohl NE, Emini EA, Schleif WA, Davis LJ, Heimbach JC, Dixon RAF, Scolnick EM, Sigal IS. Active HIV protease is required for viral infectivity. Proc Natl Acad Sci USA 1988; 85:4686–4690. Also see (b) Gottlinger HG, Sodroski JG, Haseltine WA. Proc Natl Acad Sci USA 1989; 86:5781–5785. (c) Peng C, Ho BK, Chang TW, Chang NT. J Virol 1989; 63:2550–2556.
4. Navia MA, Fitzgerald PMD, McKeever BM, Leu C–T, Heimbach JC, Herber WK, Sigal IS, Darke PL, Springer JP. Three dimensional structure of the aspartyl protease from the AIDS virus HIV-1. Nature 1989; 337:615–620. (b) Wlodawer A, Miller M, Jaskolski M, Sathyanarayana BK, Baldwin E, Weber IT, Selk LM, Clawson L, Schneider J, Kent SBH. Conserved folding in retroviral proteases: crystal structure of a synthetic HIV-1 protease. Science 1989; 245:616–621.
5. Wlodawer A, Erickson JW. Structure-based inhibitors of HIV-1 protease. Annu Rev Biochem 1993; 62:543–585.
6. Greenlee W. Renin Inhibitors. Med Res Rev 1990; 10:173–236.
7. Vacca JP, Guare JP, deSolms SJ, Sanders WM, Giuliani EA, Young SD, Darke PL, Zugay J, Sigal IS, Schleif WA, Quintero JC, Emini E, Anderson PS, Huff JR. L-687,908, a potent hydroxyethylene-containing HIV protease inhibitor. J Med Chem 1991; 34:1225–1228.
8. Lyle TA, Wiscount CM, Guare JP, Thompson WJ, Anderson PS, Darke PL, Zugay JA, Emini EA, Schleif WA, Quintero JC, Dixon RAF, Sigal IS, Huff JR. Benzocycloalkyl amines as novel C-termini for HIV protease inhibitors. J Med Chem 1991; 34:1228–1230.
9. Thompson WJ, Fitzgerald PMD, Holloway MK, Emini EA, Darke PL, McKeever BM, Schleif WA, Quintero JC, Zugay JA, Tucker TJ, Schwering JE, Homnick CF, Nunberg J, Springer JP, Huff JR. Synthesis and antiviral activity of a series of HIV-1 protease inhibitors with functionality tethered to the P1 or P1′ phenyl substituents: X-ray crystal structure assisted design. J Med Chem 1992; 35:1685–1701.
10. Lin JH, Chen I–W, King J. Dose-dependent toxicokinetics of L-689,502, a potent human immunodeficiency virus protease inhibitor, in rats and dogs. Pharmacol Exp Ther 1992; 263:105–111.
11. Roberts NA, Martin JA, Kinchington D, Broadhurtst AV, Craig JC, Duncan IB, Galpin SA, Handa BD, Kay J, Krohn A, Lambert RW, Merrett JH, Mills JS, Parkes KEB, Redshaw S, Ritchie AJ, Taylor DL, Thomas GJ, Machin PJ. Rational design of peptide-based HIV proteinase inhibitors. Science 1990; 248:358–361.
12. Vacca JP, Dorsey BD, Schleif WA, Levin RB, McDaniel SL, Darke PL, Zugay JA, Quintero JC, Blahy OM, Roth E, Sardana VV, Schlabach AJ, Graham PI, Condra JH, Gotlib L, Holloway MK, Lin J, Chen I–W, Vastag K, Ostovic D, Anderson PS, Emini EA, Huff JR. L-735,524: an orally bioavailable human immunodeficiency virus type 1 protease inhibitor. Proc Natl Acad Sci USA 1994; 91:4096–4100.

13. Dorsey BD, Levin RB, McDaniel SL, Vacca JP, Guare JP, Darke PL, Zugay JA, Emini EA, Schleif WA, Quintero JC, Lin JH, Chen I–W, Holloway MK, Fitzgerald PMD, Axel MG, Ostovic D, Anderson PS, Huff JR. L-735,524: the design of a potent and orally bioavailable HIV protease inhibitor. J Med Chem 1994; 37:3443–3451.
14. Lazdins JK, Mestan J, Goutte G, Walker MR, Bold G, Capraro HG, Klimkait T. In vitro effect of alphal-acid glycoprotein on the anti-human immunodeficiency virus (HIV) activity of the protease inhibitor CGP 61755: a comparative study with other relevant HIV protease inhibitors. J Infect Dis 1997; 175:1063–1070.
15. Felder E, Maffei S, Pietra S, Pitre D. Helv Chim Acta 1960; 888–896.
16. Bigge CF, Hays SJ, Novak PM, Drummond JT, Johnson G, Bobovski TP. Tetrahedron Lett 1989; 30:5193–5196.
17. Askin D, Eng KK, Rossen K, Purick RM, Wells KM, Volante RP, Reider PJ. Highly diastereoselective reaction of a chiral, on-racemic amide enolate with (*S*)-glycidyl tosylate. Synthesis of the orally active HIV-1 protease inhibitor L-735,524. Tetrahedron Lett 1994; 35:673–676.
18. Maligres PE, Weissman SA, Upadhyay V, Cianciosi SJ, Reamer RA, Purick RM, Sager J, Rossen K, Eng KK, Askin D, Volante RP, Reider PJ. Cyclic imidate salts in acyclic stereochemistry: diastereoselective *syn*-epoxidation of 2-alkyl-4-enamides to epoxyamides. Tetrahedron 1996; 52:3327–3338.
19. Hammer SM, Squires KE, Hughes MD, Grimes JM, Demeter LM, Currier JS, Eron JJ, Fienberg JE, Balfour HH, Deyton LR, Chodakewitz JA, Fischl MA. A controlled trial of two nucleoside analogs plus indinavir in persons with human immunodeficiency virus infection and CD4 cell counts of 200 per cubic millimeter or less. N Engl J Med 1997; 337:725–733.
20. Gulick RM, Mellors JW, Havlir D, Eron JJ, Gonzalez C, McMahon D, Richman DD, Valentine FT, Jonas L, Meibohm A, Emini EA, Chodakewitz JA. Treatment with indinavir, zidovudine, and lamivudine in adults with human immunodeficiency virus infection and prior antiretroviral therapy. N Engl J Med 1997; 337:734–739.
21. Gulick RM, Mellors JW, Havlir D, Eron JJ, Gonzales C, McMahon D, Jonas L, Meibohm A, Holder D, Schleif WA, Condra JH, Emini EA, Isaacs R, Chodakewitz JA, Richman DD. Simultaneous vs. sequential initiation of therapy with indinavir, zidovudine, and lamivudine for HIV-1 infection (100-week follow-up). JAMA 1998; 280:35–41.
22. Boden D, Markowitz M. Resistance to human immunodeficiency virus type 1 protease inhibitors. Antimicrob Agents Chemother 1998; 42:2775–2783.
23. Richman DD. News and views: Protease uninhibited. Nature 1995; 374:494; and references sited therein.
24. (a) Jacobsen H, Yasargil K, Winslow DL, Craig JC, Krohn A, Duncan IB, Mous J. Characterization of human immunodeficiency virus type 1 mutants with decreased sensitivity to proteinase inhibitor Ro 31-8959. Virology 1995; 206:527–534. (b) Markowitz M, Mo H, Kempf DJ, Norbeck DW, Bhat TN, Erickson JW, Ho DD. Selection and analysis of human immunodeficiency virus type 1 variants with increased resistance to ABT-538, a novel protease inhibitor. J Virol 1995; 69:701–706. (c) Partaledis JA, Yamaguchi K, Tisdale M, Blair EE, Falcione C, Maschera B, Myers RE, Pazhanisamy S, Byrn RA, Livingston DJ. In vitro selection and char-

acterization of human immunodeficiency virus type 1 (HIV-1) isolates with reduced sensitivity to hydroxyethylamino sulfonamide inhibitors of HIV-1 aspartyl protease. J Virol 1995; 69:5228–5235. (d) Patick AK, Rose R, Greytok J, Bechtold CM, Hermsmeier MA, Chen PT, Barrish JC, Zahler R, Colonno RJ, Lin P–F. Characterization of a human immunodeficiency virus type 1 variant with reduced sensitivity to an aminodiol protease inhibitor. J Virol 1995; 69:2148–2152. (e) Tisdale M, Myers RE, Maschera B, Parry NR, Oliver NM, Blair ED. Cross-resistance analysis of human immunodeficiency virus type 1 variants individually selected for resistance to five different protease inhibitors. Antimicrob Agents Chemother 1995; 39:1704–1710.

25. Condra JH, Schleif WA, Blahy OM, Gabryelski LJ, Graham DJ, Quintero JC, Rhodes A, Robbins HL, Roth E, Shivaprakash M, Titus D, Yang T, Teppler H, Squires KE, Deutsch PJ, Emini EA. In vivo emergence of HIV-1 variants resistant to multiple protease inhibitors. Nature 1995; 374:569–571.
26. (a) Condra JH, Holder DJ, Schleif WA, Blahy OM, Danovich RM, Gabryelski LJ, Graham DJ, Laird D, Quintero JC, Rhodes A, Robbins HL, Roth E, Shivaprakash M, Yang T, Chodakewitz JA, Deutsch PJ, Leavitt RY, Massari FE, Mellors JW, Squires KE, Steigbigel RT, Teppler H, Emini EA. Genetic correlates of in vivo viral resistance to indinavir, a human immunodeficiency virus type 1 protease inhibitor. J Virol 1996; 70:8270–8276. (b) Jacobsen H, Hanggi M, Ott M, Duncan IB, Owen S, Andreoni M, Vella S, Mous J. In vivo resistance to a human immunodeficiency virus type 1 proteinase inhibitor: mutations, kinetics, and frequencies. J Infect Dis 1996; 173:1379–1387. (c) Molla A, Korneyeva M, Gao Q, Vasavanonda S, Schipper PJ, Mo H–M, Markowitz M, Chernyavskiy T, Niu P, Lyons N, Hsu A, Granneman GR, Ho DD, Boucher CAB, Leonard JM, Norbeck DW, Kempf DJ. Ordered accumulation of mutations in HIV protease confers resistance to ritonavir. Nat Med 1996; 2:760–766.
27. Craig C, Race E, Sheldon J, Whittaker L, Gilbert S, Moffatt A, Rose J, Dissanayeke S, Chirn G–W, Duncan IB, Cammack N. HIV protease genotype and viral sensitivity to HIV protease inhibitors following saquinavir therapy. AIDS 1998; 12:1611–1618. (b) Winters MA, Schapiro JM, Lawrence J, Merigan TC. Human immunodeficiency virus type 1 protease genotypes and in vitro protease inhibitor susceptibilities of isolates from individuals who were switched to other protease inhibitors after long-term saquinavir treatment. J Virol 1998; 72:5303–5306.

5
Discovery and Development of Nelfinavir (Viracept)

Siegfried H. Reich
Agouron Pharmaceuticals, San Diego, California

I. INTRODUCTION

In the middle to late 1980s reverse transcriptase inhibitors (RTI), primarily, zidovudine (AZT), were the only available drugs for the treatment of AIDS. RTIs such as AZT produced only a limited response, owing to toxicity associated with concomitant inhibition of human enzymes; consequently, alternative strategies to HIV therapy were strongly needed. HIV protease (HIV PR) inhibitors represented a new class of therapeutic agent, with the potential for increased specificity (1). Key experiments wherein the HIV PR was either mutated or chemically inhibited, resulting in noninfectious virus, reinforced the notion that this enzyme held promise for a new therapeutic strategy in the treatment of HIV infection and AIDS (2). The unique cleavage sequence in the substrate polyprotein recognized by the HIV PR (Tyr–Pro, for example) provided some basis for expecting selectivity over human aspartyl proteases, which have a considerably different substrate specificity. Identifying a potent peptidomimetic inhibitor of HIV PR with good oral availability would prove to be a significant challenge. Up to this point, monumental efforts on the part of the drug industry on related aspartyl protease targets, such as renin, were unsuccessful in producing peptidomimetic agents with the favorable physicochemical properties and pharmacokinetic profile needed for good oral availability.

Based on this information and the knowledge that HIV PR was a small soluble protein (20 kDa) amenable to protein crystallographic analysis, research on HIV PR was initiated at Agouron in 1988. It is now well recognized that detailed three-dimensional (3D) structural information of a target protein can greatly facilitate both the identification of new lead structures and the optimization of their binding and physical properties (3). For this, it is critical that structural information be obtained within a time frame where it can influence modeling and synthesis. Furthermore, it is imperative that the analysis is iterative, for with each new inhibitor, movement of both the ligand and protein can be crucial to subsequent design. The rapid acquisition of new structural information, within a similar time frame as modeling, design, and synthesis, is a critical aspect of the iterative cycle (Fig. 1). For example, in the HIV PR project described herein, well over 50 high-resolution cocrystal structures were solved spanning a number of structural inhibitor classes.

One of the first requirements would be the preparation and purification of large amounts of pure HIV PR protein to support an aggressive structure-based design program. Production of sufficient active HIV PR for the iterative design process was accomplished in *Escherichia coli* through the use of a synthetic gene expressing an HIV PR–dihydrofolate fusion protein (4). This construct provided what, in the end, would be grams of pure HIV PR protein in support of the project. The first 2.0-Å cocrystal structure of HIV PR in complex with a peptidic statine-based inhibitor was solved in 1989 (Fig. 2). By using this structural information and that of others (5), a structure-based program in inhibitor design was initiated with the goal of identifying a potent HIV PR inhibitor and antiviral agent with good oral availability. Because of the nature of HIV infection and the requirement for

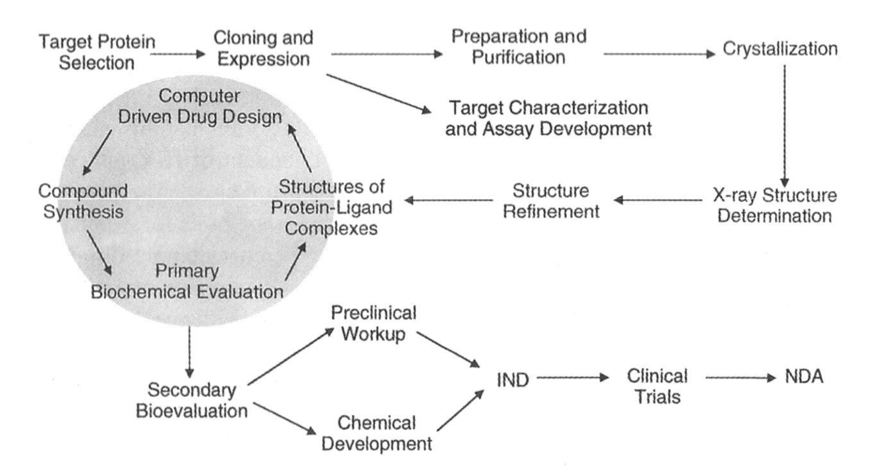

Figure 1 Drug discovery and development pathway highlighting the iterative structure-based design cycle.

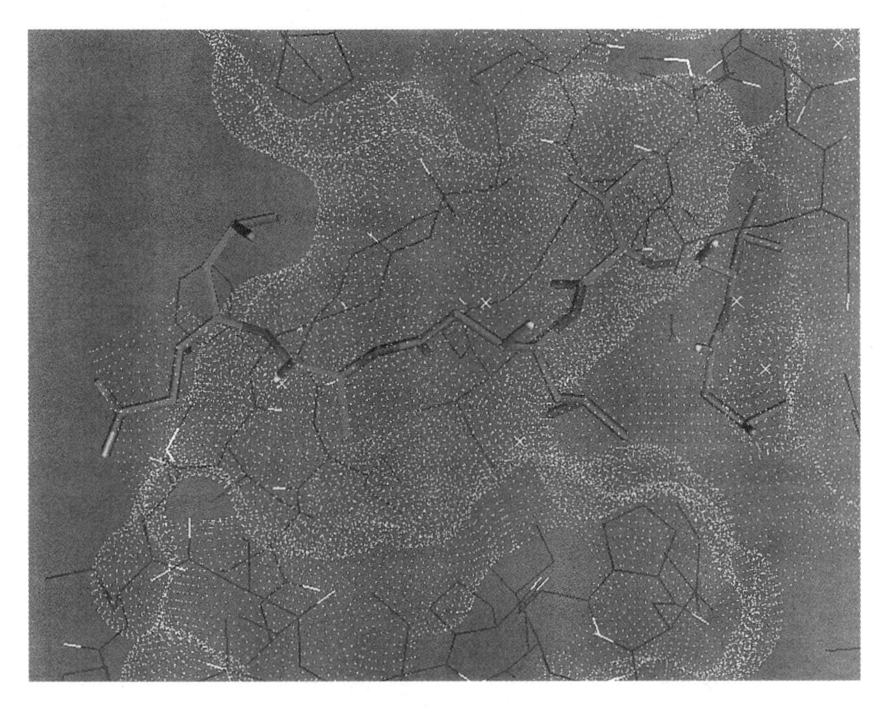

Figure 2 Cocrystal structure of peptidic statine inhibitor (Isoval-Val-Val-Sta-Ala-Sta) bound to the HIV PR. The inhibitor is shown in thick bonds and hydrogen bonds to the protein and bound water molecules are indicated with dashed lines.

prolonged therapy, a conveniently administered HIV PR inhibitor with good oral availability was desired. A combination of protein structure-based design and traditional medicinal chemistry ultimately led to the identification of nelfinavir (Viracept), a highly potent and orally available drug for the treatment of AIDS. Viracept is currently the most widely prescribed HIV PR inhibitor on the market.

II. STRUCTURE-BASED INHIBITOR DESIGN

Initial attempts at peptidomimetic inhibitors met with limited success. A plethora of de novo design ideas for nonpeptide inhibitors as well as more conventional peptidomimetic strategies were explored. A collaboration with Eli Lilly was initiated on our HIV PR project in 1990. At this time the Hoffmann–La Roche inhibitor, saquinavir, had been identified (2b) (Fig. 3). Although this was a potent inhibitor of HIV PR and a potent antiviral, it was clear that this compound had

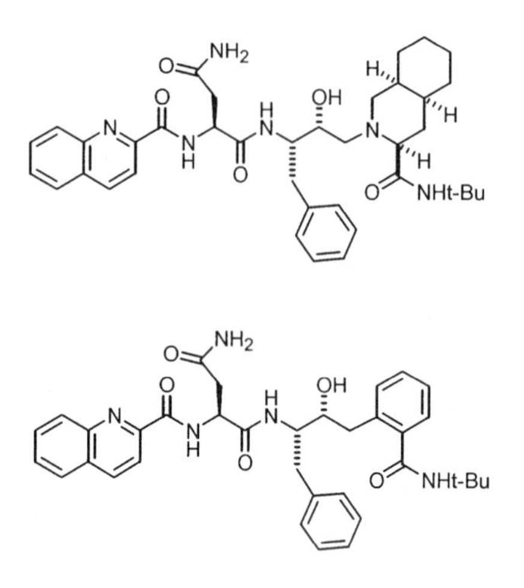

Figure 3 Structure of saquinavir and compound LY289612.

a less than optimal oral availability. The peptidic nature of this inhibitor was presumed to be the main source of the unfavorable pharmacological properties. In the collaboration, a simple modification of saquinavir was made by replacing the perhydroisoquinoline ring system with an *ortho*-substituted benzamide (see compound 2, Fig. 3) (6). This simplified the molecule and provided a potent antiviral. Unfortunately, similar to saquinavir, inhibitor **2** also suffered from poor oral availability, likely a result of a common, largely peptidic core. The cocrystal structure of inhibitor **2** was solved and further design was initiated. A radical modification of inhibitor **2** was made by removing the largely peptidic N-terminal portion and redesigning to fill the empty S1 and S2 pockets using nonpeptidic components resulting in compound **3** (Fig. 4). An initial cocrystal structure of this modest inhibitor **3** led to a nonpeptidic series of inhibitors that were optimized by using crystallographic feedback, ultimately leading to inhibitor **4** (AG1284). This research highlights the efficiency and enormous influence structural feedback can have on the optimization process. In the course of just 10 months a 10,000-fold improvement in potency was achieved, requiring the preparation of only 80 compounds, and utilizing 11 cocrystal structures. During the initial phase of this work, the inhibitors were observed to adopt an alternative binding mode, an important result that was confirmed crystallographically (Fig. 5). This alternative-binding mode was reinforced by making appropriate substitutions in the series based on 3D structural information. Compound **4** was selected for pharmacokinetic evaluation in the rat, dog, and monkey (7). Although **4** showed good oral levels after dosing in all three species, this compound was ultimately abandoned owing to its modest antiviral potency (IC_{50} 0.78, μM, racemic mixture).

From AG1132 to AG1284
A Potent, Orally Available HIV Protease Inhibitor

-Over 10,000-fold Improvement in Ki
-80 Compounds Prepared
-11 Cocrystal Structures
-10 Months

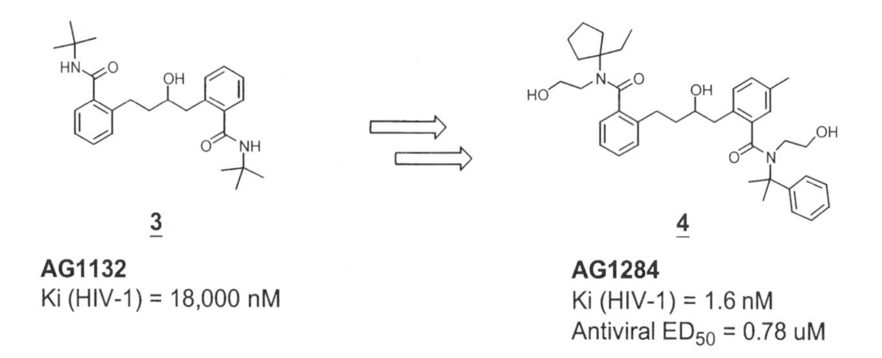

Figure 4 Optimization leading to inhibitor **4** as a preclinical lead.

III. IDENTIFICATION OF NELFINAVIR

A. A Parallel, More Conservative, Approach

Concomitant with the development process leading to compound **4**, a parallel, more conservative effort was underway to optimize inhibitor **2** by retaining the central hydroxyethylamine core and modifying the three areas depicted in Figure 6. Considering the proximity of the phenylalanine side chain at P1 and the quinoline ring at the N-terminus of **2**, we realized that by extending the P1 residue one could simultaneously reduce the size of the P2–P3 groups. Replacement of the phenylalanine side chain at P1 with a larger thioarylalanine and a smaller N-terminus was indeed tolerated. A bicyclic aromatic system was initially used to replace the more peptidic Asn–quinaldic acid portion at P2 based on modeling into the cocrystal structure of **2** in the HIV PR. This resulted in one of the more potent and considerably less peptidic inhibitors **6** (K_i = 24 nM; Fig. 7). However, **6** still had only modest antiviral activity relative to compound **2**. Further modifications of bicycle **6** led to *ortho*-substituted monocyclic anilines, which displayed

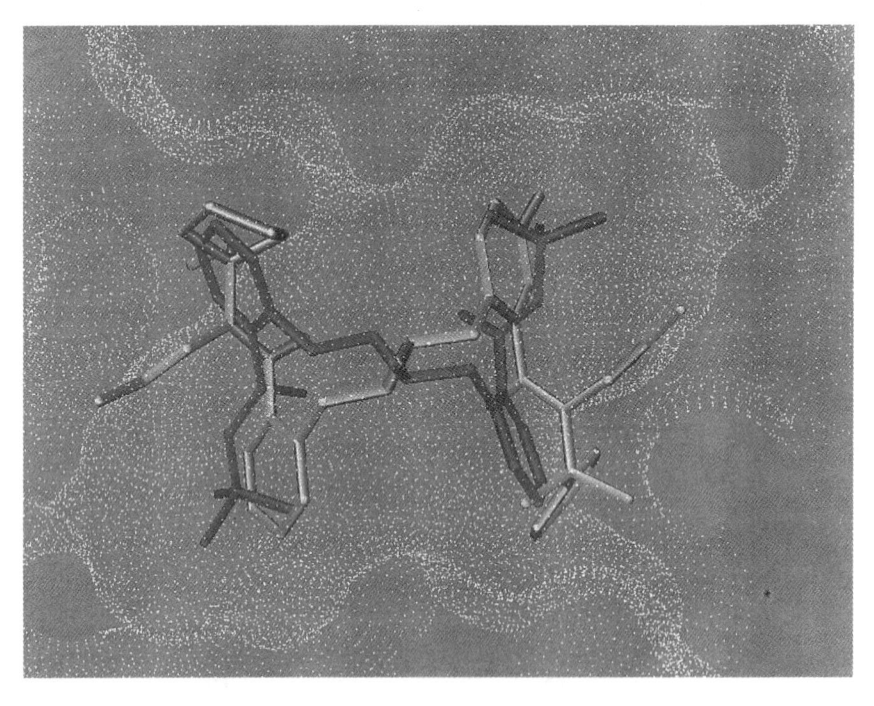

Figure 5 Comparison of the binding modes of inhibitor **4** (light gray) and the original inhibitor **3** (dark gray) from their cocrystal structures with the HIV PR.

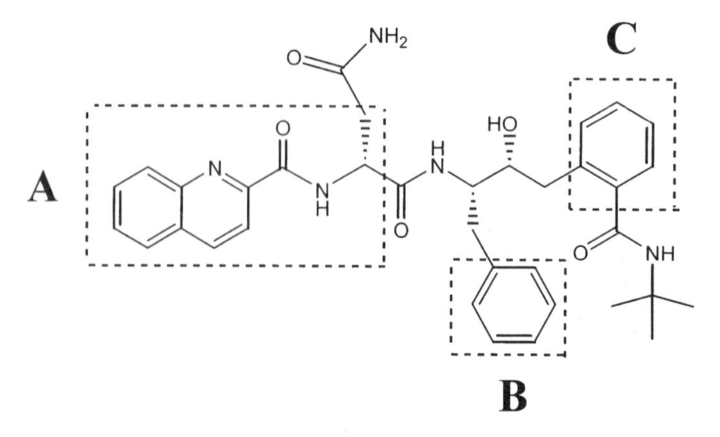

Figure 6 Areas of inhibitor **2** optimized using structure-based methods.

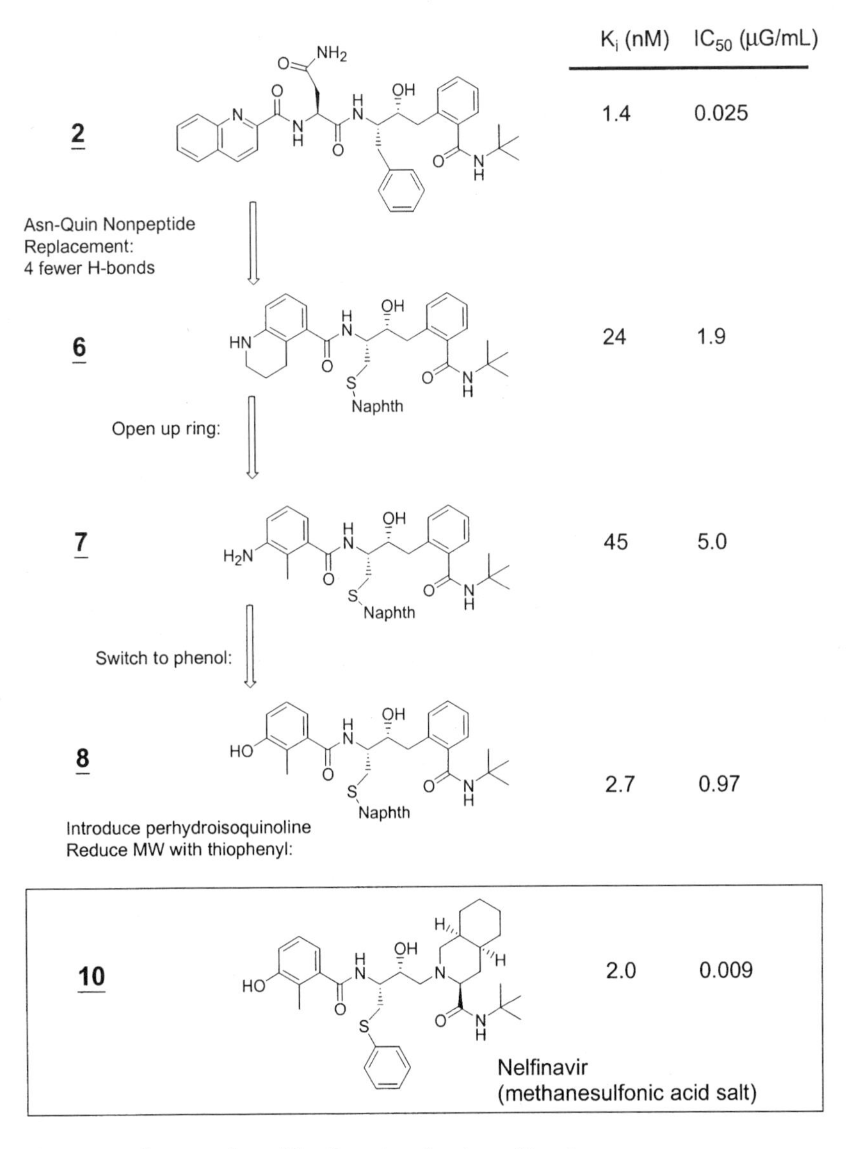

Figure 7 Structural modifications leading to nelfinavir.

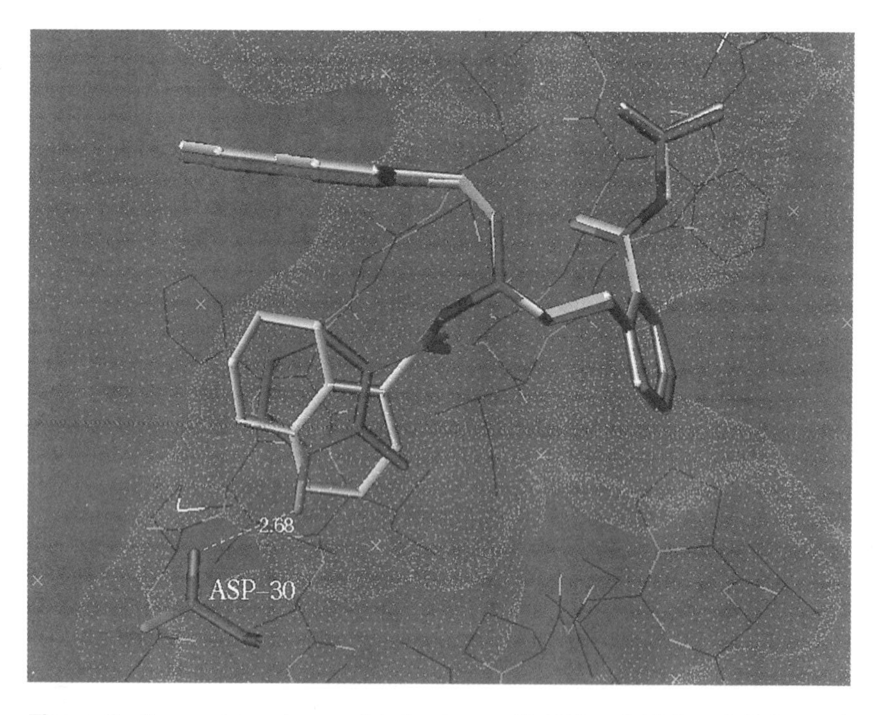

Figure 8 Comparison of cocrystal structures of inhibitors inhibitors **6** (light gray) and **8** (dark gray).

activity comparable with parent **6**. The *ortho*-methyl substitution was key to maintaining both the appropriate conformation of the aryl amide and hydrogen bonding to the protein and bound water molecule. A critical cocrystal structure of compound **7** revealed that the monocyclic compounds had undergone a 180-degree flip relative to the bicyclic inhibitors where the *ortho*-methyl group was now buried in the hydrophobic S2 pocket and the 3-amino group was hydrogen bonded to Asp30 sidechain (Fig. 8). This prompted the exploration of a 3-phenol compound **8**, expected to make a better hydrogen bond with Asp30. This compound was ten times more potent as an inhibitor of HIV PR and five times more potent as an antiviral, relative to **7**.

Information from the many cocrystal structures provided a basis for superimposing a variety of inhibitors in the same frame of reference. It was anticipated that an increase in the aqueous solubility and antiviral activity of compound **8** was needed. By superposing compound **8** and saquinavir in their respective cocrystal structures, we envisaged combining the known tertiary amine-containing perhydroisoquinoline system with the 2-methyl-3-hydroxybenzoic acid to afford inhibitor **10** (AG1343; nelfinavir). This combination was significant in that it provided a markedly improved antiviral agent, likely a result of the unique physical properties in combining the perhydroisoquinoline amide with the *o*-methyl phenol. Nelfinavir showed superior antiviral activity with a 100-fold improvement in antiviral IC_{50} relative to compound **8**. As described in the foregoing, the series of compounds containing the *t*-butyl *o*-benzamide function and saquinavir displayed universally poor oral availability. To our gratification, nelfinavir, in addition to being a significantly more potent antiviral, showed plasma levels above the ED_{90} for over 6 h when administered to rats (40 mg/kg; $n = 3$) (8).

B. Synthesis and Scale-Up of Nelfinavir

The initial synthesis of nelfinavir conducted in the medicinal chemistry group, in support of more extensive pharmacokinetic studies, is outlined in Scheme 1 (8c). Both key intermediates epoxide **A** and perhydroisoquinoline amide **B** were prepared from commercially available carbobenzyl oxy (Cbz)-protected amino acids. The unnatural amino acid *N*-Cbz-*S*-Phe-L-cysteine was obtained from

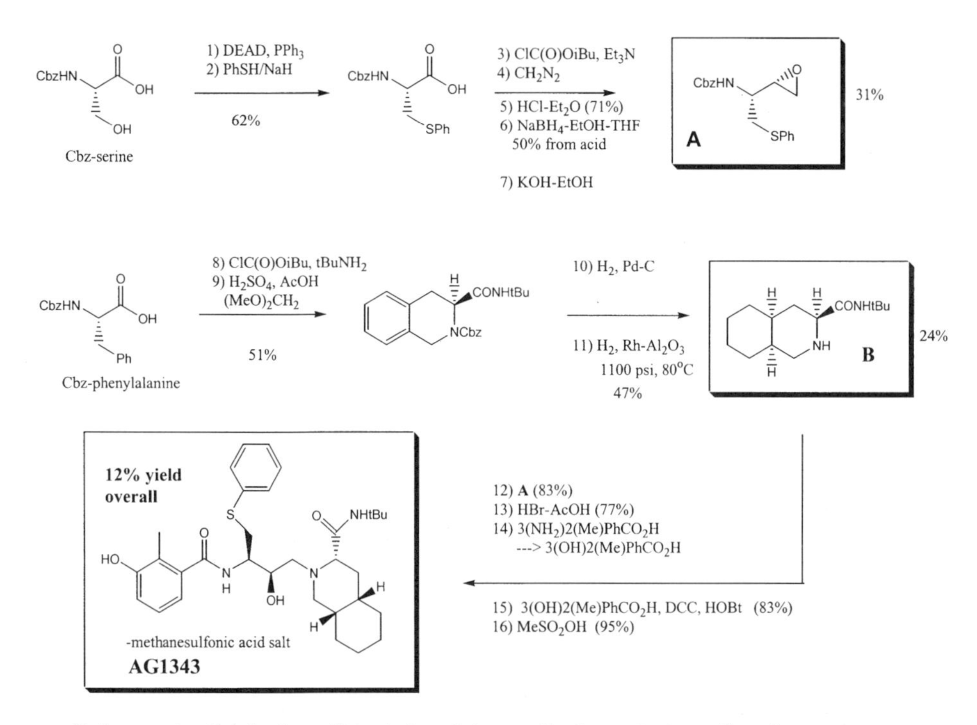

Scheme 1 Original medicinal chemistry synthetic route to nelfinavir.

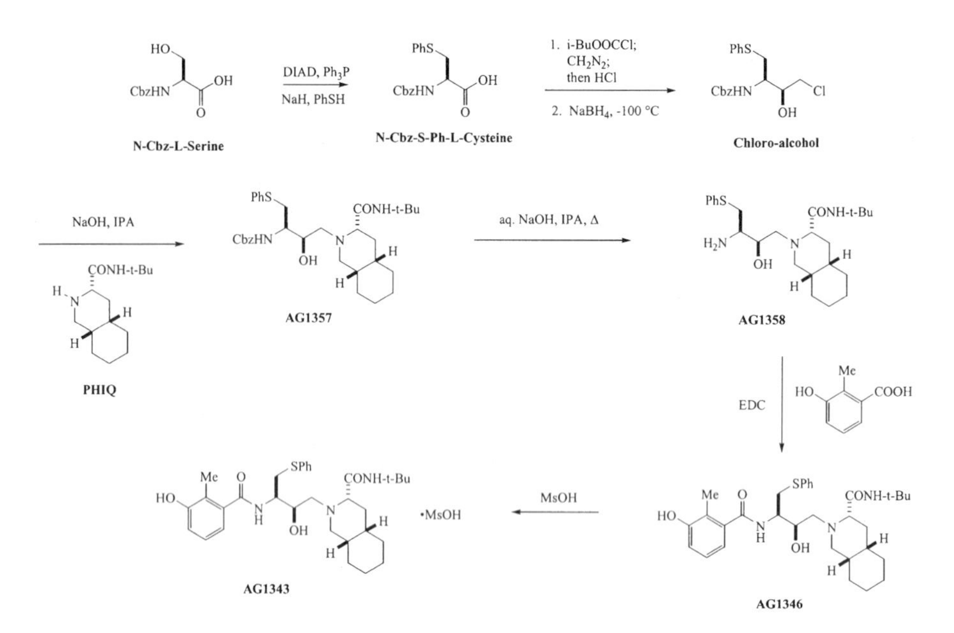

Scheme 2 Manufacturing route to nelfinavir.

L-Cbz-L-serine by formation of the corresponding β-lactone followed by in situ opening with sodium aryl thiolate. Conversion of the *N*-Cbz-*S*-Phe-L-cysteine into the diazoketone followed by treatment with HCl provided the chloroketone, which was reduced to provide the chloroalcohol. Cyclization of the chloroalcohol under basic conditions gave the desired epoxide **A** in 31% overall yield. Terminal opening of epoxide **A** with the known amine **B** (9) (prepared in 24% overall yield from Cbz–phenylalanine) and deprotection afforded the amino alcohol. Amide coupling of the amino alcohol with 2-methyl-3-hydroxybenzoic acid followed by mesylate salt formation gave nelfinavir mesylate in 16 steps and 12% overall yield.

Several improvements in the preparation of nelfinavir were made by the Agouron Process Group as outlined in Scheme 2. The introduction of DIAD in the synthesis of *N*-Cbz-*S*-Ph-cysteine gave improved yields. Lowering the temperature of the $NaBH_4$ reduction (−100°C) improved the selectivity in the formation of the chloroalcohol. Lastly, the sequence from the chloroalcohol to AG1358 was simplified significantly by using sodium hydroxide for both epoxide formation and Cbz cleavage. This new route would provide the multikilogram quantities of nelfinavir required to support clinical studies and ultimately production of drug product.

More extensive antiviral and pharmacokinetic studies were initiated on nel-

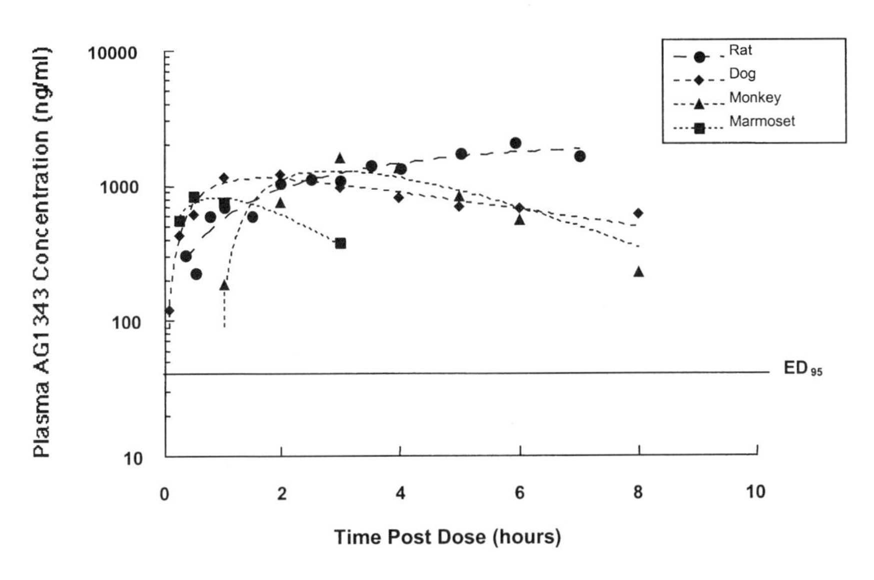

Figure 9 Pharmacokinetic data on nelfinavir in four species.

finavir owing to its superior antiviral activity. Nelfinavir was effective against the replication of several laboratory and clinical HIV-1 and HIV-2 isolates, with EC_{50} values ranging from 9 to 60 nM (vide infra) (10). Initial pharmacological studies of nelfinavir in rats indicated an oral availability of 43% in fed rats and 29% in fasted rats. Across four species, rat, dog, cynomolgus monkey, and marmosets, nelfinavir displayed significant oral availability (Fig. 9). Furthermore, a single oral dose of nelfinavir exhibited plasma levels exceeding the ED_{95} for more than 6 h in three of the four species. At this time, the preclinical results obtained with nelfinavir demonstrated potential antiviral and perhaps more importantly, pharmacokinetic advantages, when compared with most known HIV PR inhibitors. These data made it clear that further development of the compound for testing in humans was warranted.

IV. CLINICAL STUDIES

The phase 1 safety and pharmacokinetic studies of nelfinavir in humans revealed that single doses of 100–800 mg and multiple doses of 800–900 mg were well tolerated. Moreover, peak plasma levels of up to 100-fold over antiviral ED_{95} and trough levels up to 10 times the ED_{95} were achieved (Fig. 10). Consistent with earlier studies done in animals, the oral levels of nelfinavir were optimal when administered with a meal or light snack. Additional clinical studies of nelfinavir demonstrated good safety and efficacy as monotherapy and in combination

Dose (mg)	C_{max} [a] (ng/ml)	T_{max} [b] (hr)	$T_{1/2\text{-el}}$ (hr)	$AUC_{0\text{-inf.}}$ (ng hr/ml)
100	313±91	3.00	1.72±0.35	1250±230
200	439±124	3.00	2.21±0.75	1828±372
400	1577±965	4.00	2.28±0.36	7555±4175
800	3163±584	3.50	3.37±1.23	22208±6306

[a] Values shown are mean ± SD of data from 4 subjects.
[b] Median values for T_{max} are presented.

Figure 10 Mean pharmacokinetic parameters of nelfinavir (AG1343) when administered orally with food to healthy human volunteers.

with stavudine (d4T) or with zidovudine (ZDV) and lamivudine (3TC). The primary side effect observed associated with nelfinavir therapy has been mild to moderate diarrhea that can usually be controlled with nonprescription drugs, such as loperamide. More recently, a twice-daily dosing of nelfinavir with two reverse transcriptase inhibitors has been shown to significantly reduce HIV-1 RNA levels (11).

V. ANTIVIRAL PROFILE OF NELFINAVIR

Nelfinavir showed potent inhibition of the HIV PR with a K_i of 2 nM (saquinavir produced a K_i of 1 nM under the same conditions) and no significant activity against the human aspartyl proteases pepsin, renin, and gastricin at 1-μM concentrations. Antiviral efficacy of nelfinavir in both CEM-SS and MT-2 cells acutely infected with HIV-1 RF- and HIV-1 IIIB was clearly demonstrated with EC_{50}s between 30 and 40 nM (8). The corresponding cytotoxicity under these conditions was from 23 to 28 μM, providing a therapeutic index of between 500 and 900. Similar potency was achieved against a wide variety of HIV-1 strains across four different cell types (CEM-SS, MT-2, macrophage, and PBMC). Furthermore, potent inhibition of replication of HIV-2 ROD in CEM-SS cells was observed.

The HIV reverse transcriptase enzyme functions early in the virus life cycle whereas the HIV PR functions at a later stage. As might be expected, in vitro combination studies with nelfinavir and reverse transcriptase inhibitors ZDV, 3TC, zalcitabine (ddC), stavudine (d4T), and didanosine (ddI) showed additive to strong synergistic effects (12). The combination of nelfinavir with three other protease inhibitors, on the other hand, resulted in additive (saquinavir and ritonavir) to slightly antagonistic effects in vitro (indinavir). These data supported the notion of using combination therapy as a means of achieving a maximal clinical benefit.

To assess the ability of the virus to develop resistance when exposed to nelfinavir, serial passage studies were conducted in MT-2 cells. After 22 passages, a 37-fold reduced sensitivity was observed, a consequence of one primary mutation, D30N. When studies continued beyond passage 22, a different resistance pattern emerged (I84V, M46I) with an associated 30-fold reduced sensitivity. These in vitro data have been published (10). However, the more relevant resistance profile is that observed in patients who have undergone treatment with nelfinavir in completely suppressive monotherapy or combination regimens revealing a mutation different from that observed with other HIV PR inhibitors. Nucleotide sequence analysis of plasma taken from patients treated with nelfinavir showed an anomalous mutation profile. A unique aspartic acid (D) to asparagine (N) substitution at residue 30 was observed in selected patients treated with nelfinavir. From the cocrystal structure information and HIV PR inhibition discussed previously, mutation at Asp30 most likely interferes with the favorable H-bonding observed between the phenol group -OH on nelfinavir and the carboxylate of Asp30. There were other less frequent changes also observed; however, these were different from those genotypic changes associated with phenotypic resistance to other protease inhibitors. Furthermore, viral isolates that showed a high level of resistance to nelfinavir maintained susceptibility to saquinavir, indinavir, ritonavir, and amprenavir (13). The initial development of resistance to nelfinavir is unique among HIV protease inhibitors and has suggested that patients who fail nelfinavir treatment might respond to treatment with other protease inhibitors. These data have been recently substantiated both at the level of viral genotyping and phenotyping and in clinical studies and practice.

The identification and development of nelfinavir exemplifies the state of the art in drug design, discovery, and development. Starting with an aggressive structure-based program exploring a variety of inhibitor designs, a combination of structure-based methods and traditional medicinal chemistry overcame the challenges of producing a potent, well-tolerated protease inhibitor with an excellent pharmacokinetic profile. Nelfinavir was identified as a development compound in January of 1994 and phase 1 trials were initiated 10 months later. The U.S. Food and Drug Administration (FDA) approval followed just 39 months after nelfinavir's identification as a development candidate, marking the quickest development of any HIV protease inhibitor to date. Nelfinavir also became the first protease inhibitor to be cleared for marketing for use in children. Currently, Viracept (nelfinavir) is the most widely prescribed HIV PR inhibitor on the market.

ACKNOWLEDGMENTS

The discovery of Viracept involved the gifted contributions from a large number of scientists at Agouron Pharmaceuticals, Inc., and Eli Lilly. The record-breaking development of Viracept required an equally talented and committed development team (Agouron). In addition, there were extensive contributions from external collaborators, clinical investigators, and the very crucial clinical trial participants. The immense contributions from of all of the foregoing individuals who ultimately made a useful antiviral available to the many AIDS patients in need was key to the successful launch of this important product. The tremendous effort, talent, and enthusiasm of the many contributors to Viracept are respectfully acknowledged.

REFERENCES

1. Patick AK, Potts KE. Protease inhibitors as antiviral agents. Clin Microbiol Rev 1998:614–627.
2. (a) Dreyer GB, Metcalf BW, Tamaszek TA Jr, Carr TJ, Chandleer AC III, Hyland L, Fakhoury SA, Magaard VW, Moore ML, Strickler JE, Debouck C, Meek TD. Inhibition of human immunodeficiency virus 1 protease in vitro: rational design of substrate analogue inhibitors Proc Natl Acad Sci USA 1989; 86:9752–9756; (b) Roberts NA, Martin JA, Kinchington D, et al. Rational design of peptide-based HIV proteinase inhibitors. Science 1990; 248:358–361.
3. Reich SH, Webber SE. Structure-based drug design (SBDD): every picture tells a story . . . Perspect Drug Discov Design 1993; 1:371–390.
4. Hostomsky Z, Applet K, Ogden RC. High level expression of self-processed HIV-1 protease in *Escherichia coli* using a synthetic gene. Biochem Biophys Res Commun 1989; 161:1056–1063.
5. Appelt K. Crystal structures of HIV-1 protease. Perspect Drug Discov Design 1993; 1:23–48.
6. Kaldor SW, Hammond M, Dressman BA, Fritz JE, Crowell TA, Herman RA. New dipeptide isotereres useful for the inhibition of HIV-1 protease. Bioorg Med Chem Lett 1994; 4:1385–1390.
7. Reich SH, Melnick M, Davies JF II, et al. Protein structure-based design of potent orally bioavailable, nonpeptide inhibitors of human immunodeficiency virus protease. Proc Natl Acad Sci USA 1995; 92:3298.
8. [Initial disclosures on compound **10** (AG1343)] (a) Shetty B, Kaldor S, Kalish V, Reich S, Webber S. AG1343, An orally bioavailable non-peptidic HIV-1 protease inhibitor. Presented at the 10th Int AIDS Conference, Yokohama, Japan, Aug 1994; paper 321A. (b) Kalish V, Kaldor S, Shetty B, Tatlock J, Davies J, Hammond M, Dressman B, Fritz J, Appelt K, Reich S, Musick L, Wu B, Su K. Iterative protein structure-based design and synthesis of HIV protease inhibitors. Proceedings of the XIIIth International Symposium on Medicinal Chemistry, Paris, France, Sept 1994; 201s. [In early reports, AG1343 was also called LY312857.] (c) Kaldor SW, Kalish VJ, Davies JF II, Shetty BV, Fritz JE, Appelt K, Burgess JA, Campanale KM, Chirgadze NY, Clawson DK, Dressman BA, Hatch SD, Khalil DA, Kosa MB, Lubbehusen PP, Muesing MA, Patick AK, Reich SH, Su KS, Tatlock JH. Viracept (nelfinavir mesylate, AG1343): a potent, orally bioavailable inhibitor of HIV-1 protease. J Med Chem 1997; 40:3979–3985.

9. Parkes KEB, Bushnell DJ, Crackett PH, Dunsdon SJ, Freeman AC, Gunn MP, Hopkins RA, Lambert RW, Martin JA, Merrett JH, Redshaw S, Spurden WC, Thomas GJ. Studies toward the large scale synthesis of the HIV proteinase inhibitor Ro 31-8959. J Org Chem 1994; 59:3656–3664.
10. Patick AK, Mo H, Markowitz M, Appelt K, Wu B, Musick L, Kalish V, Kaldor S, Reich SH, Ho D, Webber S. Antiviral and resistance studies of AG1343, an orally bioavailable inhibitor of human immunodeficiency virus protease. Antimicrob Agents Chemother 1996; 40:292–297.
11. Sension M, Elion R, Farthing J, Currier J, Lindquist C, Richardson B, Becker M. Open label pilot studies to assess the safety and efficacy of bid dosing regimens of Viracept (nelfinavir mesylate) combined with NRTI's in HIV-infected treatment naïve patients. Abstr 5th Conf Retroviruses Opportunistic Infect 1998; p 151, abstr 387A.
12. Patick AK, Bortizki TJ, Bloom LA. Activities of human immunodeficiency virus type-1 (HIV-1) protease inhibitor nelfinavir mesylate in combination with reverse transcriptase and protease inhibitors against acute HIV-1 infection in-vitro. Antimicrob Agents Chemother 1997; 41:2159–2164.
13. Patick AK, Duran M, Cao Y, Shugarts D, Keller MR, Mazabel E, Knowles M, Chapman S, Kuritzkes KR, Markowitz M. Genotypic and phenotypic characterization of human immunodeficiency virus type 1 variants isolated from patients treated with the protease inhibitor Viracept. Antimicrob Agents Chemother 1998; 42:2637–2644.

6

Design and Synthesis of Amprenavir, A Novel HIV Protease Inhibitor

R. D. Tung, D. J. Livingston, B. G. Rao, E. E. Kim, C. T. Baker, J. S. Boger, S. P. Chambers, D. D. Deininger, M. Dwyer, L. Elsayed, J. Fulghum, B. Li, M. A. Murcko, M. A. Navia, P. Novak, S. Pazhanisamy, C. Stuver, and J. A. Thomson
Vertex Pharmaceuticals, Cambridge, Massachusetts

I. BACKGROUND AND DESIGN CONSIDERATIONS

Efforts at Vertex Pharmaceuticals toward the discovery of novel human immunodeficiency virus (HIV) aspartyl protease inhibitors began late in 1990. A substantial body of in vitro evidence had been developed by that time supporting this enzyme as a therapeutic target to suppress HIV replication, although no compounds had yet demonstrated clinical efficacy. Recombinant HIV possessing an inactive, mutated protease was known to be replication incompetent, as had been demonstrated for an analogous aspartyl protease in murine leukemia virus (1–3). Cloning and expression of the protease had been accomplished, and X-ray crystallographic structures of both apo- and inhibitor-complexed forms were available (4–6). This wealth of existing information on the HIV protease and the lack of highly effective therapeutic agents to treat HIV infection rendered the protease an attractive target for research at Vertex, which had recently been founded as a structure-based drug design company.

Potent inhibitors of the HIV protease had already been described. Most groups had based their work on chemical leads and approaches previously developed for the inhibition of other aspartyl proteases, particularly the antihypertensive target renin. Innovative work by Kempf et al. (7) utilized the C2 symmetry of the protease to develop structurally novel, tight binding agents (8). The substantial groundwork laid by those groups helped define attributes necessary for therapeutically successful HIV protease inhibitors (HIV PIs).

Early reports indicated generally poor oral bioavailability or rapid systemic clearance as a common feature of the first HIV protease inhibitors, in common with early renin inhibitors (9). Overcoming that hurdle proved one of the most difficult elements in developing therapeutically useful renin inhibitors, and still prevents commercialization of that class of agents. Consequently, from the outset, we directed our primary focus toward attaining an improved oral-dosing profile for intracellularly potent agents. As a secondary goal, because the lymphatic system and central nervous system (CNS) had been described as potential viral reservoirs, we targeted compounds that could gain access to those compartments (10). Finally, viral resistance was known to occur in response to treatment with reverse transcriptase inhibitors, and it was reasonable to believe it might also occur in response to HIV PI treatment (11). To minimize the potential for resistance development, we hoped to direct the inhibitor's binding to amide–backbone interactions to the greatest extent possible, while minimizing interactions with potentially mutable amino acyl side chains.

II. DESIGN TACTICS

Oral bioavailability still remains poorly predictable owing to the numerous and disparate biological barriers facing the candidate compound. Specific obstacles include the need for thermodynamically and kinetically relevant dissolution, appropriate partition through biological solutions and membranes, avoiding chemical and enzymatic degradation, direct elimination through renal or hepatic routes, and inappropriate sequestration (12–19). Among the few generalizations linking physicochemical characteristics with oral bioavailability, optimum pharmaceutical properties are generally associated with compounds with a balance of both aqueous and lipid solubility, within a molecular weight range of roughly 150–550 Da (15,20). Hepatic extraction of xenobiotic agents has also been related to molecular weight, both for general organic compounds and for peptides and peptidomimetics, such as renin and angiotensin-converting enzyme inhibitors (9,14,15,21). Most of the HIV protease inhibitors known at the outset of our work were in excess of 600 Da, suggesting a general strategy of reducing molecular weight.

We also chose to avoid physiologically ionizable moieties in our design process. In some instances, their presence is associated with lower oral bioavailability (22). Molecular charge, both positive and negative, is also responsible in many instances for transporter-mediated renal uptake and urinary elimination (23). Also, both the lymphatic system and CNS appear to be accessed well by lipophilic agents (24). The preferred physicochemical profile emerging from this analysis was a relatively low molecular weight compound lacking obligate charge at physiological pH, with substantial lipophilicity, but sufficient aqueous solubility to allow dissolution and absorption in the gut.

As with renin inhibitors, HIV PIs appeared to require interactions across a substantial span of the enzyme-active site to achieve potent (low to subnanomolar K_i) enzymatic inhibition. Because of both our intent to minimize molecular weight and to reduce the likelihood for viral resistance, we focused on incorporating known, favorable interactions with the invariant enzyme backbone amides. Previous work on both renin and the HIV protease indicated that hydrophobic side chain interactions, particularly in S_1 and S_1', were necessary, and S_2 was important in both renin and HIV for substrate recognition. Beyond those residues, we sought to truncate the molecule to the minimum number of ''substrate-residue equivalents'' possible, to minimize the inhibitor's molecular weight.

The specific approach that led to the discovery of amprenavir derived from examination of the substrate cleavage sites of the HIV protease. Among the eight substrate sequences present in the *gag* and *gag–pol* gene products, several possess an X-X-(Tyr/Phe)-Pro-X-X sequence (25), which indicated that a P_1' amide NH is not obligatory for binding and productive catalysis. Molecular dynamics calculations demonstrated this to be the weakest of the backbone hydrogen bonds between a (P_3–P_3') inhibitor and the enzyme (26,27), a fact underscored by the surprisingly high affinity of certain ester isostere C2 symmetric inhibitors (28,29). Secondary amide bonds have been associated with reducing passive diffusion through *in vitro* models of the gut endothelium as well as limiting blood–brain barrier penetration, so the opportunity to remove this functional group while retaining enzymatic recognition was very attractive (30–33). Another feature we took note of was the crystallographically observed structural water molecule linking the P_2 and P_1' residue carbonyls of enzyme-complexed inhibitors with Ile50 and Ile50′ on the flap regions of the HIV protease (34). Its nearly invariant presence across both reported structures and those we solved ourselves suggested a significant energetic importance of this ''flap water.'' In addition, all of the most potent aspartyl protease inhibitors have possessed a group that can productively interact with the two active-site aspartate side chain residues to mimic the transition state of substrate hydrolysis.

These observations led us to target structures possessing a moiety appropriately positioned to maximize interactions with the active-site aspartate side chains and the flap water while filling at least the central P_2–P_2' subsites of the enzyme-active site. The invariant presence of phenylalanine or tyrosine in P_1 of the natural *gag* and *gag–pol* gene product cleavage sites containing proline suggested an (*S*)-phenylalanine-derived ''transition-state'' binding unit reminiscent of various earlier aspartyl protease inhibitors (8,35,36). With this starting point, we compared distances between the flap water-binding carbonyls of other HIV protease inhibitors known in the literature with a variety of backbones that could be extended from the phenylalinol pharmacophore (Table 1).

We directed our chemistry tactics toward targets that were amenable to short, flexible synthetic pathways. To the greatest extent possible we incorporated starting materials for which a variety of analogues were commercially available, or for which short, robust syntheses were described in the literature. We also took the approach of designing multiple points of convergence for our prototype inhibitors whenever possible, to allow ready incorporation and testing of modeling and structural insights. Critically, we ensured that interactions betwen chemists, biophysicists, modelers, and enzymologists occurred on a real-time basis, to allow rapid incorporation of structural insights and biological-testing data into compound design and synthesis.

After several iterations of computational and synthetic structural explorations, we focused on the *N,N*-dialkyl sulfonamide unit to bind to the flap water and attach the P_1' and P_2' groups. That choice was based on the following consid-

Table 1

Pdb code	P_CO....P'_XO	P_CO....WAT	P'_XO....WAT
4hvp	4.02	3.04	2.48
7hvp	4.55	2.58	2.77
8hvp	4.40	2.47	2.58
9hvp	4.43	2.70	2.64
4phv	4.25	2.49	2.59
1hiv	4.65	2.62	2.68
VB-10,980	5.31	3.04	2.95
AMPRENAVIR	4.86	2.65	2.88

Inter- and intramolecular distances involving the two flap water binding groups, expressed in Ångstroms. Data for this table was derived from coordinates deposited in the Cambridge Structural Database. In columns 2 and 4, when X=carbon, the flap water binding group is an amide (or, in the case of amprenavir, a carbamate); when X=S (examples VB-10,980 and amprenavir), the group is a sulfonamide. The intermolecular distances from the inhibitor oxygens to the flap water are very similar in the amides and sulfonamides, as are as the intramolecular distances between the inhibitor oxygen atoms. This geometric consistency supports our prediction that the sulfonamides would be a reasonable choice for a flap water hydrogen bonding group. The structure of amprenavir is given in Table 4. Compound VB-10,980 is an example of general structure V in Figure 1, containing a quinaldoyl aspariginyl moiety in P_3/P_2, an isobutyl group in P_1' and a 3-acetamide, 4-fluorobenzene sulfonamide. References for the Pdb structure codes are as follow: 4HVP, reference 4; 7HVP, A. L. Swain *et al.*, *Proc. Natl. Acad. Sci. USA* **87**, 8805 (1990); 8HVP: M. Jaskolsk *et al.*, *Biochemistry*, **30**, 1600 (1991); 9HVP, Erickson *et al.* in reference 8; 4PHV, R. Bone *et al.*, *J. Am. Chem. Soc.*, **113**, 9382 (1991); 1HIV, N. Thanki *et al.*, *Protein Science,* **1**, 1061 (1992).

erations: first, molecular modeling indicated that the sulfonamide group would be able to make one or possibly two strong hydrogen bonds with the flap water. An examination of the Cambridge Structural Database showed that the two most commonly observed conformations of the *N,N*-dialkyl sulfonamide group would be both geometrically and energetically similar to the conformation that we expected to produce optimum binding to the enzyme. This presumption was later confirmed by detailed *ab initio* calculations on a series of *N,N*-dialkylbenzenesulfonamides using the 6 -13G* basis set. These studies demonstrated active–site-bound conformations (based on dihedral angles taken from amprenavir and related compounds cocomplexed with the HIV protease) within about 1–2 kcal/mol of the solution global minima. Table 2 shows a comparison of the calculated relative energies of various conformations of *N,N*-dialkylbenzenesulfonamides along with the experimental values found in the X-ray structure of amprenavir. We believed that minimizing conformational reorganization from the inhibitor's solution state to that in the enzyme's active site, thereby reducing entropic loss of binding energy, would be an important contributor to attaining potent, low molecular weight agents.

The relatively large, partial negative charge on the sulfonamide oxygens lends that group to act as a productive hydrogen bond acceptor when optimally placed. Furthermore, sulfonyl chlorides and thus sulfonamides moiety are readily accessible through well-known chemistry (23,37–39) (Fig. 1).

Compounds of this series proved to be active HIV-1 protease inhibitors, displaying K_i values of tens of nanomolar or less. X-ray diffraction analyses of these inhibitors complexes with the HIV protease showed that the bound conformation of the inhibitor backbones were substantially similar to those suggested by computational analysis. Clearly defined electron density for the flap water and good hydrogen bond distances between it and the sulfonamide oxygens were also observed in all cases, supporting our modeling prediction of strong hydrogen bond interaction between these two groups (Fig. 2). The torsional angles (χ_1 and χ_2) within the *N*-benzyl side chain, however, differed markedly from the ideal values. Further inspection revealed that only a limited range of χ_1 and χ_2 rotations are available to the CH_2-phenyl group owing to steric demands imposed by the adjacent aryl sulfonamide moiety and the side chains of the S_1' pocket. This constraint introduces considerable intramolecular strain in the enzyme-bound conformation of the inhibitor. Substituting the smaller isobutyl moiety for the *N*-benzyl P1′-binding group ameliorated that problem (see Fig. 2). The sp^3 branched isopropyl in place of the sp^2-phenyl ring allowed a broader range of χ_2 rotation, with lower intramolecular strain and better overlap with the S_1' pocket. The *N*-isobutyl-substituted compounds displayed greatly improved enzymatic inhibition relative to the parent *N*-benzyl series with reductions in K_i values of 2 log units or more in a number of directly comparable cases. The enzymatic activity of the *N*-isobutyl series was reflected by improved antiviral potency, with the better

Table 2

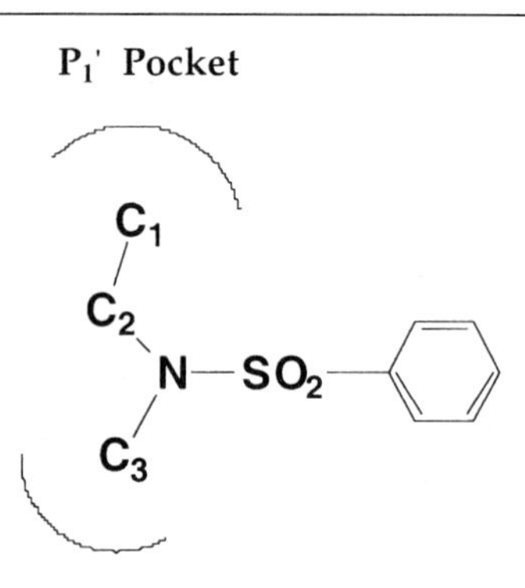

Source	C1-C2-N-S	C2-N-S-C	C3-N-S-C	N-S-C-C	
	Torsion Angles				
Cambridge[a]	110±20	-75±11	75±11	89±8	
Ab Initio[b]	130	-75	74	88	
AMPRENAVIR[c]	97	-118	82	80	
Ab Initio[d]	97	-118	82	80	ΔE ~1.5 kcal/mol

Crystallographic and theoretical data on the conformational preferences of N,N-dialkylbenzenesulfonamides. (a) The 1994 version of the Cambridge Structural Databank contains 10 entries for N,N-dimethylsulfonamide derivatives without ortho substituents. The torsional values for C_2-N-S-C, C_3-N-S-C, and N-S-C-C were derived from the data for these compounds. The database also contians 4 compounds containing the C_1-C_2-N-S moiety, which have been used to estimate the torsional preference. (b) HF/3-21G* optimizations were carried out on N-ethyl-N-methylbenzenesulfonamide. The global minimum structure had the torsion angles shown. (c) Data taken from the refined 2.0Å structure of AMPRENAVIR. (d) MP2/6-31G*//3-21G* single-point calculations shown that the global minimum conformation of N-ethyl-N-methylbenzenesulfonamide is only 1.5 kcal/mol lower in energy than the conformation in which the torsional angles are fixed at the values seen crystallographically for amprenavir.

analogues displaying cellular IC_{90} values of 100 nM or less. These inhibitors displayed low toxicity to the host T-cell line, with cellular therapeutic indices of generally more than 10^4 for the more active compounds. Similar results were subsequently reported by Vasquez et al. (40).

Further reduction in molecular size was accomplished by truncation of the P_3–P_2 portion. Small acyl substituents on the N^{α} position of the P_2 asparaginyl

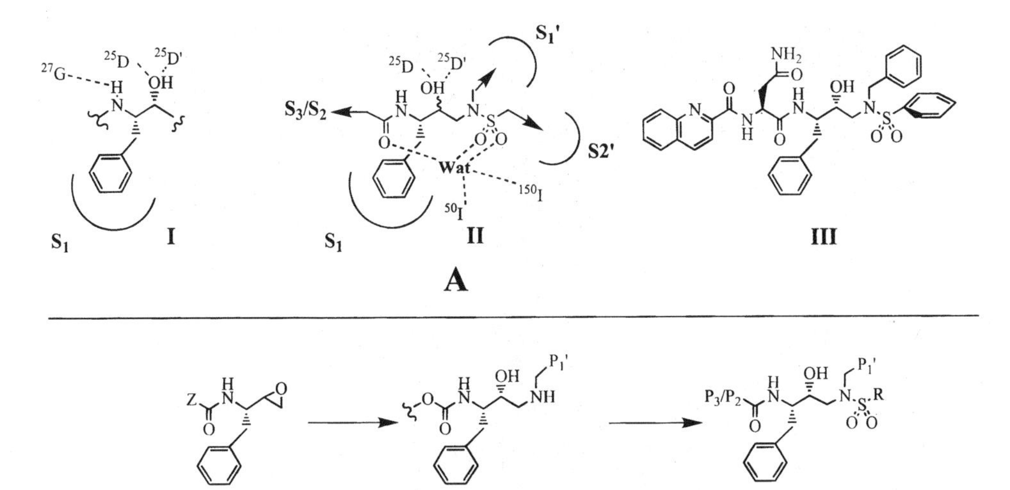

Figure 1 (A) The presence of a P_1 aromatic residue in all substrate cleavage sites containing a proline in P_1' suggested a phenylalanine-like side chain as the basis for binding to that portion of the enzyme. Addition of a transition state-mimetic alcohol group to interact with the active site aspartyl residues gave the fragment shown in structure I. Examination of a variety of functional groups designed to interact with the structural flap water yielded a combination of an amide carbonyl at P_1 and a sulfonamide at P_1' as a particularly promising combination due to the relatively shallow energetic barrier to rotation of the sulfonamide bonds, potential for strong hydrogen bonding interaction with the structural water molecule and ready synthetic access described in B. Early examples within this series utilized an N-benzyl group for P_1' binding and an aryl sulfonamide to access P_2'. The quinaldoyl asparaginyl group described by Roberts et al. (6) provided additional anchoring to provide high binding affinity in these prototypes as shown in structure III. Dashed lines represent hydrogen bonds between the enzyme and key inhibitor functional groups and bold arrows in structure II show interactions of the indicated appendages with substrate pockets in the HIV protease. (B) Synthetic access to a wide variety of analogs was available through the route indicated. The well-known phenylalanine epoxide IV may be opened by a variety of primary amines (17), and subsequent sulfonylation with commercially available or readily prepared sulfonyl chlorides allowed rapid and broad exploration of the S_1' and S_2' pockets of the enzyme. N-Deprotection of the phenylalinol moiety followed by standard amide forming condensations allowed access to the extended S_2/S_3 portions of the active site. In this scheme, "Z" represents a generic N-protecting group and "R" represents a range of available substituents.

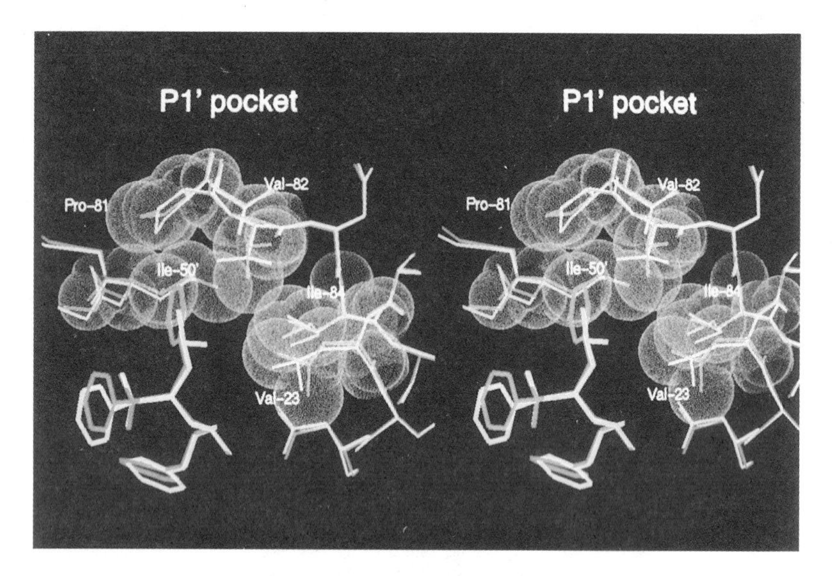

Figure 2 Stereo view of the S_1' pocket of the HIV-1 protease in complex with two inhibitors. The two inhibitors differ only at the P_1' site: one has an *N*-benzyl group whereas the other has an *N*-isobutyl group. Only parts of the enzyme (22–25, 80–85, 49′–51′) and inhibitors (P_1, P_1', and P_2' substituents) are shown for clarity. Key residues (Leu-23, Pro-81, Val-82, Ile-84, Ile-50′) of the enzyme pocket are labeled and were used to generate the van der Waals surface. Only the enzyme coordinates corresponding to the green structure were used for the generation of the surface. Inhibitor bonds are shown in thicker lines. The inhibitor with the N-benzyl side chain is involved in suboptimal intra- and intermolecular interactions in the S_1' pocket: (a) the side-chain is forced to adopt an almost eclipsed conformation of χ_2 side-chain torsion; (b) the χ_1 torsion also differs from its ideal value; (c) it does not fill part of the S_1' pocket in the vicinity of Val-32; and (d) it forces the Val-82 side-chain χ_1 angle from its preferred trans conformation to gauche. The sp^2 hybridization of the β-branched carbon and large ring size of the *N*-benzyl side chain appear to be the cause of the suboptimal binding of the *N*-benzyl sulfonamide inhibitors. Substitution of the benzyl group with an isobutyl group helped minimize most of these suboptimal interactions. The sp^3 hybridization of the β-branched carbon enables the side chain to fill the S_1' pocket while relieving the torsional strain in χ_1 and χ_2. Further, Val-82 side chain in the structure of *N*-isobutyl complex has a preferred *trans* orientation at the C1 torsion.

residue attenuated the binding efficacy of these inhibitors, consistent with the findings of Roberts et al. (41,42) on an earlier series of potent HIV protease inhibitors and did not yield significant size reductions in the resulting inhibitors. Therefore, we sought to replace the entire P_3–P_2 portion with a smaller and chemically simpler group while retaining the desirable activity and cellular therapeutic index of the parent compounds. Examination of a variety of appending groups demonstrated that simple alkyl carbamates provided excellent binding relative to a variety of other groups. Several small carbamates provided yet further enhancements of potency and were variously considered as possible P_2 groups within the lead structures (Table 3). Because of its straightforward synthesis from the commercial malic acid and combination of good activity enhancement and small size, we chose the (*S*)-tetrahydrofuryloxycarbonyl group utilized in an earlier class of HIV protease inhibitors by Ghosh et al. (43–45).

Table 3

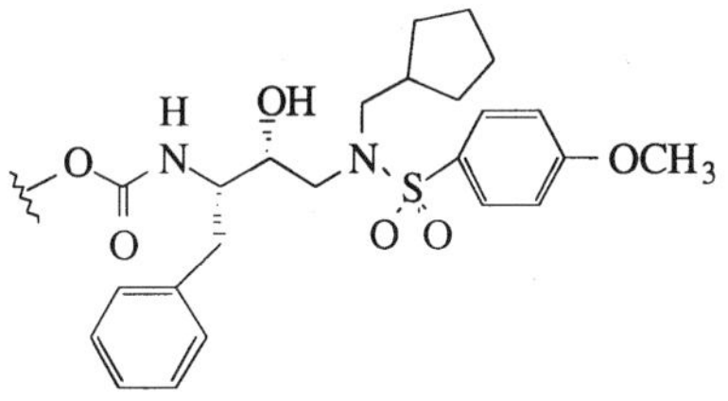

Compound	Carbamate alkyl group	K_i
VB-12,556	tert-Butyl	0.3
VB-12,784	(S)-3-hydroxy-tetrahydrothiofuranyl sulfone	0.1
VB-11,329	Allyl	0.1
VB-11,328	(S)-3-hydroxytetrahydrofuryl	< 0.1
VB-12,782	5-hydroxy-1,3-dioxanyl	< 0.1
VB-11,330	3-pyridylmethyl	< 0.1
VB-11,277	(S)-3-hydroxytetrahydrofuryl*	0.16

* Compound VB-11,351 contains an isobutyl group in place of cyclopentylmethyl interacting with P_1'

Carbamate substituents on the N,N-disubstituted hydroxyethylamine sulfonamide scaffold provide excellent inhibition of the HIV protease across a range of structures as shown. Inhibition of HIV-1 protease at varying inhibitor concentration was determined by monitoring the rate of hydrolysis of H-III (Bachem Bioscience) peptide substrate (His-Lys-Arg-Val-Leu-(p-NO_2)Phe-Glu-Ala-Nle-Ser-NH^2) using HPLC analysis. The K_i values were determined by computer fitting to equations for tight-binding competitive inhibitors (KineTic® program, BioKin, Ltd.)

As shown in Table 3, homologation of the isobutyl group in P_1' to cyclopentylmethyl also yielded further increases in enzymatic binding. However, within this now significantly smaller class of inhibitors, sufficient binding efficiency had been attained that our efforts were driven more by a desire to continuously exceed blood levels necessary for antiviral activity than by absolute inhibitory efficacy. Early compounds in this series exemplified by VB-11,328 were excellent antiviral agents, but very poorly soluble in aqueous media and displayed poor blood levels on oral administration to rats (Table 4). The high lipophilicity of these compounds reflects the mostly hydrophobic nature of the enzyme's active site and the structural complementary of the inhibitors. To enhance aqueous solubility in the inhibitors while retaining binding affinity, we examined molecular complexes with the enzyme to identify portions of the inhibitor residing in a physicochemically ''mixed'' area of the enzyme that might be easily accessible synthetically.

The phenyl ring attached to the sulfonamide nitrogen ideally met the foregoing criteria. The area distal to the sulfur is surrounded by a mixture of hy-

Table 4

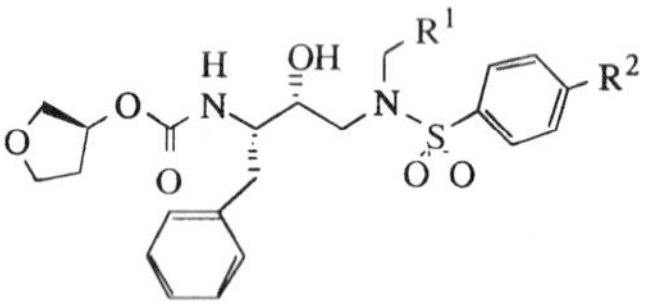

Cmpd. #	R^1	R^2	Aq. solubility (µg/mL)	K_i (nM)	IC_{90} (nM)	C_{max} (nM)
11,328	CyP	$-OCH_3$	0.3	< 0.1	13	88
11,277	^{i}Bu	$-OCH_3$	1	0.16	50	360
11,235	CyP	$NH-COCH_3$	0.7	0.3	67	107
11,103	^{i}Bu	$NH-COCH_3$	46	1.3	190	424
11,599	CyP	NH_2	33	0.2	10	287
amprenavir	^{i}Bu	NH_2	190	0.6	47	1900

Comparison of structure with activity and blood levels attained on oral dosing of selected HIV protease inhibitors. Aqueous solubility was measured by HPLC analysis of saturated phosphate-buffered saline solutions at pH 6.8. K_i values were determined as described in table 3. IC_{90} values are the concentration of compound necessary to inhibit 90% of viral replication of CCRF-CEM cells acutely infected with HIV-1IIIb as measured by extracellular p24 production. C_{max} values are the maximal blood concentration of the compound observed in adult male Sprague-Dawley rats (Charles River) drawn by retro-orbital bleed. Samples were extracted with acetonitrile: methanol and the compound concentration determined by HPLC-MS. Each time point represents the average of three animals.

drophobic side chains such as Ile50, Ile84, Ala28′, Ile32′, Val47′, and hydrophilic groups, including the main-chain amides of Asp29 and Asp30 and the side chain of Asp30. Although limited in volume, the hydrogen-bond rich environment of the surrounding pocket suggested that a variety of hydrophilic groups might be incorporated without interfering with binding to the enzyme. The commercial availability of many aryl sulfonyl chlorides and extensive literature detailing their chemistry fulfilled our synthetic requirements.

Accordingly, we synthesized a group of substituted sulfonamides incorporating various uncharged, but potentially solubilizing, functional groups and determined their solubility, enzymatic binding, intracellular antiviral activity, and blood levels achieved on oral dosing. As shown in Table 4, these compounds, as a group, possessed a wide range of aqueous solubility while retaining excellent enzymatic and viral inhibition properties. For example, amprenavir displayed more than 500-fold greater solubility in phosphate-buffered saline (PBS) than did VB-11,328, while maintaining similar antiviral activities. Compounds containing a P_1' isobutyl group often possessed much higher aqueous solubility than did their cyclopentylmethyl congeners. The combination of low molecular weight, acceptable solubility in aqueous and organic media, and excellent antiviral potency led us to examine amprenavir more closely and subsequently advance the compound to preclinical development status.

III. PRECLINICAL PROFILE

Routine antiviral analyses of our compounds was carried out by an independent commercial contractor under U.S. Food and Drug Administration (FDA)-submissible Good Laboratory Practice (GLP) conditions. We also utilized a network of secondary commercial vendors, and academic collaborators provided us with access to a wide range both of assays and virus provenance. Across these studies, we uniformly observed high antiviral potency that generally appeared to be greater against primary clinical strains than laboratory-adapted strains. These findings were also confirmed in extensive evaluation carried out by Burroughs–Wellcome, with whom we undertook a research and development collaboration in late 1993. Amprenavir has also recently been reported to have good activity against chronically infected monocytes, an important viral reservoir that is poorly suppressed by several other commercial protease inhibitors (46). No remarkable cytotoxicity or cytostatic effects were observed against a variey of cell types, including MT-2, CEM, Jurkat, Hela S3, THP-1, LB, and several lines of bone marrow progenitor cells, with TC_{50} level exceeding 20 μM in all cases, as measured by dye-uptake assay. Isolated receptor profiling and pharmacological analysis in rats (Panlabs pharmascreen and receptor screen assays) similarly demonstrated a benign profile.

Amprenavir displayed a resistance pattern on viral passaging distinct from that of other known protease inhibitors. In two separate viral-passaging experiments, carried out in one case by Vertex scientists working in the laboratory of a Harvard collaborator, and in another case by Burroughs–Wellcome researchers, the ''signature'' mutation associated with amprenavir was found to be I50V. This mutation is observed in extremely low abundance naturally and has not been observed with other approved protease inhibitors. Multiple mutations were found to be necessary for moderate to high levels (>10×) of viral resistance to amprenavir. The basis for the effect of the I50V mutant was directly observed by cocrystallizing amprenavir with that enzyme (manuscript in preparation). Little change was observed either in the enzyme or inhibitor conformations, with the result that the residue-50 side chain, changing from a methyl group to a hydrogen, reduced its hydrophobic contact with the P_2' phenyl group in amprenavir. Other commercial protease inhibitors lack a structural correlate to amprenavir's phenyl sulfonyl group, explaining the lack of effect of this mutation on their binding. Notably, the L90M mutation, which has been observed clinically in response to all other commercial PIs and has been reported to be an important contributor to protease inhibitor cross-resistance, has no apparent effect on amprenavir sensitivity as a single mutation or in addition to a background of other resistance mutations.

Oral bioavailability of amprenavir in rats was excellent (Fig. 3). Following a 10-mg/kg–oral dose, a C_{max} of 1.9 μM was observed in blood at 30 min. After 6 h, 0.45 μM or nine times the IC_{90} remained in the blood, and at 8 h the concentration was 90 nM. The elimination half-life, $T_{1/2\beta}$, was approximately 1–2 h. Following intravenous administration of 1 mg/kg, rapid distribution was followed by terminal elimination with a half-life of approximately 0.5 h. Comparison of areas under the curve (AUCs) for these two experiments, accounting for the difference in dose, yields an oral bioavailability of 70%. No evidence of acute toxicity was observed in these experiments or in others at oral doses ranging up to 3000 mg/kg. Similar observations in dogs were also noted by Burroughs–Wellcome. Negative findings in genotoxicity tests were also obtained. The composite of this promising data led us and Burroughs–Wellcome to undertake clinical development of amprenavir.

It then became necessary to produce multikilogram quantities of amprenavir under cGMP conditions to provide for longer-term animal toxicology experiments and support preparation of the Chemical, Manufacturing, and Controls section of the Investigational New Drug (IND) application. Vertex lacked the facilities to undertake such efforts internally, so we worked closely with several contract manufacturing vendors to develop appropriate manufacturing conditions and produce bulk drug substance and supporting analytical methods. Eventually, using relatively minor modifications of our original laboratory synthetic scheme, we were able to work with a vendor who produced several hundred kilos of cGMP

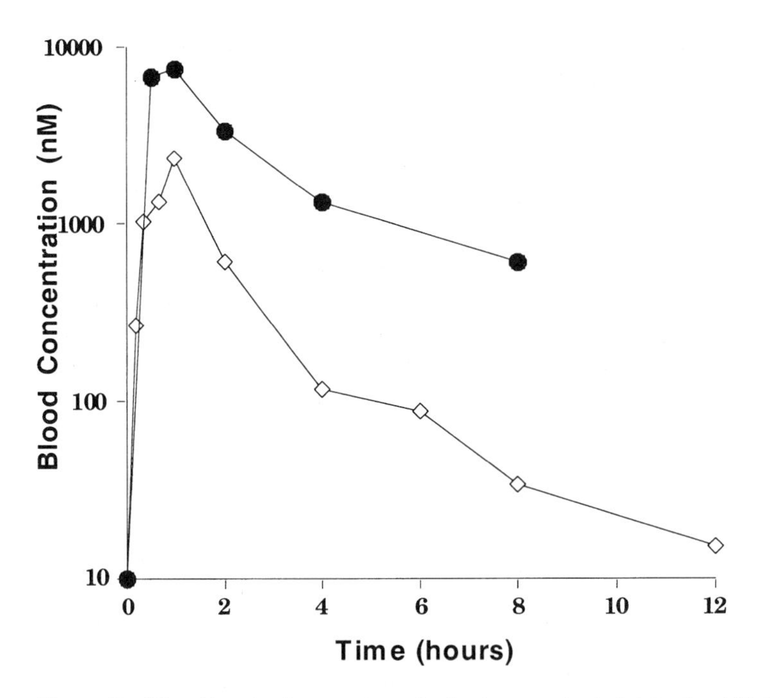

Figure 3 Blood levels observed on single-dose oral administration VX-478 to animals at 25 mg/kg. Filled circles are mean blood levels in Sprague-Dawley rats (n = 6). Open diamonds are mean blood levels in samples drawn from a peripheral vein of male and female cynomologus monkeys (n = 4). Dosing was carried out and blood levels determined as described in the legend to Table 4. The absolute bioavailability (%F) pf VX-478 at a 25 mg/kg dose is 60% in rats and 20% in monkeys.

material before transfer to Burroughs–Wellcome's process chemistry group was finalized. This material enabled the conduct of 1-month animal toxicology studies and initial human clinical evaluation.

IV. CLINICAL RESULTS

The full clinical profile of amprenavir is beyond the purview of this chapter. The following section summarizes the pharmacokinetics and viral resistance observations salient to the design considerations discussed earlier. In HIV-infected patients, amprenavir possesses a biological half-life of 7.5–10 h across a number of studies, substantially exceeding those of other currently marketed protease inhibitors, which range from approximately 1 to 5 h. Absolute bioavailability is not currently available, but pharmacokinetic modeling of the observed absorption profile suggests oral bioavailability of 70–90%. Linear pharmacokinetics are observed with the drug over the range of 150–1200 mg, the clinically approved dose. Administration concurrent with a high-fat meal (67 g fat, 967 kcal total, 62% of calories from fat) reduces C_{max} and AUC by 36 and 21%, respectively. However, owing to a delay in T_{max}, high-fat meals do not adversely affect C_{12} (C_{min}), which has been reported to be the pharmacokinetic parameter associated with clinical efficacy of HIV PIs (7,47). This bioavailability profile, allowing substantial flexibility in dosing, could be a marked advantage in terms of encouraging patient adherence, particularly in comparison with a three-times daily regimen.

The in vitro mutation profile observed on serial passaging has been largely, although not wholly, reflected in clinical experience. The majority of resistant mutants sequenced to date contain the I50V mutation, generally with secondary mutations similar to those observed in vitro. A second resistance pathway, involving mutation of residue 54, has also been observed, likely as a function of the initial genetic background of the virus. Early clinical results indicate that patients who develop resistance to amprenavir and are subsequently treated with a different protease inhibitor may have good virological responses in many, although not all, cases. Furthermore, Race et al. (48), in Clavel's laboratory have shown that, across a cross section of patients previously treated with protease inhibitors, a higher incidence of viral sensitivity to amprenavir was observed than to other commercial protease inhibitors.

Biodistribution of amprenavir is still under study. Early experiments indicate that it is present to a limited extent in cerebrospinal fluid (CSF). Preclinical studies demonstrated a substantially higher concentration in brain tissue than in CSF, and several clinical studies utilizing amprenavir indicate that both as a monotherapy and in combination therapy with nucleosides it effectively reduces viral RNA in CSF. Eron et al. also recently reported that amprenavir, both as monotherapy and as part of combination treatment with zidovudine (AZT) and lamivudine (3TC), substantially reduces viral load in seminal plasma.

V. CONCLUSIONS

Our goal in this program was to develop an effective anti-HIV therapeutic agent combining high antiviral potency with favorable pharmacological properties. To do so, we utilized multiple bound ligand–protein structures and applied a rapid modeling/synthesis/testing cycle to prototype and successively refined inhibitors. In this manner, starting from substrate-derived information, we developed a novel aryl sulfonyl diaminopropane molecular scaffold that closely mimics the enzyme active-site structure and minimizes molecular rearrangements when transiting from solution to protein-bound states, maximizing its binding energy/molecular weight ratio. The end result of these efforts, amprenavir, is currently marketed by Glaxo–Wellcome, under the trade name Agenerase, and by Kissei Pharmaceuticals in Japan, under the trade name Prozei. Amprenavir's clinical profile reflects its molecular origins, with high antiviral efficacy, excellent pharmacokinetics and oral bioavailability, a distinct viral resistance pattern, and wide biodistribution. As one of the first marketed products resulting from structure-driven drug design, amprenavir represents a landmark in the application of structural-based molecular design to pharmaceutical science, and is a potent addition to the armamentum of anti-HIV drugs.

REFERENCES

1. Crawford S, Goff SP. J Virol 53:899–907, 1985.
2. LeGrice SFJ, Mills J, Mous J. EMBO J 7:2547–2553, 1988.
3. Seelmeier S, Schmidt H, Turk V, Von der Helm K. Proc Natl Acad Sci USA 85: 6612–6616, 1988.
4. Navia MA, Fitzgerald PMD, Keever BM, Lew C–T, Heimbach JC, Herber WK, Sigal IS, Darke PL, Springer JP. Nature 337:615–620, 1989.
5. Wlodawer A, Miller M, Jaskolski M, Sathyanarayana Bk, Baldwin E, Weber IT, Selk LM, Clawson L, Schneider L, Kent SBH. Science 245:616–21, 1989.
6. Lapatto L, Blundell T, Hemmings A, et al. Nature 342:299–302, 1989.
7. Kempf DJ, Rode RA, Xu Y, Sun E, Heath–Chiozzi ME, Valdes J, Japour AJ, Danner S, Boucher C, Molla A. AIDS 12:F9–F14, 1998.
8. Norbeck DW, Kempf DJ. Annu Rep Med Chem 26:141–150, 1991.
9. Greenlee WJ. Med Res Rev 10:173–236, 1990.
10. Grant I, Atkinson JH, Hesselink JR, Kennedy CJ, Richman DD, Spector SA, McCutchan JA. Ann Intern Med 107:828–836, 1987.
11. Volberding PA, Lagakos SW, Koch MA, Pettinelli C, Myers MW, Booth DK, Balfour DH, Richman RC, Bartlett JA, Hirsch MS, Murphy RL, Hardy WD, Solero R, Fischl MA, Bartlett JG, Merigan TC, Hyslop NE, Richman DD. N Engl J Med 322: 941–949, 1990.
12. Adedoyin A, Perry PR, Wilkinson GR. Drug Metab Dispos 21:184–188, 1993.
13. Davis SS. In: Davis S, Illum L, Tomlinson E, eds. Delivery Systems for Peptide Drugs, 1986: 1.
14. Greenlee WJ. Pharm Res 4:364–374, 1987.
15. Humphrey MJ, Ringrose PS. Drug Metab Rev 17:283–310, 1986b.
16. Kararli TT. Biopharm Drug Dispos 16:351–80, 1995.
17. Lee VH. CRC Crit Rev Therap Drug Carrier Syst 5:69–97, 1988.
18. McMartin C, Peters G. Davis S, Illum L, Tomlinson E, eds. Delivery Systems for Peptide Drugs, 1986: 255.
19. Rosenberg SH, Spina KP, Condon SL, Polakowski J, Yao Z, Kovar P, Stein HH,

Cohen J, Barlow JL, Klinghofer V, Egan DA, Tricarico KA, Perun TJ, Baker WR, Kleinert HD. J Med Chem 36:460–467, 1993b.
20. Hirschmann R. Angew Chem Int Ed Engl 30;1278–1301, 1991.
21. Abou-El-Markarem MM, Millburn P, Smith RL, Williams RT. Biochem J 105: 1289–1293, 1967.
22. Humphrey MJ. In: Davis S, Illum L, Tomlinson E, eds. Delivery Systems for Peptide Drugs, 1986a: 139.
23. Somogyi A. Trends Pharmacol Sci 8:354–57, 1987.
24. Abraham MH, Kamlet MJ, Taft RW, Doherty RM, Weathersby PK. J Med Chem 28:865–870, 1985.
25. Hellen CU, Krausslich H–B, Wimmer E. Biochemistry 28:9881–9890, 1989.
26. Rao BG, Singh UC. J Am Chem Soc 114:4447, 1991a.
27. Rao BG, Murcko MA. IXth International Conference on AIDS, Berlin, 6–11 June 1999, abstr. PO-A25-0620.
28. Tung RD, Deininger DD. International patent application WO 92/16501, 1 Oct 1992; assigned to Vertex Pharmaceuticals, Incorporated.
29. Babine R, et al. Biorg Med Chem Lett 2:541–546, 1992.
30. Karls MS, Rush BD, Wilkinson KF, Burton PS, Ruwart MJ. Pharm Res 8:1477–1481, 1991.
31. Conradi RA, Hilgers RA, Ho NFH, Burton PS. Pharm Res 8:1453–1460, 1991.
32. Burton PS, Conradi RA, Ho NF, Hilgers AR, Burchardt RT. J Pharm Sci 85:1336–1340, 1996.
33. Stein WD. The Movement of Molecules Across Cell Membranes. New York: Academic Press, 1967: 65–125.
34. Miller M, Schneider J, Sathyanarayana BK, Toth MV, Marshall GR, Clawson L, Selk L, Kent SB, Wlodawer A. Science 246:1149–1152, 1989.
35. Kleinert HD, Baker WR, Stein HH. Adv Pharmacol 22:207–250, 1991.
36. Ocain TD, Abou–Gharbia M. Drugs Future 16:37–51, 1991.
37. Gilbert EE. Synthesis 1; 3–10, 1969.
38. Meyerhans A, Cheynier R, Albert J, et al. Cell 58:901–910, 1989.
39. Huang J, Widlanski TS. Tetrahedron Lett 33:2657–2660, 1992.
40. Vazquez YN, Bryant ML, Clare M, et al. J Med Chem 38:81–84, 1995.
41. Roberts NA, Redshaw B. In: Adams J, Merluzzi VJ, eds. The Search for Antiviral Drugs. Boston: Birkhauser, 1993: 129–151.
42. Roberts NA, Martin JA, Kinchington D et al. Science 248:358–361, 1990.
43. Ghosh AK, Thompson WJ, McKee SP, Duong TT, Lyle TA, Chen JC, Darke PL, Zugay JA, Emini EA, Schleif WA, Huff JR, Anderson PS. J Med Chem 36:292–94, 1993.
44. Ghosh AK, Thompson WJ, Lee HY, McKee SP, Munson PM, Duong TT, Darke PL, Zugay JA, Emini EA, Schleif WA, et al. J Med Chem 36:924–927, 1993b.
45. Wynberg H, Bantjes A. Org Synth Coll 4:534–539, 1963.
46. Perno CF, Aquaro S, Guenci T, Bagnarelli P, Clementi M, Cenci A, Balestra E, Calio R. Antiviral Res 41:A42, 1999.
47. Acosta EP, Henry K, Baken L, Page LM, Fletcher CV. Pharmacotherapy 19:708–712, 1999.

48. Race E, Dam E, Obry V, Paulous S, Clavel F. 6th Conference On Retroviruses and Opportunistic Infections, Chicago, 31 Jan–4 Feb 1999, abstr. 119.
49. Eron JE Jr., Smeaton LM, Fisens SA, Gulick RM, Currier JS, Lennox JL, D'Aquila RS, Rogers MD, Tang R, Murphy RL. J Infect Dis 182 (2000).

ADDITIONAL READINGS

Ashorn P, McQuade TJ, Thaissrivongs S, Tomasselli AG, Tarpley WG, Moss B. Proc Natl Acad Sci USA 87:7472–7476, 1990.

Bagasra P, Lavi E, Bobroski L, Khalili K, Pestaner JP, Tawadros R, Pomerantz RJ. AIDS 10:573–585, 1996.

Craig JC, Duncan IB, Hockley D, Grief C, Roberts NA, Mills JS. Antiviral Res 16:295–305, 1991.

Craig JC, Duncan IB, Whittaker L, Roberts NA. Antiviral Chem Chemother 18:161–166, 1993.

D'Aquila RT. Clin Lab Med 14:393–422, 1994; and references therein.

Erickson J, Neidhart DJ, VanDrie J, Kempf DJ, Wang XC, Norbeck DW, Plattner JJ, Rittenhouse JW, Turon M, Widenburg N, Kohlbrenner WE, Simmer R, Helfrich R, Paul DA, Knigge M. Science 249:527–533, 1990.

Foundling SI, Cooper J, Watson FE, Cleasby A, Pearl LH, Sibanda BL, Hemmings A, Wood SP, Blundell TL, Vailer MJ, Norey CG, Kay J, Boger J, Dunn BM, Leckie BJ, Jones DM, Atrash B, Hallett A, Szelke M. Nature 327:349–352, 1987.

Gray RA, Vander Velde DG, Burke CJ, Manning MC, Middaugh CR, Borchardt RT. Biochemistry 15:1323–1331, 1994.

Hahn BH, Shaw GM, Taylor ME, et al. Science 232:1548–1553, 1986.

Hosur MV, et al. J Am Chem Soc 116:847–855, 1994.

Karls MS, Rush BD, Wilkinson KF, Vidmar TJ, Burton PS, Ruwart MJ. Pharm Res 8: 1477–1481, 1991.

Kim EE, Baker CT, Dwyer MD, Murcko MA, Rao BG, Tung RD, Navia MA. J Am Chem Soc 117:1181–1182, 1995.

Kohl NE, Emini EA, Schleif WA, Davis LJ, Heimbach JC, Dixon RAF, Scolnick EM, Sigal IS. Proc Natl Acad Sci USA 85:4686–4690, 1988.

McQuade TJ, Tomasselli AG, Liu L, Karacostas V, Moss B, Sawyer TK, Heinrikson RL, Tarpley WG. Science 247:454–456, 1990.

Kohl NE, et al. Proc Natl Acad Sci USA 85:4686, 1988.

Lyle TA, Wiscount CM, Guare JP, Thompson WJ, Anderson PS, Darke PL, Zugay JA, Emini EA, Schleif WA, Quintero JC, Dixon RAF, Sigal IS, Huff JR. J Med Chem 34:1228–1230, 1991.

Meek TD, Lambert DM, Dreyer GB, Carr, TJ, Tomaszek TA Jr, Moore ML, Strickler JE, Debouk C, Hyland LJ, Matthews TJ, Metcalf BW, Tetteway SR. Nature 343: 90–92, 1990.

Moses AV, Bloom FE, Pauza CD, Nelson JA. Proc Natl Acad Sci USA 90:10474–10478, 1993.

Moree WJ, van der Marel GA, Liskamp RMJ. Tetrahedron Lett 32:409–412, 1991.

Peng C, Ho BK, Chang TW, Chang NT. J Virol 63:2550–2556, 1989.
Rao, BG, Murcko MA, IXth International Conference On AIDS, 6–11 June 1999, Berlin, abst PO-A25-0620.
Rao JKM, Erickson JW, Wlodawer A. Biochemistry 30:4663–4671, 1991.
Rich DH, Prasad JV, Sun CQ, Green J, Mueller R, Houseman K, MacKenzie D, Malkovsky M. J Med Chem 35:3803–3812, 1992.
Rosenberg SH, Spina KP, Woods KW, Polakowski J, Martin DL, Yao Z, Stein HH, Cohen J, Barlow JL, Egan DA, Tricarico KA, Baker WR, Kleinert HD. J Med Chem 36: 449–459, 1993a.
Shapiro JM, Winters MA, Stewart F, Efron B, Norris J, Kozal MJ, Merigan TC. Ann Intern Med 124:1039–1050, 1996.
Selnick HG, Bourgeois ML, Butcher JW, Radzilowski EM. Tetrahedron Lett 34:2043–2046, 1993.
Umezawa H, Aoyagi T, Morishima H, Matsuzaki M, Hamada M, Takeguchi TJ. Antibiotica 23:259–262, 1970.
Vaishnav YN, Wong–Staal F. Annu Rev Biochem 60:577–630, 1991.
Wlodawer A, Erickson JW. Annu Rev Biochem 62:543–585, 1993.

7

HIV Protease Inhibitors in Early Development

R. Alan Chrusciel and Suvit Thaisrivongs
Pharmacia & Upjohn, Kalamazoo, Michigan

Judith A. Nicholas
MDS Pharma Services, Bothell, Washington

I. INTRODUCTION

An increasing number of HIV protease inhibitors have proved successful in significantly reducing viral burden in HIV-infected patients, particularly when used in combination regimens. Substantial durability has also been demonstrated, validating the notion that HIV protease is an appropriate target for an aggressive therapeutic intervention of the disease state. The availability of this new class of drugs has added notable survival benefits and quality-of-life improvements to a sizable population of HIV-infected patients. Moreover, the initially insurmountable research challenge of transforming peptide mimetic enzyme inhibitors into orally deliverable products was met with resounding successes. It has also been a well-documented testament of the significant contribution of a structure-based paradigm of drug discovery that would convince even the most skeptical researchers. As increasing number of licensed HIV protease inhibitors appeared, accelerated usage of this class of medication is now widespread in increasingly enormous number of HIV-infected patients. As with the earlier class of anti-HIV agents—the HIV reverse transcriptase inhibitors—the increasing volumes of data, accumulated from the treatment of HIV-infected patients with regimens containing HIV protease inhibitors, have identified a sizable number of cases for whom the therapeutic effect, as typically measured by the reduction in viral load, was much less than satisfactory. Several complicated issues are associated with this treatment failure, including the conundrum of patient compliance to a smorgasbord of combination regimens and the emergence of viral strains that have shown alarming resistance to HIV protease inhibitors. Many of the observed clinical isolates exhibit phenotypical cross-resistance against various HIV protease inhibitors, both licensed and under development. Genotypic analyses of these resistant strains continue to provide informative characteristics of these mutated viruses and clues to the reasons for their cross-resistant patterns. For a whole host of logical reasons then, there continues to be a need for new HIV protease inhibitors that would further extend the scope of usefulness of the existing licensed therapeutic agents. This chapter will highlight a few selected HIV protease inhibitors under early development, many of which seek to provide significant improvements over the currently available agents and to address notable unmet medical need, especially for patients who have failed or will fail currently available anti-HIV treatment regimens.

II. HIV PROTEASE INHIBITORS IN EARLY CLINICAL OR LATE PRECLINICAL DEVELOPMENT

A. PNU-140690

Phenprocoumon was identified as a competitive inhibitor ($K_i = 1\ \mu M$) of the HIV-1 protease from a broad screening program. Its already proven oral bioavailability in humans and a timely access to its X-ray crystal structure in an HIV protease complex, made it a very attractive template to initiate a structure–activity relation (SAR) program (1). An iterative cycle of structure-based design initially identified PNU-96988 ($K_i = 38$ nM; in vitro antiviral $IC_{50} = 3\ \mu M$) as Pharmacia & Upjohn's (P&U's) first orally bioavailable clinical candidate (2). Because of its relatively weak antiviral activity, the blood levels in early phase I studies in healthy volunteers did not achieve the needed efficacious compound concentrations. Further structure-based design effort led to the discovery of PNU-103017 ($K_i < 1$ nM, in vitro antiviral $IC_{50} = 1$–$2\ \mu M$) as the second clinical candidate (3). Some liver enzyme elevations were observed in a few healthy volunteers after multiple dosing in early phase I studies, and the development of this compound was also discontinued.

The third-generation clinical candidate, PNU-140690 (Fig. 1), was a result of another parallel structure-based design effort. With the engineered tandemly linked HIV-1 protease, PNU-140690 showed potent enzyme inhibitory activity with a K_i value of 8 pM (4). The crystal structure of the HIV-1 protease triple mutant (Q7K/L33I/L63I) complexed with PNU-140690 was determined. This particular protease construct provided an enzyme with significantly improved stability for structural studies. The 4-hydroxyl group of PNU-140690 showed nearly symmetrical hydrogen bonding to the two catalytic aspartic acids. The carbonyl oxygen at C-2 also made nearly symmetrical hydrogen-bonding interactions with the NH groups of the flap residues Ile50A and Ile50B, and replaced

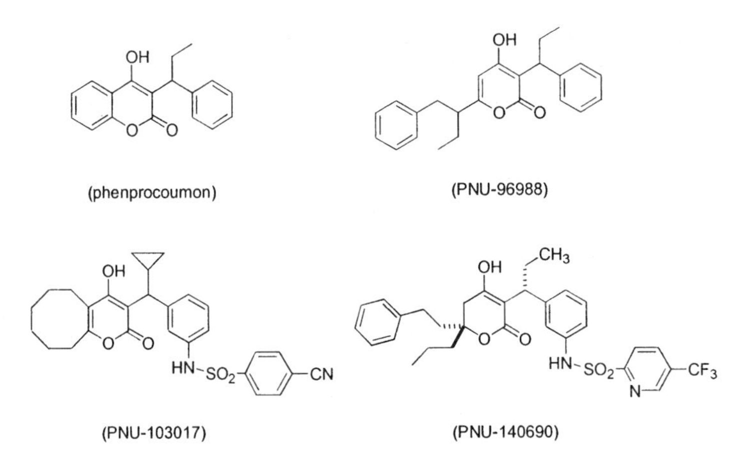

Figure 1 HIV protease inhibitors from Pharmacia & Upjohn.

the ubiquitous water molecule found in complexes of peptide-derived inhibitors. The sulfonamide functionality showed strong hydrogen bonding to a few residues at the active site. The phenethyl and the propyl groups at C-6 extended into the S_1' and S_2' subsites, respectively; the ethyl and the phenyl groups at C-3α occupied the S_1 and S_2 subsites, respectively. The 5-(trifluoromethyl)-2-pyridyl group curled into the S_3 subsite.

PNU-140690 was highly selective for HIV proteases (K_i <1 nM against HIV-2 protease, V82A, V82F/I84V mutant proteases), for it only weakly inhibited (K_i > 1μM) human pepsin, human cathepsin D, and human cathepsin E. This compound potently suppressed viral replication of laboratory and clinical HIV-1 strains. In HIV-1_{IIIB}-infected H9 cells and in HIV-$1_{JR\text{-}CSF}$-infected peripheral blood monocyte cells (PBMC), PNU-140690 showed IC_{90} values of 0.16 and 0.18 μM, respectively (5). When ten diverse HIV-1 isolates from patients, many of which were AZT-resistant isolates, were used to infect PBMC, PNU-140690 also demonstrated consistently potent activity with an average IC_{90} value of 0.16 μM. Many HIV protease inhibitors have significantly reduced in vitro antiviral activity as a result of high-affinity binding to plasma proteins, in particular to α_1-acid glycoprotein (AAG). The effects of added AAG, human plasma albumin, and human plasma on the antiviral activity of PNU-140690 were assessed with HIV-1_{IIIB}-infected H9 cells and HIV-$1_{JR\text{-}CSF}$-infected PBMC. The increases in the IC_{90} values resulting from the presence of the added proteins ranged from two- to sixfold. The trough blood level in humans of 1 μM was then targeted for expected antiviral activity in HIV-infected patients.

The single-dose safety, tolerance, and pharmacokinetics of the capsules of the disodium salt of PNU-140690 was studied in normal healthy volunteers with a dose range of 100–2000 mg (6). PNU-140690 was generally well-tolerated at all dose levels, and gastrointestinal effects, mostly mild or occasionally moderate, were the most commonly observed adverse events. Drug concentrations in excess of 1 μM, the predicted IC_{90} value based on in vitro virology and protein-binding data, were observed for longer than than 8 h after a single dose of 500 mg and higher. Preliminary data from a multiple-dose study in healthy volunteers suggest that PNU-140690 was well-tolerated for up to the 10-day duration of studies, with the primary adverse events related to gastrointestinal effects. On average, trough PNU-140690 concentrations of more than 1 μM have been achieved at doses of 900 mg every 8 h.

To gain some information on the degree of cross-resistance, the laboratory-resistant variant, HIV-1 NL4-3 (p37)—resulted from 37 passages of HIV-1 NL4-3 in the presence of ritonavir—was studied. This viral strain, with several coding changes after genotypic analysis, was 80-fold more resistant to ritonavir, 47-fold more resistant to indinavir, and at least 125-fold more resistant to nelfinavir and saquinavir than the wild-type strain. PNU-140690 retained good activity (IC_{90} = 0.45 μM) against this HIV-1 NL4-3 (p37) strain compared with the activity of PNU-140690 against the wild-type HIV-1 NL4-3 strain (IC_{90} = 0.07 μM). The mutations in HIV-1 NL4-3 (p37) that confer very high-level resistance to several peptide mimetic protease inhibitors caused only a sixfold increase in resistance to PNU-140690. Additionally, paired clinical isolates from four patients treated with ritonavir (pre- and posttreatment) were used to infect PBMC, and genotypic analysis revealed a variety of coding changes that resulted in resistant variants. Up to 67-fold and 33-fold increases in the IC_{90} values for ritonavir and indinavir, respectively, between pre- and post-ritonavir therapy isolates were observed. However, PNU-140690 showed IC_{90} values of 0.2 μM for the pretherapy isolates and no greater than 0.6 μM in IC_{90} values for the post-ritonavir therapy isolates. Thus, there was an increase of only threefold resistance against PNU-140690 (5). It is anticipated that clinical viral strains displaying cross-resistance to several peptide mimetic HIV protease inhibitors might be expected to be sensitive to PNU-140690.

B. BMS-234475

CGP-61755 (recently licensed by Bristol–Myers Squibb [BMS] and redesignated BMS-234475) is structurally characteristic of a rationally designed peptide-derived inhibitor in which the scissile bond of the natural substrate is replaced by a transition-state isostere (Fig. 2). The challenge in developing peptide mimetic inhibitors is the need to reduce the inherent peptide character of these molecules to achieve good antiviral potency and an adequate pharmacokinetic profile (7). The successful strategies to date have included minimizing molecular weight,

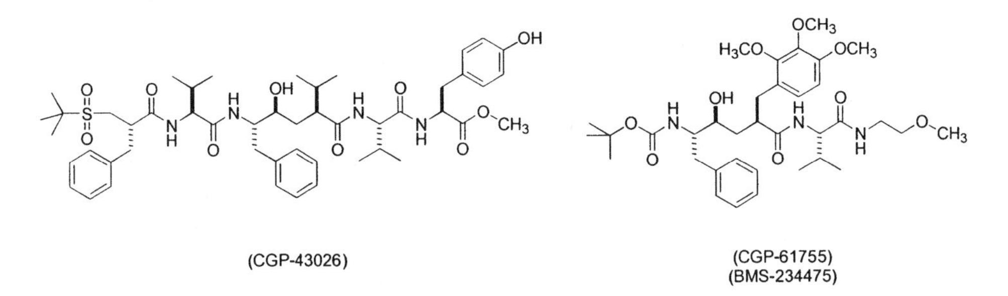

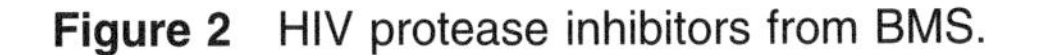
Figure 2 HIV protease inhibitors from BMS.

improving water solubility, and removing as many peptide bonds as possible. CGP-61755 is the result of an optimization program that began by the identification of lead structures, such as the renin inhibitor CGP-43026, which was found by screening to have HIV-1 protease inhibitory activity (8). CGP-61755 is a 2,3,4-trimethoxy-substituted phenylalanyl-derived compound, incorporating a hydroxyethylene transition-state isostere, that demonstrates excellent enzyme inhibitory activity against HIV-1 protease (IC_{50} <0.5 nM) with selectivity over cathepsin-D, pepsin, and renin (IC_{50}s of 100, 8000, and >670,000 nM, respectively) (9). Good antiviral cell culture activity is displayed by this peptide-mimetic, HIV-1 protease inhibitor, with an IC_{90} of 30–100 nM, in an acute infection of HIV-1 in MT-2, PBLs and macrophages. No cytotoxicity is observed with CGP-61755 in MT-2 cells with concentrations up to 100 μM. Against chronically infected macrophages, the IC_{90} is 1 μM, which is comparable with the IC_{90}s determined for indinavir (10 μM) and saquinavir (1 μM) (10).

CGP-61755 has been evaluated in cell culture assays in combination with the reverse transcriptase (RT) inhibitors, zalcitabine (ddC) and didanosin (ddI), and the protease inhibitor, indinavir; it has additive to synergistic effects with these antiviral agents (11). Synergistic effects were observed when used in combination with zidovudine (AZT) or saquinavir. CGP-61755 has been reported to be fully active against the saquinavir-resistant G48V/L90M double-mutant virus.

When dissolved in plasma, a high proportion of CGP-61755 was protein-bound. To evaluate the influence of protein binding on antiviral activity, the cell culture activity of CGP-61755 was studied in the absence or presence of 2 mg/mL AAG (chosen to represent the upper concentration limit of this plasma protein seen in AIDS patients) and compared with similar studies with saquinavir, ritonavir, indinavir, and SC-52151 (which failed to show antiviral activity in vivo, presumably owing to binding to AAG) (12). In MT-2 cells, indinavir activity was least affected by the addition of plasma protein (twofold increase of IC_{90}), whereas ritonavir, saquinavir, VX-478 (141W94), and telinavir (SC-52151) were affected significantly (greater than a tenfold increase in IC_{90} in MT-2 cells). CGP-61755 showed an intermediate fourfold increase in IC_{90} owing to added AAG. Similar effects were observed using human lymphocytes.

When CGP-61755 was dosed orally at 120 mg/kg in mice, using a sesame oil–based formulation, a C_{max} of 10.2 μM was achieved (Table 1). The AUC_{0-8h} was reported as 54.49 μMh. After 8 h, the plasma concentration was 4.9 μM, exceeding the antiviral IC_{90} (30 nM) more than 100-fold. In dogs, a C_{max} of 4.8 μM was achieved with an AUC_{0-8h} of 21.06 μM h^{-1} when dosed at 1.2 g/dog.

BMS is conducting phase II clinical evaluation of BMS-234475 (lasinavir), with an azapeptide inhibitor (CGP-73547; BMS-232623) in preclinical studies.

C. ABT-378

The C_2-symmetry found in the active site of HIV protease is an important design element that has been successfully utilized by workers from Abbott for the preparation of potent inhibitors. Ritonavir, although not a true C_2-symmetric molecule, evolved from the study of very active symmetric and pseudosymmetric HIV protease inhibitors such as A-77003.

Recently, workers from Abbott have disclosed information on their second-generation HIV protease inhibitor, ABT-378 (Fig. 3). A brief description of the preclinical data on ABT-378 is also provided in Chapter 3. Two important properties that their second-generation candidate is intended to possess are (1) lower serum protein binding, which compromises the antiviral activity of ritonavir; and (2) high potency against the initial mutants selected by ritonavir in vivo (most notably Val82) (13). ABT-378 is an approximately eightfold more potent inhibitor of HIV protease than ritonavir, with improved antiviral activity in cell culture (see Table 1). Addition of 50% human serum increased the IC_{50} for ABT-378 against HIV-1 replication in cell culture five- to eightfold (14). Detailed studies of the protein binding of ritonavir and ABT-378 to the serum proteins AAG and human serum albumin (HSA) indicated that the in vitro activity of ABT-378 was significantly affected by the presence of AAG but not HSA. In contrast, in vitro activity of ritonavir was significantly affected by the presence of both proteins.

Examination of the activity of ABT-378 and ritonavir against HIV proteases containing V82A, V82F, V82S, and V82T mutations expressed and purified from *Escherichia coli*, demonstrated that its affinity was only moderately affected compared with wild-type enzyme (3- to 21-fold increase in K_i), (15). In contrast, ritonavir displayed a 10- to 260-fold reduction in affinity when examined with these mutant enzymes. These in vitro results could be rationalized, based on X-ray crystallographic studies of the complex of ABT-378 and HIV protease, which demonstrated that the hydrophobic interactions of the inhibitor with Val82 had been successfully minimized (16). Cross-resistance effects of ABT-378 were studied by the serial passage of HIV-1 (NLA4-3) in MT-4 cells in the presence

Table 1 HIV Protease Inhibitors in Development

	K_i (nM)	IC_{50} (nM)[a]	IC_{90} (nM)	Species	IV dose (mg/kg)	Oral dose (mg/kg)	$t_{1/2}$ (h)[b]	C_{max} (μM)[a]	%F	Plasma protein effects[d]	Ref.
PNU-140690	0.008	30	100–200	Rat	5	10[m]	5.4	3–10	30–45	6-fold[f]	4, 5, 6
				Dog	5	10	1	4–10	20		
				Human		900[e]		20			
CGP-61755	0.5		30–100	Mice	—	120		10.2	20–80	4-fold[h]	9, 38
(BMS 234475)				Dog		1200[g]		4.8	50–80		
ABT 378	0.0013	17		Rat	—	10	(<1)	1.53	25	6-fold[i], 1.3-fold[j], 13-fold[k]	13–19
KNI 272	0.0055	20–140		Rat	10	10[l]	2.86	1.35	42	3 to 5-fold[m]	20, 21, 25, 28, 39
				Dog	50	100[m]	2.16	0.78–4.8	7–26	15 to 25-fold[p]	31, 40
DMP 450	0.28		130	Rat	10	10	1.3	2.25	71	4.5 to 8.4-fold[p]	
				Dog	10	10	3.6	11.2	79		
				Monkey	10	10	0.8	1.60	26		
DMP 850	0.040		71	Dog	—	10	(3)	10.5	50–90	8-fold	41
DMP 851	0.016		57	Dog	—	10	(8)	12.6	63	17-fold	41
PD 169277	0.17	600		Mice	—	25	(1.6)	17.3	96		36, 42
				Dog	5	10	(0.5)	38.0	31		
PD 171277	0.43	500		Mice	—	25	(2.2)	23	80		36, 42
				Dog	10	10	(1.3)	168	58		
PD 178390	0.11	210								5-fold[g]	37, 43

[a] IC_{50}: 50% effective concentration, HIV-1 acute infection assay (cell lines or PBMC).
[b] Terminal phase plasma half-life after i.v. (for po) dose.
[c] C_{max}: maximum plasma concentration following oral dose.
[d] Increase in IC_{50} (or IC_{90}) in the presence of added serum protein.
[e] Single dose of 900 mg.
[f] H9 cell assay containing 10% fetal bovine serum and 75% human plasma.
[g] 1.2 g/dog as a single dose.
[h] Increase in IC_{50} in the presence of 2 mg/mL AAG.
[i] Increase in IC_{50} on addition of 50% HS.
[j] Increase in IC_{50} on addition of 25 mg/mL HSA.
[k] Increase in IC_{50} on addition of 1 mg/mL AAG.
[l] i.d. administration.
[m] 100 mg/dog.
[n] Increase in antiviral IC_{50} in the presence of 50% FCS compared with 15% FCS.
[o] Increase in antiviral IC_{50} in the presence of 80% FCS compared to 15% FCS.
[p] AAG and HSA added to tissue-culture media (2 different endpoints measured).
[q] H9 cells containing 30% HS.

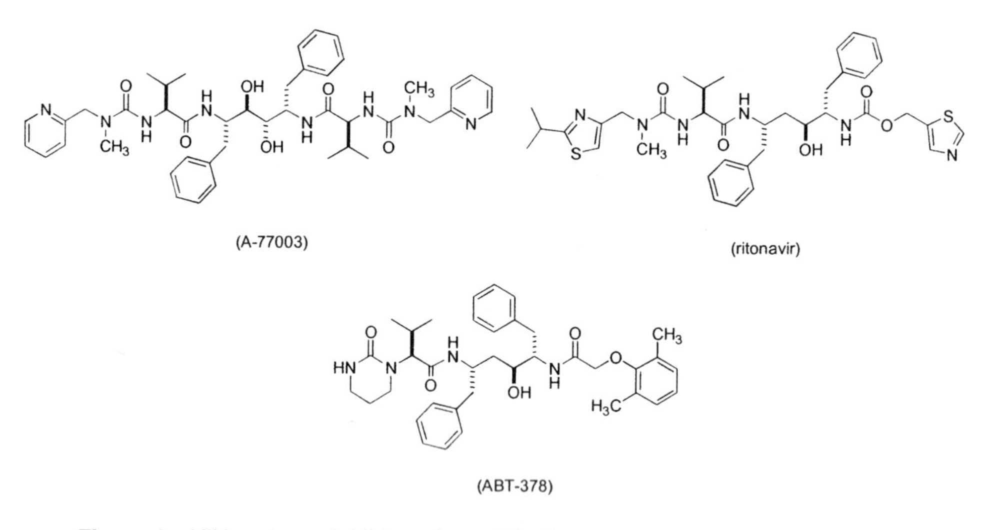

Figure 3 HIV protease inhibitors from Abbott.

of this novel inhibitor (17). After 17 passages, a virus that is greater than 300-fold resistant to ABT-378 was selected. Genotypic analysis revealed that the virus contained the I84V, L10F, M46I, T91S, and V47A multations, suggesting that partial cross-resistance to ritonavir and indinavir is likely, but sensitivity to saquinavir remains.

Following oral dosing of 10 mg/kg in rats, the pharmacokinetic profile of ABT-378 was characterized by a maximum plasma concentration of 1.53 μM and 25% bioavailability (see Table 1) (18). In monkeys and dogs, however, plasma concentrations of ABT-378 could not be detected after a single oral dose. Administration of a 10-mg/kg dose of ABT-378 in combination with a 10-mg/kg dose of ritonavir provided a dramatic and sustained increase in plasma levels of ABT-378 in rats and monkeys, presumably the consequence of ritonavir's inhibition of the P450 metabolism of ABT-378. When coadministered with ritonavir in rats, ABT-378 plasma concentrations 8 h after dosing were fivefold higher than the antiviral IC_{50} adjusted for the presence of human serum. In a phase I drug interaction study in healthy volunteers, coadministration of ritonavir (up to 300 mg) and ABT-378 (up to 600 mg) increased the AUC of ABT-378 100- to 300-fold (19). The combination achieved 24-h plasma levels 86-fold higher than the IC_{50} adjusted for protein binding.

D. KNI-272

Similar to the successful strategy employed by workers from Hoffmann–La Roche in the development of saquinavir, Kiso and co-workers considered that the unusual recognition of HIV protease for the Phe–Pro and Tyr–Pro sequences afforded a useful starting point for the rational design of substrate-based HIV protease inhibitors. The ensuing synthetic program identified the pseudohexapeptide inhibitor KNI-93 (IC_{50}, nM) that utilized the novel transition state isostere allophenylnorstatine [APNS; (2*S*,3*S*)-3-amino-2-hydroxy-4-phenylbutyric acid] (20). Initial optimization aimed at reducing molecular weight and minimizing the number of natural peptide bonds led to the novel pseudotripeptide KNI-102 (IC_{50}, 89 nM). Further synthetic efforts directed at the modification of the individual residues to achieve a balance in lipophilicity/hydrophilicity, considered essential for good membrane permeability and thus good antiviral activity, and reduction in the molecular size led to the discovery of KNI-272 (Fig. 4), an extremely active (IC_{50}, 6.5 nM; K_i, 5.5 pM) and selective inhibitor of HIV protease (IC_{50}, human plasma renin $>$ 100 μM, porcine pepsin $>$ 10 μM, bovine cathepsin D $>$ 100 μM) (21), with good antiviral activity (see Table 1).

One of the key synthetic modifications contributing to the discovery of KNI-272 was the replacement of the pyrrolidine ring of the proline residue with a thiazolidine ring, which led to beneficial effects on enzyme inhibitory activity and antiviral potency. A subsequent crystal structure of KNI-272 complexed with HIV protease determined to 2.0 Å resolution revealed a unique inhibitor-binding conformation with a number of interesting interactions between the inhibitor and enzyme. Perhaps the most notable feature was that the *trans*-conformation of the amide bond between the APNS and thiazolidine-4-carboxylic acid residues results in an unexpectedly symmetric mode of binding for the hydroxymethylcarbo-

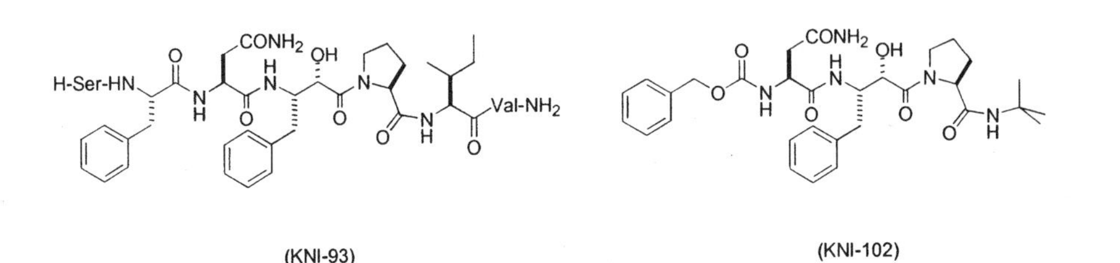

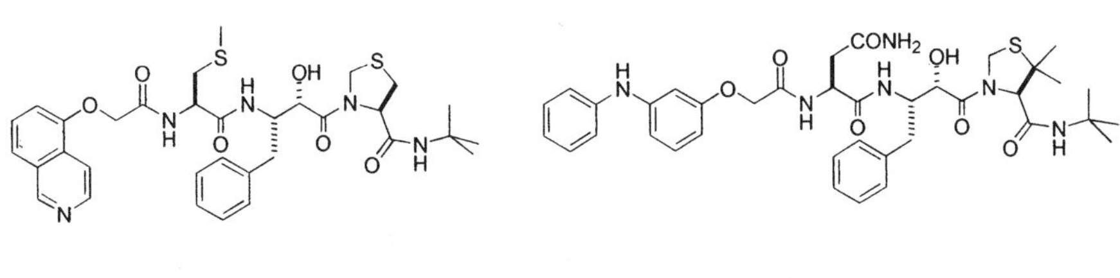

Figure 4 HIV protease inhibitors from Kyoto Pharmaceutical University.

nyl portion of the transition-state isostere (22). In vitro drug selection experiments using KNI-272 have resulted in mutant viruses with amino acid substitutions, including V32I, K45I, F53L, A71V, I84V, and L89M, that exhibit 5- to 258-fold decrease in sensitivity to the inhibitor (23). Use of the refined model of the KNI-272–HIV protease complex suggests that the structural basis for the key I84V resistance may result from the loss of the hydrophobic interaction between the thiazolidine ring and the Ile84 Cδ1 methyl group. Screening of other APNS-containing inhibitors identified KNI-241, which displays good activity against wild-type and KNI-272-resistant viruses (IC_{50}s, 0.01–0.04 μM) (24).

When tested against a variety of HIV strains, including HIV-1_{LAV}, HIV-1_{RF}, HIV-1_{MN}, and HIV-2_{ROD} in $CD4^+$ ATH8 cells, potent antiviral activity of KNI-272 was demonstrated (IC_{50}s, 0.02–0.14 μM), with negligible cytotoxicity (25). Replication of AZT-resistant HIV-1 was also suppressed by KNI-272 (IC_{50}, 0.02 μM). KNI-272 is highly protein bound, with antiviral activity decreasing approximately 15- to 25-fold when fetal calf serum (FCS) concentration was increased from 15 to 80% (26). Although the binding of KNI-272 in human plasma was largely attributed to AAG, subsequent experiments demonstrated that this lipophilic inhibitor also interacts with the warfarin-binding site of HSA (27).

Bioavailability of KNI-272 has been examined by intraduodenal administration in rats and orally in dogs (see Table 1). Oral administration to dogs has been examined using four different dosage forms, including enteric capsules. The highest bioavailability observed (26%) was provided with conventional gelatin capsules, suggesting that delivery of KNI-272 to the stomach was preferred to release of the drug in the upper or middle of the small intestine (28).

Preliminary results from a phase 1 dose-escalation clinical study of drug administered orally for 4–12 weeks demonstrated an average decrease in viral load of up to 0.27 $\log_{10}$ with the seven of nine patients who were able to be evaluated on a dose of 6.6 mg/kg four times daily (29). A dose-limiting transient elevation of hepatic transaminase levels was observed at this dose, which could be partially ameliorated by escalating to the target dose over 4 weeks. Oral bioavailability was determined to be 25–32%.

E. DMP-850/851

Workers from Dupont–Merck pioneered the development of cyclic ureas as a novel class of C_2-symmetric inhibitors. From X-ray crystallographic data from several hydroxyethylene-based inhibitor HIV protease complexes, and analysis of the extensive structure–activity relations established for C_2-symmmetric diol-protease inhibitors (of which no crystal complexes had yet been published) a three-dimensional (3D) pharmacophore model was generated that suggested that the cyclic urea structure would provide a useful template to construct HIV protease inhibitors (30). On entropic grounds, it was anticipated that good enzyme affinity could be realized using this small heterocycle, as the urea carbonyl was expected to displace the structural water molecule that normally bridges the P2 and P1′ carbonyl groups of peptide substrates or peptide mimetic inhibitors to the flap regions of the enzyme. Because this bridging water is a structural feature unique to retroviral proteases, good enzyme selectivity among the other mammalian aspartic proteases was also anticipated. Subsequent synthetic efforts led to extremely potent and selective inhibitors of HIV protease, typified by DMP-450 (Fig. 5; see Table 1). DMP-450 is a weakly basic cyclic urea with improved aqueous solubility (>170 mg/mL as bismesylate salt) over DMP-323, Dupont–Merck's first clinical candidate in this structural class, which was dropped from clinical trials owing to variable pharmacokinetics largely attributed to limited solubility (6 μg/mL) (31). DMP 450 is a potent inhibitor of the replication of laboratory strains and clinical isolates of HIV and exhibits substantial oral bioavailability in all species examined (see Table 1).

Under selective pressure of cyclic urea-based HIV protease inhibitors, key mutations of Val82 and Ile84 are observed, and they are considered to be among the major factors leading to the development of resistance to this class of inhibitors. Under selective *in vitro* pressure from DMP-450, significant loss of sensitivity was observed (45-fold) associated with the presence of two to five mutations, including the I84V mutation. A 100-fold decrease in potency was observed when recombinant virus selected *in vitro* against DMP-323 was examined, suggesting the likelihood of cross-resistance within this class. Comparison of the X-ray crys-

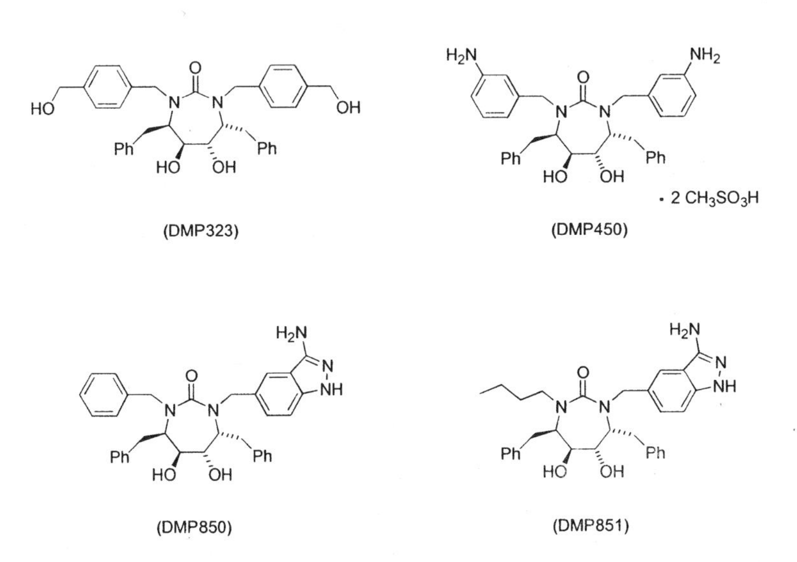

Figure 5 HIV protease inhibitors from DuPont Merck.

tal complexes of DMP-450 with wild and recombinant mutant proteases (I84V, V82F, V82F/I84V) suggests that the difference in K_i observed (1000-fold increase in K_i determined for DMP-450 and the V82F/I84V mutant protease) is due to the loss of key van der Waals (vdw) interactions between the inhibitor and protease (32). Although DMP-450 is highly protein-bound (90–93% as determined by dialysis experiments) the effects of protein binding on DMP-450's antiviral activity is moderate, with only a 4.5- to 8.4-fold increase in IC_{90} observed in the presence of added AAG and HSA.

In a phase I dose-escalation study in 56 healthy volunteers, oral doses of 60–1250 mg yielded a mean C_{max} of 0.28–11.75 μM at 0.75–2 h postdosing. Half-life ($t_{1/2\beta}$) ranged from 1.3–3.9 h at low dose to 5.2–8.2 h at high dose. Clearance decreased with increasing dose and food reduced the rate, but not the extent of absorption. DMP-450 was licensed to Avid Corporation and Avid was later acquired by Triangle Pharmaceuticals, who will continue phase I/II clinical development of DMP-450.

Recently, workers from DuPont–Merck reported on two new *asymmetric* cyclic urea HIV protease inhibitors, DMP-850 and DMP-851. These ureas are potent inhibitors of HIV protease and demonstrate antiviral activity similar to the two previous clinical candidates DMP-323 and 450 (see Table 1) (33). It is anticipated that an asymmetric cyclic urea may have advantages over a symmetric inhibitor relative to resistance, as a single mutation produces two symmetrically related changes in the enzyme, one in each of the monomers. It is also expected that the introduction of a flexible alkyl chain into an asymmetric urea will provide better solubility, without a significant loss of activity. The poor solubility associated with the symmetric cyclic ureas has partly been attributed to the *intermolecular* diol–urea bidentate hydrogen bonds observed in the crystal lattice and symmetry-related high crystal-packing energy.

Compared with DMP-450, sensitivity for the key I84V mutation is less with DMP-850 and DMP-851 (fourfold increase in IC_{90} as compared with sevenfold for DMP-450), suggesting that these new ureas are, in fact, less susceptible to this key mutation. Addition of serum proteins causes an 8- and 17-fold increase in sensitivity to the antiviral activities of DMP-850 and DMP-851, respectively. Excellent pharmacokinetic properties are displayed by DMP-851 after oral administration to dogs (see Table 1). In single-dose pharmacokinetic studies at 10 mg/kg, plasma levels remained above the IC_{90} (wild-type and adjusted for protein binding and sensitivity to clinical isolates) for longer than 20 h. DMP-850 and 851 were slated for clinical trial entry by the end of 1997.

F. Parke–Davis Dihydropyrones

Researchers from Parke–Davis have also independently identified 5,6-dihydro-4-hydroxy-2-pyrones as an interesting class of HIV protease inhibitors (34). High-volume screening of the Parke–Davis compound collection resulted in the identification of the 4-hydroxy-2-pyrone **1** (Fig. 6) as a inhibitor of HIV protease with low micromolar affinity (K_i, 1.1 μM) (35). As part of the subsequent synthetic program based on this lead, it was determined that potent inhibitors of the enzyme were readily prepared using a 5,6-dihydro-4-hydroxy-2-pyrone template (e.g., Fig. 6 **2** has an enzyme inhibitory IC_{50} of 6.2 nM). Subsequent X-ray crystallographic studies of the complex of **2** with HIV-1 protease demonstrated that the inhibitor fulfilled requisite hydrogen bond interactions with the enzyme and effectively occupied the four central substrate-binding sites. Disubsitution at the 6-position of the heterocycle provided effective interaction of the inhibitor with the S_1 and S_2 subsites of the enzyme. Unfortunately, poor antiviral activities were associated with this initial series of dihydropyrone protease inhibitors. Recent reports, however, disclose that dramatic improvements in antiviral activity could be achieved in the absence of significant cellular toxicity (TC > 100 μM) by incorporating appropriate polar functional groups in the phenethyl ring at C-6 and in *S*-aryl ring at C-3 as exemplified by PD-169277 (see Table 1) (36).

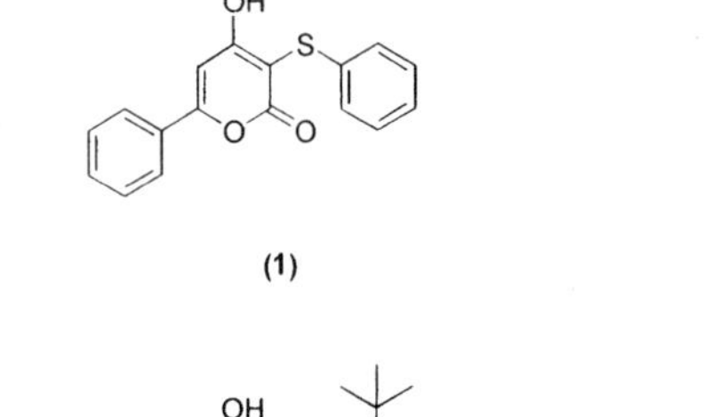

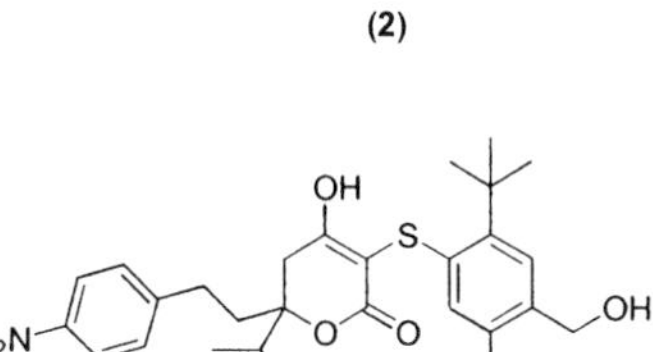

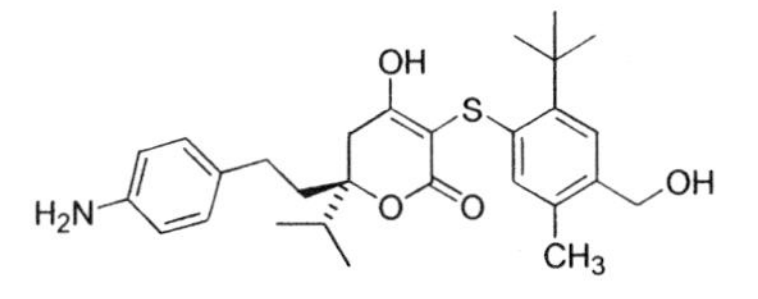

Figure 6 HIV protease inhibitors from Parke-Davis.

Good pharmacokinetic properties were observed with PD-169277 in mice and dogs (see Table 1). In dogs, an improved pharmacokinetic profile was observed for the closely related amino-substituted dihydropyrone, PD171277. Of interest, a decrease in systemic clearance was observed for PD-171277 (CLs of 0.65 mL min^{-1} kg^{-1}) when compared with the closely related hydroxy compound PD-169277 (CLs of 2.57 mL min^{-1} kg^{-1}). Improved solubility of PD-171277 at the low pH of the upper gastrointestinal tract and a reduction in the number of sites for conjugation were suggested to contribute to the improvement in the pharmacokinetic properties observed.

In those cases thus far reported by Parke–Davis, the *S* stereochemistry at the 6 position of the dihydropyrone ring provides better activity than the *R* stereoisomer. For example, the *S* enantiomer, PD-178390, has an approximately 220-fold better K_i and 94-fold better cellular antiviral activity than its *R* antipode. When tested against several resistant strains of HIV that contained mutations commonly associated with protease inhibitors (e.g., M46I, I54V, L63P, V82T, V82F, I84V), very modest increases in antiviral IC_{50}s were observed for PD-178390 (less than twofold) (37). In H-9 cells, the antiviral activity of PD-178390 was essentially unchanged by the addition of 2 mg/mL AAG; however, the addition of 30% human serum led to an approximately fivefold increase in the IC_{50} when compared with the presence of 10% FCS.

III. CONCLUDING REMARKS

The HIV protease inhibitors described in this chapter are at various stages of development. They generally meet the basic criteria of having very impressive in vitro enzymatic inhibitory potency against HIV-1 protease, with high selectivity against other aspartyl proteases; displaying high in vitro antiviral activities, as assessed against a panel of cells infected by laboratory and clinical HIV isolates with moderately adverse protein-binding effect; and demonstrating sufficient oral bioavailability and pharmacokinetic properties in animal species. The available data from the clinical development for a number of compounds have already reported initial acceptable safety profile and preliminary promising efficacious results on viral burden reduction in HIV-infected patients. As experience grows in a larger population of patients using a variety of combination drug therapies, it is anticipated that a few of these newer agents will emerge as effective anti-HIV chemotherapeutic products that would extend the scope of successful drug treatment available with currently licensed HIV protease inhibitors. While limited information is available on the pattern of resistance and cross-resistance from in vitro studies, one anxiously awaits the hopeful outcomes of the studies of many of these newer compounds in previously treatment-failure patients, especially in HIV protease inhibitor-experienced patients. On the fundamental front of basic scientific pursuits, the fruitful attempts to develop long-term effective anti-HIV therapy continue to be dependent upon the detailed elucidation and understanding of the critical steps involved in the virus–host relation, from infection to the development of AIDS symptoms. A more complete scientific picture of synergistic approaches to disease prevention, intervention, and treatment is a prerequisite for the successful management of the HIV pandemic.

REFERENCES

1. Romines KR, Chrusciel RA. 4-Hydroxypyrones and related templates as nonpeptidic HIV protease inhibitors. Curr Med Chem 1995; 2:825–838.
2. Thaisrivongs S, Tomich PK, Watenpaugh KD, Chong K–T, Howe JW, Yang C–P, Strohbach JW, Turner SR, McGrath JP, Bohanon MJ, Lynn JC, Mulichak AM, Spinelli PA, Hinshaw RR, Pagano PJ, Moon JB, Ruwart MJ, Wilkinson KF, Rush BD, Zipp GL, Dalga RJ, Schwende FJ, Howard GM, Padbury GE, Toth LN, Zhao Z, Koeplinger KA, Kakuk TJ, Cole SL, Zaya RM, Piper RC, Jeffrey P. Structure-based design of HIV protease inhibitors: 4-hydroxycoumarins and 4-hydroxy-2-pyrones as non-peptidic inhibitors. J Med Chem 1994; 37:3200–3204.
3. Skulnick HI, Johnson PD, Howe WJ, Tomich PK, Chong K–T, Watenpaugh KD, Janakiraman MN, Dolak LA, McGrath JP, Lynn JC, Horng M–M, Hinshaw RR, Zipp GL, Ruwart MJ, Schwende FJ, Zhong W–Z, Padbury GE, Dalga RJ, Shiou L, Possert PL, Rush BD, Wilkinson KF, Howard GM, Toth LN, Williams MG, Kakuk TJ, Cole SL, Zaya RM, Thaisrivongs S, Aristoff PA. Structure-based design of sulfonamide substituted non-peptidic HIV protease inhibitors. J Med Chem 1995; 38:4968–4971.
4. Thaisrivongs S, Skulnick HI, Turner SR, Strohbach JW, Tommasi RA, Johnson PD, Aristoff PA, Judge TM, Gammill RB, Morris JK, Romines KR, Chrusciel RA, Hinshaw RR, Chong K–T, Tarpley WG, Lynn JC, Horng M–M, Tomich PK, Seest EP, Dolak LA, Howe WJ, Howard GM, Schwende FJ, Toth LN, Padbury GE, Wilson GJ, Shiou L, Zipp GL, Wilkinson KF, Rush BD, Ruwart MJ, Koeplinger KA, Zhao Z, Cole S, Zaya RM, Kakul TJ, Janakiraman MN, Watenpaugh KD. Structure-based design of HIV protease inhibitors: sulfonamide-containing 5,6-dihydro-4-hydroxy-2-pyrones as nonpeptidic inhibitors. J Med Chem 1996; 39:4349–4353.
5. Poppe SM, Slade DE, Chong K–T, Hinshaw RR, Pagano PJ, Markowitz M, Ho D, Mo H, Gorman RR, Dueweke TJ, Thaisrivongs S, Tarpley WG. Antiviral activity of the dihydropyrone PNU-140690, a new non-peptidic HIV protease inhibitor. Antimicrob Agents Chemother 1997; 41:1058–1063.
6. Borin MT, Carlson GF, Wang Y, Brewer JE, Daenzer CL, Baldwin JR, Li H. Single-dose safety, tolerance, and pharmacokinetics of PNU-140690, a new HIV protease inhibitor, in healthy volunteers. 37th ICAAC, Toronto, Ontario, Sept 28–Oct 1, 1997.
7. Kempf DJ. Progress in the discovery of orally bioavailable inhibitors of HIV protease. Perspect Drug Disc Design 1994; 2:427–436.
8. Capraro H–G, Bold G, Fässler A, Cozens R, Klimkait T, Lazdins J, Mestan J, Ponci-

oni B, Rösel JL, Stover D, Lang M. Synthesis of potent and orally active HIV-protease inhibitors. Arch Pharm Pharmacol Med Chem 1996; 329:273–278.

9. Cozens RM, Bold G, Capraro H–G, Fässler A, Mestan J, Lang, Poncioni B, Stover D, Rösel JL. Synthesis and pharmacological evaluation of CGP 57813 and CGP 61755, HIV-1 protease inhibitors from the Phe-c-Phe peptidomimetic class. Antiviral Chem Chemother 1996; 7:294–299.
10. Lazdins JK. CGP 61755 a novel and potent HIV-1 protease inhibitor with a suitable profile for clinical development. 3rd Conference on Retroviruses and Opportunistic Infection, Washington, DC, Jan 28–Feb 1, 1996.
11. Klimkait T. CGP 61755, a potential combination partner for saquinavir: a synergistic HIV-1 protease inhibitor with high antiviral potency. Antiviral Res 1996; 30: A33.
12. Lazdins JK, Mestan J, Goutte G, Walker MR, Bold G, Capraro G, Klimkait T. In vitro effect of α_1-acid glycoprotein on the anti-human immunodeficiency virus (HIV) activity of the protease inhibitor CGP 61755: a comparative study with other relevant HIV protease inhibitors. J Infect Dis 1997; 175:1063–1070.
13. Sham H, Kempf D, Molla A, Marsh K, Betebenner D, Chen X, Rosenbrook W, Wideburg N, Chen C, Kati W, Kumar G, Korneyeva M, Vasavanonda S, McDonald E, Saldivar A, Chernyavskiy T, Carillo A, Lyons N, Park C, Stewart K, Plattner J, Norbeck D. Design, synthesis and biological properties of ABT-378, a highly potent HIV protease inhibitor. 4th Conference on Retroviruses and Opportunistic Infection, Washington, DC, Jan 22–26, 1997.
14. Molla A, Vasavanonda S, Denissen J, Kumar G, Grabowski B, Sham H, Norbeck D, Kohlbrenner W, Plattner J, Kempf D, Leonard J. Effect of human serum proteins on the antiretroviral activity of ritonavir and ABT-378, potent inhibitors of HIV protease. 4th Conference on Retroviruses and Opportunistic Infection, Washington, DC, Jan 22–26, 1997.
15. Chen C, Niu P, Kati W, Norbeck D, Sham H, Kempf D, Kohlbrenner W, Plattner J, Leonard J, Molla A. Activity of ABT-378 against HIV protease containing mutations conferring resistance to ritonavir. 4th Conference on Retroviruses and Opportunistic Infection, Washington, DC, Jan 22–26, 1997.
16. Stewart K, Park C, Nienaber V, Sham H, Betebenner D, Chen C, Wideburg N, Saldivar A, Kati W, Rosenbrook W, Flentge C, Chen X, Molla A, Kempf D, Norbeck D. Molecular modeling and X-ray crystallographic studies of ritonavir, ABT378, and their analogs bound to HIV protease. 4th Conference on Retroviruses and Opportunistic Infection, Washington, DC, Jan 22–26, 1997.
17. Carrillo A, Sham H, Norbeck D, Kempf D, Kohlbrenner W, Plattner J, Leonard J, Molla A. Selection and analysis of HIV-1 variants with increased resistance to ABT-378, a novel protease inhibitor. 4th Conference on Retroviruses and Opportunistic Infection, Washington, DC, Jan 22–26, 1997.
18. Marsh K, McDonald E, Sham H, Kempf D, Norbeck D. Enhancement of ABT-378 pharmacokinetics when administered in combination with ritonavir. 4th Conference on Retroviruses and Opportunistic Infection, Washington, DC, Jan 22–26, 1997.
19. Lal R, Hsu A, Chen P, Dennis S, Elshourbagy T, Locke C, Lam W, Japour A, Leonard J, Granneman GR, Sun E. Single dose pharmacokinetics of ABT-378

in combination with ritonavir. 37th ICAAC, Toronto, Ontario, Sept 28–Oct 1, 1997.
20. Kiso Y. Biopolymers 1996; 40:235–244.
21. Mimoto T, Imai J, Kisanuki S, Enomoto H, Hattori N, Akaji K, Kiso Y. Kynostatin (KNI)-227 and-272, highly potent anti-HIV agents: conformationally constrained tripeptide inhibitors of HIV protease containing allophenylnorstatine. Chem Pharm Bull 1992; 40:2251–2253.
22. Baldwin ET, Narayana B, Gulnik S, Liu B, Topol IA, Kiso Y, Mimoto T, Mitsuya H, Erickson JW. Structure of HIV-1 protease with KNI-272, a tight-binding transition-state analog containing allophenylnorstatine. Structure 1995; 3:581–590.
23. Gulnik SV, Suvorov LI, Liu B, Yu B, Anderson B, Mitsuya H, Erickson JW. Kinetic characterization and cross-resistance patterns of HIV-1 protease mutants selected under drug pressure. Biochemistry 1995; 34:9282–9287.
24. Tanaka M, Mimoto T, Anderson B, Gulnik S, Bhat TN, Yusa K, Hayashi H, Kiso Y, Erickson JW, Mitsuya H. Identification of protease inhibitors containing allophenylnorstatine active against both wild-type and KNI-272-resistant HIV-1 variants. XIth Int Conf AIDS, Vancouver, Canada, July 7–12, 1996.
25. Kageyama S, Mimoto T, Murakawa Y, Nomizu M, Ford H, Shirasaka T, Gulnik S, Erickson J, Takada K, Hayashi H, Broder S, Kiso Y, Mitsuya H. In vitro anti-human immunodeficiency virus (HIV) activities of transition state mimetic HIV protease inhibitors containing allophenylnorstatine. Antimicrob Agents Chemother 1993; 37: 810–817.
26. Kageyama S, Anderson BD, Hoesterey BL, Hayashi H, Kiso Y, Flora KP, Mitsuya H. Protein binding of human immunodeficiency virus protease inhibitor KNI-272 and alteration of its in vitro antiretroviral activity in the presence of high concentrations of proteins. Antimicrob Agents Chemother 1994; 38:1107–1111.
27. Kiriyama A, Nishiura T, Ishino M, Yamamoto Y, Ogita I, Kiso Y, Takada K. Binding characteristics of KNI-272 to plasma proteins, a new potent tripeptide HIV protease inhibitor. Biopharm Drug Dispos 1996; 17:739–751.
28. Kiriyama A, Sugahara M, Yoshikawa Y, Kiso Y, Takada K. The bioavailability of oral dosage forms of a new HIV-1 protease inhibitor, KNI-272 in beagle dogs. Biopharm Drug Dispos 1996; 17:125–134.
29. Humphrey RW, Nguyen B–Y, Wyvill KM, Shay LE, Lietzau J, Ueno T, Fukasawa T, Hayashi H, Mitsuya H, Yarchoan R. A phase I trial of HIV protease inhibitor KNI-272 in patients with AIDS or symptomatic HIV infection. XIth Int Conf AIDS, Vancouver, Canada, July 7–12, 1996.
30. Lam PYS, Jadhav PK, Eyermann CJ, Hodge CN, Ru Y, Bacheler LT, Meek JL, Otto MJ, Rayner MM, Wong YN, Chang C–H, Weber PC, Jackson DA, Sharpe TR, Erickson–Viitanen S. Rational design of potent, bioavailable, nonpeptide cyclic ureas as HIV protease inhibitors. Science 1994; 263:380–384.
31. Lam PYS, Ru Y, Jadhav PK, Aldrich PE, DeLucca GV, Eyermann CJ, Chang C–H, Emmet G, Holler ER, Daneker WF, Li L, Confalone PN, McHugh RJ, Han Q, Li R, Markwalder JA, Seitz SP, Sharpe TR, Bacheler LT, Rayner MM, Klabe RM, Shum L, Winslow DL, Kornhauser DM, Jackson DA, Erickson–Viitanen S, Hodge CN. Cyclic HIV protease inhibitors: synthesis, conformational analysis, P2/

P2′ structure–activity relationship, and molecular recognition of cyclic ureas. J Med Chem 1996; 39:3514–3525.
32. Ala PJ, Huston EE, Klabe RM, McCabe DD, Duke JL, Rizzo CJ, Korant BD, DeLoskey RJ, Lam PYS, Hodge CN, Chang C–H. Molecular basis of HIV-a protease drug resistance: structural analysis of mutant protease complexed with cyclic urea inhibitors. Biochemistry 1997; 36:1573–1580.
33. Lam PYS, Rodgers JD, Li R, Ru Y, Jadhav PK, Chang C–H, Clark CG, Seitz SP, Bacheler LT, Lam GN, Erickson–Viitanen S, Trainor GV, Anderson PS. Cyclic HIV protease inhibitors: nonsymmetric cyclic ureas with excellent oral bioavailability. 213th ACS National Meeting, MEDI-278, Washington, DC, Apr 13–17, 1997.
34. Tummino PJ, Vara Prasad JVN, Ferguson D, Nouhan C, Graham N, Domagala JM, Ellsworth E, Gajda C, Hagen SE, Lunney EA, Para KS, Tait BD, Pavlovsky A, Erickson JW, Gracheck S, McQuade TJ, Hupe DJ. Discovery and optimization of nonpeptide HIV-1 protease inhibitors. Bioorg Med Chem 1996; 4:1401–1410.
35. Vara Prasad JVN, Para KS, Lunney EA, Ortwine DF, Dunbar JB, Ferguson D, Tummino PJ, Hupe D, Tait BD, Domagala JM, Humblet C, Bhat TN, Liu B, Guerin DMA, Baldwin ET, Erickson JW, Sawyer TK. Novel series of achiral, low molecular weight and potent HIV-1 protease inhibitors. J Am Chem Soc 1994; 116:6989–6990.
36. Hagen SE, Vara Prasad JVN, Boyer FE, Domagala JM, Ellsworth EL, Gajda C, Hamilton HW, Markoski LJ, Steinbaugh BA, Tait BD, Lunney EA, Tummino PJ, Ferguson D, Hupe D, Nouhan C, Gracheck SJ, Saunders JM, VanderRoest S. Synthesis of 5,6-dihydro-4-hydroxy-2-pyrones as HIV-1 protease inhibitors: the profound effect of polarity on antiviral activity. J Med Chem 1997; 40:3707–3711.
37. Sharmeen L, Heldsinger A, McQuade T, Domagala J, Gracheck S. Antiviral activity of PD173606, PD177298, PD178390, and PD178392: non-peptidic HIV-1 protease inhibitors. 37th ICAAC, Toronto, Ontario, Sept 28–Oct 1, 1997.
38. Lazdins, JK, Bold G, Capraro HG, Cozens R, Fässler A, Flesch G, Klimkait T, Lang M, Mestan J, Poncioni B, Rösel J, Stover D, Tintelnot–Blomley M, Walker MR, Woods–Cook K. Profile of CGP-61755: a novel and potent HIV-1 protease inhibitor that shows enhanced anti-HIV activity when combined with other antiretroviral agents in vitro. Schweiz Med Wochenschr 1996; 126:1849–1851.
39. Kiriyama A, Mimoto T, Kisanuki S, Kiso Y, Takada K. Comparison of a new orally potent tripeptide HIV-1 protease inhibitor (anti-AIDS drug) based on pharmacokinetic characteristics in rats after intravenous and intraduodenal administrations. Biopharm Drug Dispos 1993; 14:697–707.
40. Hodge CN, Aldrich PE, Bacheler LT, Chang CH, Eyermann CJ, Garber S, Grubb M, Jackson DA, Jadhav PK, Korant B, Lam PYS, Maurin MB, Meek JL, Otto MJ, Rayner MM, Reid C, Sharpe TR, Shum L, Winslow DL, Erickson–Viitanaen, S. Improved cyclic urea inhibitors of the HIV-1 protease: synthesis, potency, resistance profile, human pharmacokinetics and X-ray crystal structure of DMP-450. Chem Biol 1996; 3:301–314.
41. Lam PYS, Rodgers JD, Li R, Ru Y, Jadhav PK, Chang C–H, Clark CG, Seitz SP, Bacheler LT, Lam GN, Erickson–Viitanen S, Trainor GV, Anderson PS. Cyclic HIV protease inhibitors: nonsymmetric cyclic ureas with excellent oral bioavailability. 213th ACS National Meeting, MEDI-278, Washington, DC, Apr 13–17, 1997.

42. Guttendorf RJ, Middlebrook AM. Pharmacokinetics of a series of *S*-aryl-4-hydroxy-5,6-dihydropyrone HIV-1 protease inhibitors in dogs. 37th ICAAC, Toronto, Ontario, Sept 28–Oct 1, 1997.
43. Gajda C, Boyer FE, Ellsworth EL, Hagen SE, Hamilton HW, Kibbey CE, Lunney EA, Markowki LJ, Pavlovsky A, Vara Prasad JVN, Rubin J, Steinbaugh B, Tait BD, Tummino PJ, Urumov A, Zeikus E. An efficient asymmetric synthesis of lead dihydropyrone HIV protease inhibitors. 37th ICAAC, Toronto, Ontario, Sept 28–Oct 1, 1997.

8
Pharmacology and Drug Interactions of HIV Protease Inhibitors

Charles W. Flexner
The Johns Hopkins University School of Medicine and School of Hygiene and Public Health, Baltimore, Maryland

I. INTRODUCTION

Pharmacokinetic properties and drug interactions have played a larger role in the development and clinical use of protease inhibitors than for any other class of antiretroviral drugs. Maintaining adequate drug concentrations, avoiding and managing drug interactions, and minimizing drug toxicity all require a detailed understanding of drug disposition.

After oral administration, drug concentrations reach a peak (C_{max}) in the circulation, and then follow log-linear (first-order) elimination from the plasma (Fig. 1). Drug concentrations reach their trough (C_{min}) at the end of a dosing interval. Because some threshold concentration of drug is probably required to suppress HIV replication, efforts are made to maintain measurable drug concentrations throughout a dosing interval. The preferred target for drug concentrations at the end of a dosing interval is not well-defined for any of the five HIV protease inhibitors in current use. However, the concentration of drug required to inhibit HIV replication by 50% (IC_{50}) or 90% (IC_{90}) is often used as a reference point (see Fig. 1).

In general, higher drug concentrations are better at suppressing virus replication, but are also more prone to produce drug toxicity. The virological benefits of increased drug concentrations must be weighed against the risks of short-term and long-term toxicity. Further problems include that in vitro determination of

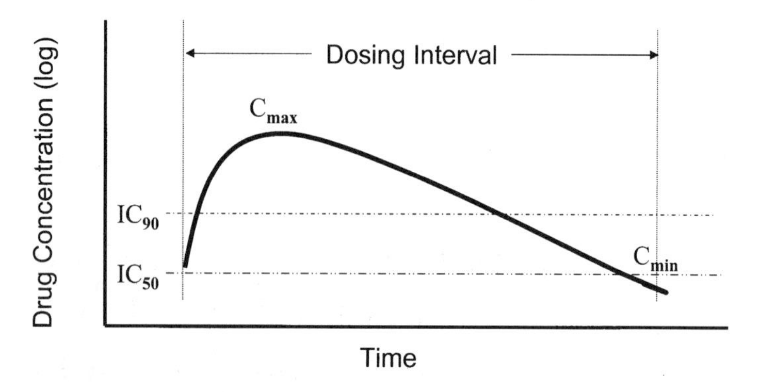

Figure 1 Representative concentration–time curve for an oral drug: The solid bold line indicates the drug concentration displayed on a logarithmic scale. IC_{90}, drug concentration required to inhibit virus replication by 90%; IC_{50}, drug concentration required to inhibit virus replication by 50%; C_{max}, maximal drug concentration during a dosing interval; C_{min}, minimal drug concentration during a dosing interval.

IC_{50} or IC_{90} is somewhat arbitrary, and depends on cell type and virus type, length of culture incubation, percentage of animal or human serum present in the culture, and other factors; reported IC_{50} values from different laboratories may vary by a factor of 10 or more (1).

Finally, all protease inhibitors are used in combination with other antiretroviral drugs; efficacy and tolerability are bound to depend on contributions of the other drugs in the regimen. Despite the goal of maintaining drug concentrations above some virologically relevant level, such as an IC_{90}, some drugs, such as saquinavir, were clinically efficacious using regimens in which trough concentrations were close to or equal to zero (2).

A general principle applied in the design of clinical drug regimens has been to achieve average concentration–time profiles associated with maximal anti-HIV effects and acceptable toxicity profiles in large clinical trials (3). The care provider and patient must carefully consider factors that might perturb this situation: for example, drug–drug or drug–food interactions.

II. BASIC PHARMACOKINETIC PROPERTIES

A. Metabolism, Absorption, and Distribution

Table 1 provides the pharmacokinetic properties of the five HIV protease inhibitors approved by the U. S. Food and Drug Administration (FDA) for use in HIV-infected patients as of this writing. These drugs are: amprenavir (Agenerase;

Table 1 Pharmacokinetics of Approved HIV Protease Inhibitors[a]

Drug	Dose (mg)	% Bioavailability (approx. oral F)	Food effect (%)	C_{max} (μg/mL)	T_{max} (h)	$T_{1/2}$ (h)	Variability (CV %AUC)	Protein binding (%)	V_d (L/kg)	CSF (%)	Clearance route	P450 induction	P450 inhibition
Amprenavir	1200 bid	NR	−21	5.4	1.9	7.1–10.6	63	90	6.1	1	Hepatic (75%) 3A4	No	Yes (3A4)
Indinavir	800 q8h	60–65	−77[b]	7.7	0.8	1.8	22–47	60–65	NR	2.2–76	Hepatic (88–90%) 3A4	No	Yes (3A4)
Nelfinavir	750 tid	> 78	+200–300	3.0–4.0	2.0–4.0	3.5–5.0	NR	> 98	2.0–7.0	< 1	Hepatic (> 78%) 2C19	Yes	Yes (3A4)
Ritonavir	600 bid	66–75	−7/+15	11.2	2.0–4.0	3.0–5.0	30–36	98–99	0.4	1	Hepatic (> 95%) 3A4	Yes	Yes (3A4≫ 2D6)
Saquinavir	600–1200 tid[b]	< 4–12	+670	0.2	NR	NR	46–84	98	10.0	< 1	Hepatic (> 97%) 3A4	No	Yes (3A4)

[a] Published mean values and ranges from studies in adults without hepatic or renal dysfunction: AUC, area under the concentration–time curve during an average dosing interval; bid, twice-daily; C_{max}, maximal concentration during a dosing interval; CSF, cerebrospinal fluid; CV, coefficient of variation; F, bioavailability; NR, not reported; P450, cytochrome P450 drug-metabolizing enzymes; tid, thrice-daily; T_{max} time to maximal concentration; $T_{1/2}$, half-life of the principal elimination (β) phase; V_d volume of distribution.

[b] Dose of saquinavir depends on formulation: 600 mg t.i.d. for the hard-gel, and 1200 mg t.i.d. for the soft-gel formulation.

Source: Ref. 4.

Glaxo Wellcome), indinavir (Crixivan; Merck & Co., Inc.), nelfinavir (Viracept; Agouron Pharmaceuticals), ritonavir (Norvir; Abbott Laboratories), and saquinavir (Fortovase or Invirase; Roche Pharmaceuticals).

The approved protease inhibitors are each metabolized by cytochrome P450 (CYP450) enzymes. Nelfinavir is converted to its major metabolite by the CYP450 2C19 isoform (5), but all other protease inhibitors are principally metabolized by CYP450 3A4 (6). In vitro, CYP3As from human intestine and liver equally metabolize saquinavir, suggesting that both sources of enzymes are factors in the extensive first-pass clearance of this drug (7). The effects of first-pass metabolism contribute to limited oral bioavailability and pharmacokinetic variability for saquinavir (3), and to a lesser extent, for the other drugs in this category.

With oral administration, peak absorption occurs within 1–3 h (see Table 1). Plasma elimination half-lives vary from 1.8 h for indinavir to 7–10 h for amprenavir (see Table 1). Elimination studies of ^{14}C-labeled drugs demonstrate fecal recovery of parent drug and metabolites ranging from a low of 75% (amprenavir) to 88% (saquinavir), and urine levels between 1% (saquinavir) to 19% (indinavir), confirming the liver as the major source of elimination.

Interindividual variability in pharmacokinetics is large, as indicated by a coefficient of variation (CV) for the mean area under the concentration–time curve (AUC) of 30% or more in all cases (see Table 1). Several factors contribute to pharmacokinetic variability, including the effects of first-past metabolism and food (3,4,6). In general, pharmacokinetic parameters obtained from studies of HIV-infected patients show greater variability than those taken from studies of healthy volunteers. This likely reflects the effects of intercurrent disease, variability in weight, and concurrent medications, rather than a pharmacokinetic effect of HIV infection per se.

All of the approved HIV protease inhibitors are highly protein-bound, with the exception of indinavir (see Table 1). Amprenavir, nelfinavir, and ritonavir bind extensively to α_1-acid glycoprotein (AAG); the estimated association constant (K_a) for nelfinavir and saquinavir is close to 1 μM (8). The addition of physiological concentrations of AAG increases the IC_{90} (that is, reduces the anti-HIV potency) of several peptidic protease inhibitors by a factor of ten or more (9,10). Binding affinities for albumin are generally much lower than for AAG, and this protein has less effect on drug activity in vitro (11).

B. Food Effects

Food has an important effect on the oral absorption of these drugs, although the impact varies by drug, composition of the meal, and even formulation of the drug (3,6). A high-fat meal substantially increases the bioavailability of saquinavir and nelfinavir, but reduces the bioavailability of indinavir and amprenavir (see

Table 1). The same high-fat meal increases the bioavailability of ritonavir capsules, but decreases the bioavailability of ritonavir liquid formulation (see Table 1). Nelfinavir and saquinavir should be taken with a moderate-fat meal; indinavir should be given in the fasted state or with a light, low-fat snack. Amprenavir and ritonavir may be taken with or without food, but amprenavir should not be given with a high-fat meal (see Table 1).

C. Formulation Differences

In addition to the aforementioned difference in food effect on ritonavir capsules versus liquid, amprenavir as an oral solution is 14% less bioavailable than the capsule formulation (12,13). Saquinavir was developed as two successive formulations; the newer Fortovase formulation has approximately threefold greater oral bioavailability than the older Invirase formulation (14). The prescriber needs to be aware that different formulations are not necessarily interchangeable for the same parent drug.

D. Penetration into Central Nervous System

The penetration of protease inhibitors into body compartments, such as brain and genital tract, are important considerations; these compartments may act as reservoirs for HIV, and genital secretions serve as the source of infecting virus in most naturally acquired cases. For all approved HIV protease inhibitors, except indinavir, fractional penetration into the central nervous system (CNS) is low (see Table 1). This may be related to the fact that most of these agents are highly protein-bound in the plasma, and only the free drug fraction is available to diffuse into the cerebrospinal fluid (CSF). The reported ratio of CSF to plasma concentrations is 1% or less for amprenavir (12,13), nelfinavir (15), ritonavir (16), and saquinavir (16), whereas the comparable ratio for indinavir is at least 12% (17). This roughly parallels the free fraction of each drug available in human plasma (see Table 1).

Drug transporters such as P-glycoprotein (P-gp) may also play an important role in CSF penetration. Transgenic mice lacking P-gp had up to 30-fold greater CSF penetration of HIV protease inhibitors than control animals (18) (see following discussion).

Whether or not this has a bearing on clinical drug activity is controversial; in most patient studies, reductions in HIV RNA in the plasma are accompanied by similar reductions in the CSF (16,19). In almost all cases, if the plasma HIV RNA becomes undetectable, the CSF viral RNA also becomes undetectable. This degree of viral load suppression in the CSF is achieved with agents, the fractional penetration of which is as low as 0.2% for nelfinavir (15,20). Only one of 17 patients receiving amprenavir had a higher CSF viral load than plasma viral load in one study, and the magnitude of this difference was small (21). These findings suggest that the relatively low concentrations of free drug in the CSF are adequate to suppress HIV replication. Alternatively, CSF viral RNA may not adequately reflect virus replication in the brain parenchyma.

E. Penetration into the Genital Tract

Because genital secretions are a major source of sexual transmission of HIV, drugs that penetrate into these fluids more efficiently may be able to better inhibit replication and transmission of the virus. In general, antiretroviral concentrations in seminal plasma are much higher than in the CSF, and some antiretrovirals actually appear to be concentrated in the semen as compared with the blood (22). Indinavir concentrations in the semen were equal to plasma concentrations (100% fractional penetration), whereas ritonavir and saquinavir fractional penetration into seminal plasma was only 2–5% (23).

Despite this, all three drugs were associated with equivalent suppression of viral load in the semen of treated patients (23). This could be because only low concentrations of saquinavir or ritonavir are adequate to suppress HIV in the genital tract, or could indicate that the HIV measured in semen comes directly from and is reflective of the blood. Penetration of protease inhibitors into vaginal fluid is more difficult to quantify, and little information is available in the literature.

III. PHARMACOKINETICS IN SPECIAL POPULATIONS

A. Hepatic Insufficiency

Liver disease may affect the clearance of all approved HIV protease inhibitors, for elimination depends on hepatic CYP450 activity. In general, drug doses need to be reduced only in those patients with moderate to severe hepatic impairment, as measured by Child–Pugh classification. Dose adjustment should be based on published dosing guidelines, whenever possible.

Amprenavir and indinavir have been studied in patients with impaired hepatic function (12,13,24). In patients with moderate cirrhosis, the AUC of amprenavir was increased by more than twofold, compared with healthy volunteers; in patients with severe cirrhosis, the amprenavir AUC was increased more than threefold, resulting in recommendations for dosage adjustment in these situations (12,13). Likewise, the AUC and half-life of indinavir were modestly altered in patients with mild to moderate hepatic insufficiency and cirrhosis; patients with severe hepatic impairment have not been evaluated for pharmacokinetic changes (24).

In five patients with liver disease, nelfinavir pharmacokinetics were highly variable compared with patients without liver disease (25). Elimination half-life was greatly prolonged (up to 20 h, compared with 3.5–5 h in patients without liver disease). Also, formation of the nelfinavir M8 metabolite was reduced (M8 AUC 2–5% of NFV AUC, as compared with 33% in patients without liver disease) (25). Dose adjustment was successfully applied in these patients to produce nelfinavir concentrations that were within the range of those seen in patients without liver disease.

Amprenavir oral solution should be used with caution in patients with hepatic impairment. This solution contains propylene glycol, and patients with impaired liver function may be at increased risk of propylene glycol toxicity (12).

B. Renal Insufficiency

None of these compounds have been evaluated pharmacokinetically in patients with renal insufficiency. Because renal clearance is minimal for this class, there is little reason to expect a need for dosage adjustment in patients with renal disease. The one exception is indinavir, which should be used with caution in patients with preexisting kidney problems, given its propensity for nephrolithiasis and nephrotoxicity.

Amprenavir oral solution is contraindicated in patients with renal failure. Because this solution contains propylene glycol, patients with renal failure are more susceptible to propylene glycol toxicity and acidosis (12).

C. Pediatrics

Most of these compounds have now been evaluated in pediatric patients. In general, pharmacokinetics are predictable and dose-proportional. For example, at single doses between 5 and 20 mg/kg, the AUC of amprenavir increased proportionally, whereas the C_{max} increased less than proportionally (12,13). One recent study suggested that dosing indinavir based on body surface area, as is often done in pediatric populations, may overdose some children, especially those with a small body surface area (26). However, in this study, indinavir trough concentrations were also lower, on average, than those in adults (26). Indinavir may need to be dosed more frequently in children, for example, every 6 h, to achieve concentrations comparable with those in adults (26,27). Nelfinavir concentrations (AUC) achieved in children using the currently recommended regimen are comparable with those in adults (28).

D. Race, Gender, and Age

There is no evidence for racial or gender-based differences in the pharmacokinetic properties of the HIV protease inhibitors. Although variability in pharmacokinetic parameters is generally higher in patients than in healthy volunteers, most of the differences seen in protease inhibitor pharmacokinetics in women, as compared with men in the same study, can be attributed to differences in body size and the use of fixed drug doses. The pharmacokinetics of amprenavir, indinavir, and saquinavir are not different between male and female patients (12–14,24). The pharmacokinetics of amprenavir were no different in black than in nonblack racial groups (12,13).

There are no reported studies of the pharmacokinetics of HIV protease inhibitors in elderly patients. Because the main pathway for protease inhibitor metabolism is phase I (CYP450) enzymes, and because phase I enzyme activity is diminished with advancing age, the clearance of HIV protease inhibitors may be reduced in the elderly, especially those older than 65, as compared with younger patients. Therefore, protease inhibitor concentrations could increase, on average, with advancing age.

E. Therapeutic Drug Monitoring

The results of several clinical studies suggest a relation between protease inhibitor concentration in the patient and anti-HIV effect. In three dose-ranging monotherapy trials, rapid emergence of drug resistance was observed in subjects receiving lower than the currently recommended doses of ritonavir (29,30) or indinavir (31). Reductions in plasma HIV RNA and risk of treatment failure have been correlated with systemic concentrations of indinavir (32–37), saquinavir (38,39), and ritonavir (40,41). Several studies have suggested that indinavir trough concentrations of 100 ng/mL or less were correlated with suboptimal virological response and increased risk of drug resistance (36,37).

Given these associations and the propensity of HIV protease inhibitors (and other antiretrovirals) to select drug-resistant HIV variants, it might seem advisable to measure drug concentrations in patients and compensate for low concentrations by increasing drug dose. Unfortunately, no prospective study has convincingly demonstrated the value of therapeutic drug monitoring (TDM) and dose-individualization for HIV protease inhibitors. Although drug concentrations are certainly one component of treatment response, there are many others: adherence, tolerability, and intrasubject variability in pharmacokinetics owing to the effects of food and intercurrent medications.

Reflecting the complexity of concentration–response relation in clinical trials, several studies have found no correlation between plasma concentrations of protease inhibitors and virological outcome (42–44). Concentration–response relation may be difficult or impossible to sort out in clinical trials. For example, if a large number of patients in a cohort are nonadherent, these patients can have ''normal'' (meaning average) or even high drug concentrations on the day of measurement, but may fail therapy because they often do not take their medication. If drug concentrations vary from day to day in the same patient, or if the drug is difficult to measure, this will confound attempts to define the concentration–effect relation. As antiretrovirals are always given in combination with other anti-HIV drugs, patients with low concentrations of one drug can still have a good outcome if they are responding to other drugs in the regimen.

Several obstacles must be overcome if TDM is to become a valuable tool for HIV protease inhibitors. First, target concentrations that take into account both efficacy and toxicity must be established; these target concentrations must be validated in prospective clinical trials. Then, prospective randomized trials need to demonstrate that TDM benefits patients independent of other potentially confounding variables. It should be pointed out that second- and third-generation protease inhibitors—with longer half-lives, better tolerance, more convenience, and less frequent dosing—may be associated with a greatly reduced need for TDM as compared with currently available drugs in this class.

IV. DRUG INTERACTIONS

A. Metabolic Pathways

All approved HIV protease inhibitors are substrates for CYP450 enzymes, and they are susceptible to drug interactions involving P450 inhibitors or inducers (6,45). All five approved HIV protease inhibitors can inhibit the metabolism of CYP450 3A4 at clinically achieved concentrations; ritonavir is also a weak inhibitor of CYP 2D6 (46). Nelfinavir and ritonavir are moderately potent inducers of hepatic drug-metabolizing enzymes, including various CYP450 isoforms and glucuronyl transferases (6,45).

Of the five approved HIV protease inhibitors, ritonavir is by far the most potent inhibitor of CYP450 3A4. Ritonavir's CYP450 inhibition has mixed competitive and noncompetitive features in vitro (46), but for practical purposes, it can be considered a reversible inhibitor in vivo because of the rapid turnover of these enzymes in the liver. Amprenavir, indinavir, and nelfinavir are less potent inhibitors, and saquinavir is the least potent (6,45). The K_i for inhibition of terfenadine metabolism in vitro is 0.017 μM for ritonavir (46), but 0.7 μM for saquinavir (7), which indicates a 40-fold difference in potency.

Ritonavir metabolism is characterized by autoinduction; steady-state trough concentrations fall two- to threefold during the first 2 weeks of monotherapy using a fixed dose (47). Ritonavir and nelfinavir can accelerate the clearance of other metabolized drugs because they are inducers of hepatic drug-metabolizing enzymes. Concurrent nelfinavir reduces the zidovudine AUC by 35%, and ritonavir by 25%, presumably as a consequence of induction of glucuronyl transferases (6). Nelfinavir decreases the ethinyl estradiol AUC by 47% and ritonavir by 40%; these protease inhibitors, therefore, are contraindicated in women taking oral contraceptives that contain the combination of norethindrone and ethinyl estradiol (6).

B. Drug Transporters

Drug transporters such as P-glycoprotein (P-gp), which is the product of the multidrug resistance *mdr1* gene originally described in cells resistant to certain anticancer agents, represent an additional pathway for drug interactions. Several HIV protease inhibitors are P-gp substrates in vitro (48,49). Some HIV protease inhibitors can also inhibit the activity of P-gp (50,51). Transgenic mice deficient in P-gp had up to 30-fold higher CSF concentrations of indinavir, nelfinavir, and saquinavir than control animals (18), suggesting a role of P-gp in exclusion of these drugs from the CSF. However, the in vitro anti-HIV activity of indinavir, nelfinavir, saquinavir, and ritonavir was not affected by P-gp expression (52). This suggests that HIV protease inhibitor penetration into cells expressing a high level of P-gp activity is still adequate to suppress virus replication.

Because P-gp is expressed in the intestinal tract, drug interactions involving orally administered drugs may involve this transport system. Selective P-gp blockade could reduce first-pass metabolism of drugs such as saquinavir, and could also be used to increase CNS penetration of HIV protease inhibitors.

C. Interactions Between HIV Protease Inhibitors

As a consequence of P450 inhibition and induction, all approved HIV protease inhibitors have significant pharmacokinetic drug interactions when given together (Table 2). In several cases, the magnitude of the pharmacokinetic interaction between protease inhibitors is substantial. For example, ritonavir inhibits the first-pass metabolism of saquinavir and increases the saquinavir AUC by up to 50-fold in single-dose studies, and 20- to 30-fold at steady-state (56). Nelfinavir increases the saquinavir AUC by 392% and the indinavir AUC by 51% in single-dose studies (57), although the magnitude of this interaction is much less at steady state, presumably because of P450 induction.

The magnitude of the pharmacokinetic interaction between protease inhibitors has encouraged the clinical development of dual protease inhibitor combinations to allow reduced drug doses, less frequent dosing, and higher trough concentrations at the end of a dosing interval, as compared with using only one protease inhibitor at a time (4). These interactions are discussed in more detail in Chapter 9.

Table 2 Selected Pharmacokinetic Drug Interactions: Impact of Concurrent Medications on Approved HIV Protease Inhibitors[a]

Drug	Amprenavir[b]	Indinavir	Nelfinavir	Ritonavir	Saquinavir[c]
HIV protease inhibitors					
Amprenavir	—	−38	+15	NR	+21
Indinavir	+33	—	+83	NR	+364/+620
Nelfinavir	NC	+51	—	+9	+392
Ritonavir	+127/+143[d]	+200/+500[e]	+152	—	+1587
Saquinavir	−32	NR	+18	NC	—
Other P450 inhibitors					
Ketoconazole	+31	+62	+35	NR	+130/+300
Clarithromycin	+18	+29	NR	+12	+187
Fluconazole	NR	−19	NR	+12	NR
Fluoxetine	NR	NR	NR	+19	NR
P450 inducers					
Rifabutin	−15	−32	−32	NR	−43
Rifampin	−82	−92	−82	−35	−84
Nucleoside analogues					
Didanosine	NR	NR	NC	NC	NR
Lamivudine	NC	NC	NR	NR	NR
Stavudine	NR	NC	NR	NR	NR
Zidovudine	+13	+13	NC	NC	NC
Non-nucleoside reverse transcriptase inhibitors					
Delavirdine	NR	+72	NR	+2	+520
Efavirenz[f]	−25[d]	−31	+20	+18	−62
Nevirapine	NR	−28	NR	NC	−27

Values shown are the reported mean percentage change in AUC (area under the concentration–time curve) during a dosing interval; + indicates an increase in AUC; − indicates a decrease in AUC; NC, no statistically significant change; NR, not reported.

Source: Ref 6; additional data;[b] 12, 13;[c] 14;[d] 53;[e] 54;[f] 55.

D. Interactions with Other Antiretroviral Drugs

Because protease inhibitors are used as part of combination therapy with other antiretroviral drugs, their potential to interact with these agents is of great interest. As most of the antiretroviral nucleosides are renally cleared, no pharmacokinetic interaction with protease inhibitors is expected, and such is true (see Tables 2 and 3). However, zidovudine and abacavir are cleared by glucuronyl transferases, which might be induced by nelfinavir and ritonavir. In single-dose studies, concurrent nelfinavir reduced the zidovudine AUC by 35%, and ritonavir reduced the zidovudine AUC by 25%, presumably by induction of glucuronyl transferases (57,58). Because the intracellular concentration of zidovudine triphosphate (the active anabolite) is not significantly affected by such a reduction in plasma concentrations of zidovudine (59), no dose adjustment is recommended when these drugs are given together.

The three non-nucleoside reverse transcriptase inhibitors approved in the United States—efavirenz (Sustiva), nevirapine (Viramune), and delavirdine (Rescriptor)—are all substrates for CPP450 enzymes (45,55). Furthermore, efavirenz and nevirapine are modest CPP450 inducers, and efavirenz and delavirdine are moderate CPP450 3A4 inhibitors, producing important opportunities for interaction with protease inhibitors (45,55). Concurrent delavirdine increased the saquinavir AUC by 520%, and could be used to boost saquinavir concentrations. Efavirenz, however, reduced the AUC of amprenavir by 25%, indinavir by 31%, and saquinavir by 62% (see Table 2); coadministration of efavirenz with these three drugs, therefore, is not recommended (55).

Efavirenz does not reduce concentrations of nelfinavir or ritonavir (see Table 2). In fact, 200 or 500 mg of ritonavir twice daily boosted the amprenavir AUC by nearly 150% in one clinical study (53). Adding 600 mg of efavirenz daily in these subjects then had no further effect on the amprenavir AUC (53). This suggests that the inhibitory effect of ritonavir on amprenavir clearance supercedes the CPP450 induction of efavirenz. These three drugs might be used together to overcome the detrimental effect of efavirenz on the amprenavir AUC, and improve amprenavir's overall pharmacokinetic profile.

E. Interactions with Antimicrobial Agents

Patients with HIV infection often require antimicrobial drugs to treat or prevent intercurrent opportunistic infections. Antimicrobials that are potent CPP450 inhibitors could increase plasma concentrations of protease inhibitors. Concurrent ketoconazole, for example, increased the indinavir AUC by 62%, the nelfinavir AUC by 35%, and the saquinavir AUC by 300% (see Table 2). An interaction of this magnitude could be beneficial for the clinical use of saquinavir, but is substantially less than the interaction produced by ritonavir (56).

Table 3 Selected Pharmacokinetic Drug Interactions: Impact of Approved Protease Inhibitors on Concurrent Medications[a]

Drug	Amprenavir[b]	Indinavir	Nelfinavir	Ritonavir	Saquinavir[c]
Anti-infectives					
Clarithromycin	NC	+53	NR	+77	+45
Isoniazid	NR	+13	NR	NR	NR
Ketoconazole	+44	+68	NR	NR	NC
Rifabutin	+193	+204	+207	+350	NR
Sulfamethoxazole	NR	NC	NR	NR	NR
Nucleoside analogues					
Didanosine	NR	NR	NR	−13	NR
Lamivudine	NC	−6	+10	NR	NR
Stavudine	NR	+25	NC	NR	NR
Zalcitabine	NR	NR	NR	NR	NC
Zidovudine	+31	+17/+36	−35	−25	NC
Other agents					
Desipramine	NR	NR	NR	+145	NR
Ethinyl estradiol	NR	+24	−47	−40	NR
Norethindrone	NR	+26	−18	NR	NR
Sildenafil	NR	NR	NR	+1000	+210
Theophylline	NR	NR	NR	−43	NR
Trimethoprim	NR	+19	NR	+20	NR

[a] Values shown are the reported mean percentage change in AUC (area under the concentration–time curve) during a dosing interval; + indicates an increase in AUC; − indicates a decrease in AUC; NC, no statistically significant change; NR, not reported.

Source: Ref. 6; additional data [b]12, 13; [c]14.

More concerning is the potential effect of concurrent CPP450 inducers, such as rifampin or rifabutin. These antimycobacterial drugs may increase the clearance and decrease the plasma concentrations of CPP450-metabolized drugs such as HIV protease inhibitors. This might decrease drug effectiveness, and promote the development of drug resistance. Concurrent rifabutin decreased the amprenavir AUC by 15%, the indinavir AUC by 32%, the nelfinavir AUC by 32%, and the saquinavir AUC by 43% (see Table 2). For patients taking indinavir or nelfinavir who require concurrent therapy with rifabutin for mycobacterial infections, the rifabutin dose should be decreased to 150 mg daily (60,61). Concurrent rifabutin is not recommended with ritonavir (58) or saquinavir (14).

Rifampin is a more potent CYP450 inducer than rifabutin (62), and has a more profound effect on protease inhibitor pharmacokinetics. Concurrent rifampin reduced amprenavir's AUC by 82%, indinavir's AUC by 92%, nelfinavir's by 82%, ritonavir's by 35%, and saquinavir's by 84% (see Table 2). Therefore, rifampin contraindicated for treatment of tuberculosis in patients requiring HIV protease inhibitors (61).

HIV protease inhibitors may also affect the pharmacokinetics of concurrent antimicrobial agents that are CYP450 substrates. For example, amprenavir increased the rifabutin AUC by 193%, indinavir increased the rifabutin AUC by 204%, nelfinavir by 207%, and ritonavir by 350% (see Table 3). Ritonavir increased concentrations (AUC) of the 25-desacetyl metabolite of rifabutin by more than 35-fold (58). Concurrent administration of ritonavir and rifabutin has been associated with an increased incidence of rifabutin toxicity, mainly uveitis (63). Ritonavir and rifabutin should not be coadministered except under close supervision, with a substantial reduction in the rifabutin dose, probably to 150 mg three times weekly or 300 mg once weekly (64).

F. Other Clinically Important Interactions

Several CYP450 3A4 substrates with narrow therapeutic indices are absolutely contraindicated for coadministration with protease inhibitors, owing to potentially toxic drug interactions. Physicians should avoid coadministration of drugs such as terfenadine (now withdrawn from the U.S. market), astemizole, cisapride, ergotamines, and some benzodiazepines (45). Induction interactions caused by nelfinavir or ritonavir may lead to clinically significant reductions in the concentrations of drugs such as warfarin, oral contraceptives, and methadone. These combinations should be avoided altogether (in the case of oral contraceptives) or used with caution and close monitoring (45).

Sildenafil (Viagra) is a CYP450 3A4 substrate that may have cardiovascular side effects. In a single-dose study, indinavir increased the sildenafil AUC by 4.4-fold (65). Saquinavir increased the sildenafil AUC by 3-fold, and ritonavir increased the AUC by 11-fold in healthy volunteers (66). Ritonavir also substantially increased concentrations of the sildenafil active metabolite (66). A 47-year-old man who had been taking ritonavir and saquinavir for more than a year died of a myocardial infarction after taking sildenafil (67). Although this report is worrisome, the patient had other myocardial risk factors. Patients taking ritonavir who wish to use sildenafil should take half the recommended dose and not repeat dosing within 48 h (66).

Methadone is also metabolized by CYP450, and its concentrations could be affected by CYP450 inhibitors or inducers. Many methadone users are HIV-infected, and the potential interaction of methadone with protease inhibitors needs to be considered. Methadone is a racemic mixture, and in a single small clinical study, nelfinavir decreased the methadone AUC for the *R*-enantiomer by 47% and for the *S*-enantiomer by 39% (68). Despite this, there were no reports of clinical methadone withdrawal and no need to increase methadone dose in patients participating in this study.

In another trial, patients on methadone maintenance were treated with ritonavir–saquinavir combination therapy at a dose of 400/400 mg twice daily. At steady state, the methadone AUC decreased 27% for the *R*-enantiomer, and 33% for the *S*-enantiomer (69). Ritonavir–saquinavir also had a minor effect on displacing methadone from its plasma protein-binding sites, and when AUC was expressed as free methadone, the changes secondary to ritonavir–saquinavir treatment were statistically significant only for the inactive *S*-enantiomer (69). Not surprisingly, the 12 subjects in this study showed no signs of either narcotic withdrawal or excessive narcotic intoxication. Nelfinavir or ritonavir–saquinavir can be used in patients on a methadone maintenance regimen, and in most cases there should be no need for dose adjustment.

G. Interactions with Foods and Health Foods

Concurrent food, especially high- to moderate-fat meals, can significantly alter the bioavailability of protease inhibitors, as discussed earlier. Food effect on systemic drug concentrations is generally in the positive direction, but food may negatively affect the bioavailability of indinavir (6).

It is now apparent that certain foods and health foods contain substances that can alter activity or expression of CYP450 enzymes and produce significant pharmacokinetic drug interactions with CYP450 substrates such as HIV protease inhibitors. Grapefruit juice, for example, contains CYP450 3A4 inhibitory activity. Coadministration of grapefruit juice and the saquinavir hard gelatin capsule formulation (Invirase) resulted in a two- to threefold increase in the saquinavir AUC (70). This most likely reflects inhibition of the first-pass metabolism of saquinavir.

More dire consequences would be expected from coadministration of a CYP450 inducer, which would increase drug clearance, reduce systemic drug concentrations, and increase the likelihood of treatment failure or emergence of resistant virus. There have been case reports suggesting that St. John's wort might contain a substance that could act as a CYP450 enzyme inducer. In a small study in healthy volunteers, St. John's wort (300 mg three times daily) reduced the indinavir AUC by 57% and reduced the indinavir trough (C_{min}) by 81% (71). St. John's wort reduced cyclosporine A concentrations by $> 50\%$ in two heart transplant recipients, and was associated with heart transplant rejection (72). Both of these studies indicate that St. John's wort can act as an inducer of CYP450 3A4.

Physicians need to be aware that some foods and health foods may cause undesirable drug interactions with HIV protease inhibitors and other prescribed drugs. HIV-infected patients taking protease inhibitors should avoid St. John's wort, and should be cautioned about the use of health foods in general.

V. CONCLUSION

Understanding the clinical pharmacology and pharmacokinetics of HIV protease inhibitors is essential for optimal clinical benefit. The available HIV protease inhibitors are susceptible to resistance if drug exposure is inadequate. Food, CYP450 inhibitors or inducers, and liver disease may significantly alter concentrations of these drugs. All of the approved HIV protease inhibitors can affect the metabolism of CPP450 substrates, and should be used with caution in patients taking other metabolized drugs.

ACKNOWLEDGMENT

The author wishes to thank Laura Rocco for assistance in manuscript preparation.

REFERENCES

1. Speck RS, Flexner C, Tian C–J, Yu X–F. Comparison of human immunodeficiency virus type 1 Pr55^{Gag} and Pr160$^{Gag\text{-}Pol}$ processing intermediates that accumulate in primary and transformed cells treated with peptidic and nonpeptidic protease inhibitors. Antimicrob Agents Chemother 2000; 44:1397–1403.
2. Salgo MP, Beattie D, Bragman K, Donatacci L, Jones M, Montgomery L. Saquinavir (Invirase) vs. HIVID (zalcitabine) vs. combination as treatment for advanced HIV infection in patients discontinuing/unable to take Retrovir (zidovudine). Program and Abstracts of the Eleventh International Conference on AIDS. Vancouver, BC: 1996: abstr. Mo.B.410.
3. Flexner C. Pharmacokinetics and pharmacodynamics of HIV protease inhibitors. *Infect Med* 1996; 13(suppl F): 16–23.

4. Flexner C. Dual protease inhibitor therapy in HIV-infected patients: pharmacologic rationale and clinical benefits. Annu Rev Pharmacol Toxicol 2000; 40:651–676.
5. Lillibridge JH, Kerr BM, Shetty BV, Lee CA. Prediction and interpretation of P450 isoform selective drug interactions for the hydroxy-*t*-butylamide metabolite of the HIV protease inhibitor nelfinavir mesylate [abstr]. Program and Abstracts of the 5th International ISSX Meeting, Cairns, Australia, 1998.
6. Flexner C. HIV protease inhibitors. N Engl J Med 1998; 338:1281–1292.
7. Fitzsimmons ME, Collins JM. Selective biotransformation of the human immunodeficiency virus protease inhibitor saquinavir by human small-intestinal cytochrome P4503A4. Drug Metab Dispos 1997; 25:256–666.
8. Bakker J, Tazartes D, Flexner C. A fluorescence quenching assay for determining the binding affinity (*K*a) of HIV protease inhibitors to alpha-1 acid glycoprotein. Program and Abstracts of the 12th World AIDS Conference, Geneva, Switzerland, 1998: abstr 42268.
9. Flexner C, Richman DD, Bryant M, Karim A, Yeramian P, Meehan P, Haubrich R, Para MF, Fischl MA. Effect of protein binding on the pharmacodynamics of an HIV protease inhibitor [abstr]. Antiviral Res 1995; 26:A282.
10. Lazdins JK, Mestan J, Goutte G, Walker MR, Bold G, Capraro HG. In vitro effect of a_1-acid glycoprotein on the anti-human immunodeficiency virus activity of the protease inhibitor CGP 61755: a comparative study with other relevant HIV protease inhibitors. J Infect Dis 1997; 175:1063–1070.
11. Molla A, Chernyavskiy T, Vasavanonda S, et al. Synergistic anti-HIV activity of ritonavir and other protease inhibitors in the presence of human serum [abstr]. Program and Abstracts of the 12th World AIDS Conference. Geneva, Switzerland, 1998.
12. Glaxo Wellcome, Inc. Agenerase (amprenavir) oral solution product monograph. Research Triangle Park, NC: Glaxo Wellcome, Inc, 2000.
13. Glaxo Wellcome, Inc. Agenerase (amprenavir) capsules product monograph. Glaxo Research Triangle Park, NC: Wellcome, Inc, 2000.
14. Roche Pharmaceuticals, Inc. Fortovase (saquinavir) soft gelatin capsules product monograph. Nutley, NJ: Roche Pharmaceuticals, Inc, 2000.
15. Aweeka F, Jayewardene A, Staprans S, Bellibas SE, Kearney B, Lizak P, Novakovic–Agopian T, Price RW. Failure to detect nelfinavir in the cerebrospinal fluid of HIV-1-infected patients with and without AIDS dementia complex. J Acquir Immune Defic Syndr 1999; 20:39–43.
16. Kravcik S, Gallicano K, Roth V, Cassol S, Hawley–Foss N, Badley A, Cameron DW. Cerebrospinal fluid HIV RNA and drug levels with combination ritonavir and saquinavir. J Acquir Immune Defic Syndr 1999; 21:371–375.
17. Collier AC, Marra C, Coombs RW, Zhong L, Stone J, Nguyen B. Cerebrospinal fluid indinavir and HIV RNA levels in patients on chronic indinavir therapy. Program and Abstracts of the Infectious Diseases Society of America 35th Annual Meeting, Washington, DC, 1997: abstr 22.
18. Kim RB, Fromm MF, Wandell C, et al. The drug transporter P-glycoprotein limits oral absorption and brain entry of HIV-1 protease inhibitors. J Clin Invest 1998; 101:289–294.
19. Letendre SL, Caparelli E, Ellis RJ, Dur D, McCutchan JA. Levels of serum and

cerebrospinal fluid (CSF) indinavir (IDV) and HIV RNA in HIV-infected individuals [abstr]. Program and Abstracts of 6th Conference on Human Retroviruses and Opportunistic Infections, Chicago, IL, 1999.

20. Haas D, Clough L, Johnson B, Bermingham K, McKinsey J, Donlon R, Spearman P, Wilkinson G, Harris V, Shoup R, Farnsworth A, Lillibridge J. Quantification of nelfinavir and its active metabolite (M8) in CSF and plasma, and correlation with CSF HIV-1 RNA response. Program and abstracts of the 7th Conference on Retroviruses and Opportunistic Infections, San Francisco, CA, 2000: abstr 313.
21. Murphy R, Currier J, Gerber J, D'Aquila R, Smeaton L, Sommadossi JP, Tung R, Gulick R. Antiviral activity and pharmacokinetics of amprenavir with and without zidovudine/3TC in the cerebral spinal fluid of HIV-infected adults. Program and Abstracts of the 7th Conference on Retroviruses and Opportunistic Infections, San Francisco, CA, 2000: abstr 314.
22. Kashuba AD, Dyer JR, Kramer LM, Raasch RH, Eron JJ, Cohen MS. Antiretroviral drug concentrations in semen: implications for sexual transmission of human immunodeficiency virus type 1. Antimicrob Agents Chemother 1999; 43:1817–1826.
23. Taylor S, Back D, Drake S, Gibbons S, Reynolds H, White D, Pillay D. Antiretroviral drug concentrations in semen of HIV-infected men: differential penetration of indinavir (IDV), ritonavir (RTV), and saquinavir (SQV). Program and Abstracts of the 7th Conference on Retroviruses and Opportunistic Infections, San Francisco, CA, 2000: abstr 318.
24. Merck & Company. Crixivan (indinavir) capsules product monograph. West Point, PA: Merck & Company, 2000.
25. Khaliq Y, Gallicano K, Seguin I, Fyke K, Carignan G, Badley A, Cameron DW. Therapeutic drug monitoring of nelfinavir in HIV patients with liver disease. Program and Abstracts of the 6th Conference on Retroviruses and Opportunistic Infections, Chicago, IL, 1999: abstr 369.
26. Gatti G, Vigano' A, Sala N, Vella S, Bassetti M, Bassetti D, Principi N. Indinavir pharmacokinetics and pharmacodynamics in children with human immunodeficiency virus infection. Antimicrob Agents Chemother 2000; 44:752–755.
27. Fletcher CV, Brundage RC, Remmel RP, Page LM, Weller D, Calles NR, Simon, C, Kline MW. Pharmacologic characteristics of indinavir, didanosine, and stavudine in human immunodeficiency virus-infected children receiving combination therapy. Antimicrob Agents Chemother 2000; 44:1029–1034.
28. Starr SE, Fletcher CV, Spector SA, Yong FH, Fenton T, Brundage RC, Manion D, Ruiz N, Gersten M, Becker M, McNamara J, Mofenson LM, Purdue L, Siminski S, Graham B, Kornhauser DM, Fiske W, Vincent C, Lischner HW, Dankner WM, Flynn PM. Combination therapy with efavirenz, nelfinavir, and nucleoside reverse-transcriptase inhibitors in children infected with human immunodeficiency virus type 1. N Engl J Med 1999; 341:1874–1881.
29. Danner SA, Carr A, Leonard JM, et al. A short-term study of the safety, pharmacokinetics and efficacy of ritonavir, an inhibitor of HIV-1 protease. N Engl J Med 1995; 333:1528–1533.
30. Markowitz M, Saag M, Powderly WG, et al. A preliminary study of ritonavir, an inhibitor of HIV-1 protease, to treat HIV-1 infection. N Eng J Med 1995; 333:1534–1539.

31. Emini EA. Resistance to anti-human immunodeficiency virus therapeutic agents. Adv Exp Med Biol 1995; 390:87–95.
32. Stein DS, Fish DG, Billelo JA, et al. A 24-week open-label phase I/II evaluation of the HIV protease inhibitor MK-689 (indinavir). AIDS 1996; 10:485–492.
33. Burger DM, Hoetelmans RMW, Hugen PWH, et al. Low plasma concentrations of indinavir are related to virological treatment failure in HIV-1 infected patients on indinavir-containing triple therapy. Antiviral Ther 1998; 3:215–220.
34. Hoetelmans RMW, Reijers MHE, Weverling GJ, et al. The effect of plasma drug concentrations on HIV-1 clearance rate during quadruple therapy. AIDS 1998; 12: F111–F115.
35. Descamps D, Flandre P, Calvez V, et al. Mechanisms of virologic failure in previously untreated HIV-infected patients from a trial of induction-maintenance therapy. JAMA 2000; 283:205–211.
36. Acosta EP, Henry K, Baken L, Page LM, Fletcher CV. Indinavir concentrations and antiviral effect. Pharmacotherapy 1999; 19:708–712.
37. Acosta EP, Havlir DV, Richman DD, Zhou XJ, Hirsch M, Collier AC, Tebas P, Sommadossi J–P. Pharmacodynamics (PD) of indinavir (IDV) in protease-naïve HIV-infected patients receiving ZDV and 3TC. Program and Abstracts of the 7th Conference on Retroviruses and Opportunistic Infections. San Francisco, CA, 2000: abstr 455.
38. Schapiro JM, Winters MA, Stewart F, et al. The effect of high-dose saquinavir on viral load and $CD4^+$ T-cell counts in HIV-infected patients. Ann Intern Med 1996; 124:1039–1050.
39. Gieschke R, Fotteler B, Buss N, Steiner J–L. Relationship between exposure to saquinavir monotherapy and antiviral response in HIV-positive patients. Clin Pharmacokinet 1999; 37:75–86.
40. Molla M, Korneyeva M, Gao Q, et al. Ordered accumulation of mutations in HIV protease confers resistance to ritonavir. Nat Med 1996; 2:760–766.
41. Lorenzi P, Yerly S, Abderrakim K, et al. Toxicity, efficacy, plasma drug concentrations and protease mutations in patients with advanced HIV infection treated with ritonavir and saquinavir. AIDS 1997; 11:F95–F99.
42. Chodakewitz J, Deutsch P, Leavitt R, McCrea J, Nessly M, Sterrett A, Winchell G. Relationships between indinavir (IDV) pharmacokinetics and antiviral activity in phase I/II trials. Program and Abstracts of the 12th World AIDS Conference, Geneva, Switzerland, 1998: abstr 42266.
43. Perello L, Gougard C, Delfraissy JF, Taburet AM. Therapeutic drug monitoring in HIV-infected patients receiving indinavir. Program and Abstracts of the 12th World AIDS Conference, Geneva, Switzerland, 1998: abstr 42272.
44. Dalmau D, Ochoa De Euchaguen AOE, Martinez–Lacasa JML, Sanchez Rodriguez CSR, Vidal JV. Indinavir Pharmacokinetics and their correlation with virologic and immunologic parameters. Program and Abstracts of the 12th World AIDS Conference, Geneva, Switzerland, 1998: abstr 42268.
45. Piscitelli SC, Flexner C, Minor JR, Polis MA, Masur H. Drug interactions in patients infected with human immunodeficiency virus. Clin Infect Dis 1996; 23:685–693.
46. Kumar GN, Rodrigues AD, Buko AM, Denissen JF. Cytochrome P450-mediated

metabolism of the HIV-1 protease inhibitor ritonavir (ABT-538) in human liver microsomes. J Pharmacol Exp Ther 1996; 277:423–431.

47. Hsu A, Granneman GR, Witt G, Locke C, Denissen J, et al. Multiple-dose pharmacokinetics of ritonavir in human immunodeficiency virus-infected subjects. Antimicrob Agents Chemother 1997; 41:898–905.
48. Alsenz J, Steffen H, Alex R. Active apical secretory efflux of the HIV protease inhibitors saquinavir and ritonavir in Caco-2 cell monolayers. Pharm Res 1998; 15: 423–428.
49. Kim AE, Dintaman JM, Waddell DS, Silverman JA. Saquinavir, an HIV protease inhibitor, is transported by P-glycoprotein. J Pharmacol Exp Ther 1998; 286:1439–1445.
50. Drewe J, Gutmann H, Fricker G, Torok M, Beglinger C, Huwyler J. HIV protease inhibitor ritonavir: a more potent inhibitor of P-glycoprotein than the cyclosporine analog SDZ PSC 833. Biochem Pharmacol 1999; 57:1147–1152.
51. Washington CB, Duran GE, Man MC, Sikic BI, Blaschke T. Interaction of anti-HIV protease inhibitors with the multidrug transporter P-glycoprotein (P-gp) in human cultured cells. J Acquir Immune Defic Syndr 1998; 19:203–209.
52. Srinivas RV, Middlemas D, Flynn P, Fridland A. Human immunodeficiency virus protease inhibitors serve as substrates for multidrug transporter proteins MDR1 and MRP1 but retain antiviral efficacy in cell lines expressing these transporters. Antimicrob Agents Chemother 1998; 42:3157–3162.
53. Piscitelli S, Bechtel C, Sadler B, Falloon J. The addition of a second protease inhibitor eliminates amprenavir–efavirenz drug interactions and increases amprenavir plasma concentrations. Program and Abstracts of the 7th Conference on Retroviruses and Opportunistic Infections. San Francisco, CA, 2000: abstr 78.
54. Hsu A, Granneman GR, Cao G, et al. Pharmacokinetic interaction between ritonavir and indinavir in healthy volunteers. Antimicrob Agents Chemother 1998; 42:2784–2791.
55. Dupont Pharma. Sustiva (efavirenz) capsules product monograph. Wilmington, DE: Dupont Pharma, Inc, 2000.
56. Hsu A, Granneman GR, Cao G, et al. Pharmacokinetic interactions between two human immunodeficiency virus protease inhibitors, ritonavir and saquinavir. Clin Pharmacol Ther 1998; 63:453–463.
57. Agouron Pharmaceuticals. Viracept (nelfinavir mesylate) tablets and oral powder product monograph. Torrey Pines, CA: Agouron Pharmaceuticals, 1999.
58. Abbott Laboratories. Norvir (ritonavir) capsules product monograph. North Chicago, IL: Abbott Laboratories, 1999.
59. Flexner C, Hendrix CW. Pharmacology of antiretroviral agents. In: DeVita VT, Hellman S, Rosenberg SA, eds. AIDS: Etiology, Diagnosis, Treatment, and Prevention, 4th ed. Philadelphia: Lippincott–Raven, 1997:479–493.
60. Hamzeh F, Benson C, Gerber J, Currier J, McCrea J, Deutsch P, Ruan P, Wu H, Flexner C. Steady-state pharmacokinetic interaction of modified-dose indinavir and rifabutin. Program and Abstracts of the 7th Conference on Retroviruses and Opportunistic Infections. San Francisco, CA, 2000: abstr 90.
61. Centers for Disease Control and Prevention. Impact of protease inhibitors on the

treatment of HIV-infected tuberculosis patients with rifampin. MMWR 1996; 45: 921–925.

62. Barditch–Crovo P, Trapnell CB, Ette E, Zacur H, Coresh J, Rocco LE, Hendrix C, Flexner C. Comparison of the effects of rifampin and rifabutin on the pharmacokinetics and pharmacologic effects of a combined oral contraceptive. Clin Pharmacol Ther 1999; 65:428–438.
63. Sun E, Heath–Chiozzi M, Cameron DW, et al. Concurrent ritonavir and rifabutin increases risk of rifabutin-associated adverse events. Program and Abstracts of the Eleventh International Conference on AIDS, Vancouver, BC, 1996: abstr 18.
64. Gallicano K, Khaliq Y, Seguin I, Fyke K, Carignan G, Tseng A, Walmsey S, Cameron DW. A pharmacokinetic (PK) study of intermittent rifabutin (RB) dosing with a combination of ritonavir (RTV) and saquinavir (SQ) in HIV patients. Program and Abstracts of the 7th Conference on Retroviruses and Opportunistic Infections. San Francisco, CA, 2000: abstr 313.
65. Merry C, Barry M, Ryan M, et al. Clinically significant interaction between indinavir and sildenafil. Program and Abstracts of the Seventh European Conference on Clinical Aspects and Treatment of HIV-Infection, Lisbon, Portugal, 1999: abstr 780.
66. Muirhead GJ, Wulff, MB, Fielding A, et al. Pharmacokinetic (PK) interactions between the HIV protease inhibitors ritonavir (RTV) and saquinavir (SQV) and Viagra (sildenafil citrate). Program and Abstracts of the 39th Interscience Conference on Antimicrobial Agents and Chemotherapy. San Francisco, CA, 1999: abstr 659.
67. Hall MC, Ahmad S. Interaction between sildenafil and HIV-1 combination therapy. Lancet 1999; 353:2071–2072.
68. Hsyu PH, Lillibridge JH, Maroldo L, Weiss WR, Kerr BM. Pharmacokinetic (PK) and pharmacodynamic (PD) interactions between nelfinavir and methadone. Program and Abstracts of the 7th Conference on Retroviruses and Opportunistic Infections. San Francisco, CA, 2000: abstr 87.
69. Gerber JG, Gal J, Rosenkranz S, Mildvan D, Gulick RM, Aberg J, Flexner C, Aweeka F, Hsu A for the ACTG 401 Study Team. The effect of ritonavir (RTV)/ saquinavir (SQV) on the stereoselective pharmacokinetics (PK) of methadone (M)[abstr]. Program and Abstracts of the 13th World AIDS Conference. Durban, South Africa, 2000.
70. Roche Laboratories. Invirase (saquinavir mesylate) capsules product monograph. Nutley, NJ: Roche Laboratories, 1997.
71. Piscitelli SC, Burstein AH, Chaitt D, Alfaro RM, Falloon J. Indinavir concentrations and St John's wort. Lancet 2000; 355:547–548.
72. Ruschitzka F, Meier PJ, Turina M, Luscher TF, Noll G. Acute heart transplant rejection due to Saint John's wort. Lancet 2000; 355:548–549.

9
Beneficial Pharmacokinetic Interactions: Are Two Protease Inhibitors Better Than One?

Charles W. Flexner
The Johns Hopkins University School of Medicine and School of Hygiene and Public Health, Baltimore, Maryland

I. INTRODUCTION

Although combination chemotherapy with protease inhibitors has had a major influence on the morbidity and mortality of human immunodeficiency virus (HIV) infection, these new drug regimens are associated with several problems (1). First, many patients cannot manage the large pill burden and strict dietary requirements, and they have difficulty adhering to the prescribed regimen. Consequently, treatment failure with initial combination regimens may be as high as 50% in some settings (2).

Various potentially beneficial metabolic drug interactions exist for combinations of two HIV protease inhibitors. One drug is used to inhibit the metabolism of the second agent, producing increased bioavailability, decreased clearance, or both (1). Two-way interactions also exist, in which the pharmacokinetic profile of each drug benefits (1).

Combining drugs to take advantage of beneficial pharmacokinetic interactions is a strategy that dates back to coadministration of probenecid and penicillin. Additional examples include imipenem–cilastatin and cyclosporine–ketoconazole. These regimens use an inhibitor of drug clearance to allow reduced dose and reduced dosing frequency, with substantial improvement in cost and convenience for the patient. Dual protease inhibitor regimens are unique in that both drugs are active for the disease being treated and attack the same pharmacological target.

Improving the convenience, tolerability, and cost of available regimens, and promoting long-term adherence to effective regimens, are priorities for clinical and preclinical drug development. Simultaneous administration of two protease inhibitors takes advantage of beneficial pharmacokinetic interactions, and may circumvent many of the drugs' undesirable pharmacological properties. For example, ritonavir increases saquinavir concentrations at steady state by up to 30-fold (3), allowing reduction of saquinavir dose and dosing frequency. The magnitude of the pharmacokinetic interaction between ritonavir and saquinavir is one of the largest ever described in human subjects.

Ritonavir decreases the systemic clearance of indinavir, and overcomes the deleterious effect of food on indinavir bioavailability (4). These benefits reflect inhibition of presystemic clearance and first-pass metabolism, as well as inhibition of systemic clearance mediated by hepatic cytochrome P450 (CYP) 3A4. Combining low doses of ritonavir with lopinavir (ABT-378) into the same formulation takes advantage of a similar interaction to develop a novel antiretroviral regimen (5). Dual protease inhibitor combinations with lesser pharmacokinetic effects have been developed to improve concentration–time profiles and reduce the risk of treatment failure. Several dual protease inhibitor combination regimens have shown great promise in clinical trials, and they are now recommended as components of therapy for HIV-infected patients.

II. RITONAVIR–SAQUINAVIR COMBINATIONS

That ritonavir is an extraordinarily potent inhibitor of the in vitro metabolism of saquinavir and other peptidic HIV protease inhibitors was known as early as 1995. Ritonavir inhibited the metabolism of 3.8 μg/mL of saquinavir in hepatic microsomes with a 50% inhibitory concentrations (IC_{50}) of 0.029 μg/mL (3). At the same time, saquinavir had no effect on the in vitro metabolism of ritonavir. Animal studies showed that 10 mg/kg of ritonavir increased the saquinavir area under the concentration–time curve (AUC) by up to 38-fold (6).

Because the primary resistance mutations seen in patients treated with either of these drugs (Val → Phe at position 82 of the HIV protease for ritonavir; mutations at position 48 and 90 for saquinavir) do not overlap (7), combination chemotherapy may provide a further benefit in terms of delaying or preventing resistance. A complete list of possible benefits from dual protease inhibitor regimens is provided in Table 1.

In a single-dose, crossover study in healthy volunteers, ritonavir increased the saquinavir AUC by 50- to 132-fold, and increased the saquinavir C_{max} by 23- to 35-fold (3) (Fig. 1). For a fixed dose of ritonavir, saquinavir concentrations

Table 1 Potential Clinical Advantages of Dual Protease Inhibitor Therapy[a]

Pharmacokinetic effects	Clinical consequences	Other potential benefits
Increased bioavailability	Reduced dose	Decreased pill burden
Decreased systemic clearance	Reduced-dosing frequency	Decreased cost of therapy
Increased AUC	Increased antiretroviral activity	Improved convenience
Increased trough (C_{min})	Less likelihood of resistance	Dual agents lacking cross-resistance
Decreased peak (C_{max})	Reduced drug toxicity	Improved adherence
Reduced pharmacokinetic variability	More predictable drug concentrations	
Increased formation of active metabolites		
Decreased clearance of active metabolites		

[a] AUC, area under the concentration–time curve; C_{max}, peak concentration; C_{min}, trough concentration.
Source: Ref. 41.

were proportional to saquinavir dose. However, when the saquinavir dose was held fixed, the relation between ritonavir dose and saquinavir pharmacokinetics was nonlinear; saquinavir AUC increased in proportion to ritonavir AUC until the ritonavir AUC exceeded 100 μg hr^{-1} mL^{-1}, at which point the increase in saquinavir AUC became less than proportional (3). Saquinavir had a small, but statistically significant, effect on the ritonavir AUC (6.4% mean increase) in this study (3).

The authors point out that systemic plasma clearance of saquinavir may be at least ten times higher than hepatic blood flow. This could be attributed to its administration with food, or to significant prehepatic clearance by intestinal cytochrome P450 or P-glycoprotein. These results suggest that the poor oral bioavailability of saquinavir (1–12%, depending on formulation and conditions) reflects extensive first-pass metabolism, rather than poor absorption. The increase in saquinavir concentrations with ritonavir is the result of improved bioavailability, perhaps to as much as 100%, with little effect on postabsorptive systemic clearance. Estimates that ritonavir reduces saquinavir's first-pass metabolism by 33-fold (3) correspond remarkably well with the increase in saquinavir C_{max} seen in single-dose studies. That the saquinavir AUC ratio, with or without ritonavir, was 50–400 suggests that the postabsorptive contribution of ritonavir (presumably owing to inhibition of P450 3A4) was only a four- to fivefold further increase in the AUC.

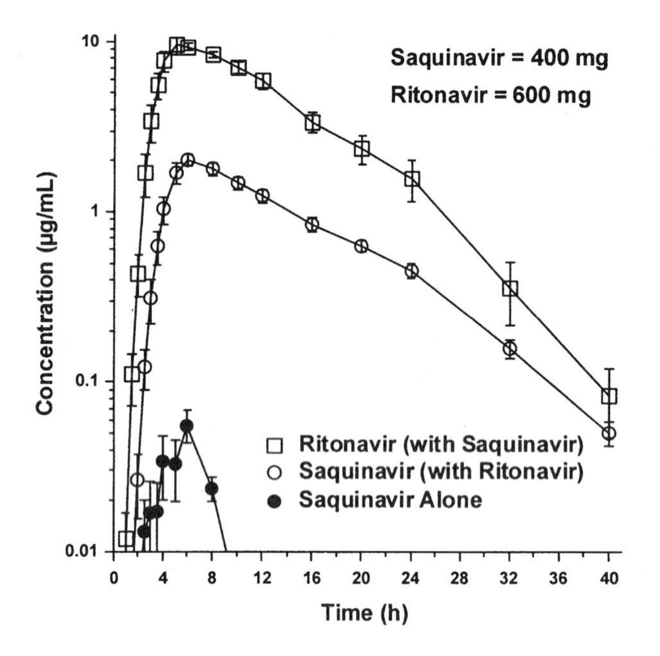

Figure 1 Effect of ritonavir on the pharmacokinetics of saquinavir: plasma concentration–time profiles in human subjects (mean ± standard error of the mean) for oral saquinavir 400 mg alone (closed circles), saquinavir 400 mg plus 600 mg ritonavir (open circles), and ritonavir 600 mg alone (open squares). (From Ref. 3.)

Because a high-fat meal increases saquinavir plasma concentrations by three- to fourfold, and because these studies imply that saquinavir's poor bioavailability is a consequence of presystemic clearance, rather than poor absorption, one must wonder how a fatty meal enhances saquinavir bioavailability. These data suggest that a high-fat meal may contain substances that specifically interfere with saquinavir metabolism by CYP3A, or that block intestinal drug transporters such as P-glycoprotein, or both.

In 1997, Hoffmann–La Roche, Inc., the manufacturer of saquinavir, made available a new oral formulation that had a two- to threefold improvement in bioavailability. This decreased the relative pharmacokinetic benefit of ritonavir, but did not alter the pharmacokinetic profile produced per milligram dose of saquinavir. That is because the presystemic effect of ritonavir is the same regardless of formulation, making the apparent bioavailability of a given oral dose of saquinavir 100% whether its inherent bioavailability is 4 or 12%.

An added pharmacokinetic benefit of combining ritonavir with saquinavir is a reduction in intersubject variance. Ritonavir reduced the percentage coefficient of variability (CV%) for saquinavir pharmacokinetic parameters from about 70 to about 30% (3). Differential expression of intestinal CYP 3A contributes to high intersubject variability in the pharmacokinetics of drugs such as saquinavir that undergo extensive first-pass metabolism (8). Eliminating this pathway as a significant contributor to saquinavir clearance would be expected to reduce pharmacokinetic variance, as is true. This makes drug concentrations more predictable in the clinical setting.

Other known inhibitors of CYP 3A4 increase the steady-state AUC of saquinavir by no more than fivefold, and inhibitors of intestinal cytochrome P450, such as grapefruit juice, increase the saquinavir AUC by no more than threefold (1,3). Ritonavir does not affect the pharmacokinetics of most other P450 substrates, even those with extensive first-pass metabolism, to nearly the same extent as it affects saquinavir, usually increasing the AUC by no more than fivefold (1). This suggests a unique chemical specificity for the interaction between ritonavir and saquinavir. It is likely that ritonavir inhibits intestinal P450 3A4, and recent data suggest that ritonavir may also be a potent inhibitor of P-glycoprotein (9,10). Selective interaction with one or both of these pathways may account for the surprising magnitude of ritonavir's effect on saquinavir oral bioavailability.

Ritonavir (200 mg twice daily) did increase the AUC of triazolam (0.125 mg) by 20-fold in a two-day study in healthy volunteers, and decreased triazolam clearance by 27-fold (11). Triazolam is a pure CYPC 3A4 substrate, and is not known to be metabolized by other P450 isoforms. In contrast, ritonavir's effect on zolpidem metabolism, which is only partly through CYP 3A4, was substantially less (11). Therefore, the magnitude of ritonavir's inhibition of saquinavir metabolism could also indicate that saquinavir is a pure 3A4 substrate and has no alternative routes of elimination.

Because ritonavir is a P450 inducer and undergoes autoinduction during the first 10–14 days of therapy (12), steady-state concentrations of saquinavir should be lower when these two drugs are combined. Multidose pharmacokinetic interaction studies found that the steady-state saquinavir AUC was increased only 20- to 30-fold (13). This is still a substantial increase, but lower than that seen in single-dose studies.

The current clinical recommendation is to combine 400 mg ritonavir with 400 mg saquinavir twice daily. Although ritonavir's approved dose is 600 mg twice daily (b.i.d.), the lower dose was chosen to account for the plateau in ritonavir's pharmacokinetic benefit with increasing doses, and to compensate for gastrointestinal toxicity seen with the 600-mg dose. This regimen has been well-tolerated and highly effective in long-term clinical trials.

In antiretroviral-naive patients, ritonavir plus saquinavir used as sole therapy suppressed HIV viral load to fewer than 400 copies/mL in most subjects after 48 weeks of treatment; overall dropout rates were 10–15%, often due to elevated liver enzymes in subjects with preexisting hepatitis virus infections (14).

Ninety percent of subjects continuing on this regimen, some of whom added nucleoside analogues, had viral loads suppressed to fewer than 400 copies/mL after 60 weeks of therapy (14). Success rates in treatment-experienced patients have not been as favorable (15–17), presumably because of cross-resistance from prior protease inhibitor use.

Adding ritonavir plus saquinavir to zidovudine–lamivudine therapy produced durable suppression of HIV viral load to fewer than 200 copies/mL in 10 of 16 patients taking these four drugs for 48 weeks (18). Furthermore, 10 of 16 patients who had failed nelfinavir- or indinavir-containing regimens had viral loads suppressed to fewer than 400 copies/mL 24 weeks after switching to ritonavir plus saquinavir plus nucleoside analogues (19).

An important question is whether lower doses of ritonavir will have as much pharmacokinetic benefit as the 400-mg dose. One single-dose study in healthy volunteers found that combining 200 mg of ritonavir with 600 mg saquinavir increased the saquinavir AUC by an average of 74-fold (20), an effect similar to that seen with 400 mg of ritonavir. Combining 100 mg of ritonavir with 600 mg of saquinavir increased the saquinavir AUC by an average of nearly 30-fold (20). Thus, 200 or 100 mg b.i.d. of ritonavir appears to provide as much, or nearly as much, pharmacokinetic benefit as 400 mg b.i.d. Because the lower ritonavir doses are likely to be better tolerated, these regimens may become increasingly popular.

III. RITONAVIR–INDINAVIR COMBINATIONS

The K_i for inhibition of indinavir metabolism in human hepatic microsomes is 0.085 μg/mL (4). Thus, the magnitude of the pharmacokinetic interaction between ritonavir and indinavir is not as great as that seen with ritonavir and saquinavir. When rats were given a single dose of 10 mg/kg of each drug, ritonavir increased the indinavir AUC by eightfold (6). There are several features of indinavir pharmacokinetics that would benefit from ritonavir coadministration. These include indinavir's rapid hepatic metabolism with a half-life of 1.8 h, an every 8 h–dosing regimen, food restrictions, hydration requirements, and large interindividual pharmacokinetic variability (4).

The combination of 200 or 400 mg ritonavir with 400 or 600 mg indinavir increased the indinavir AUC by three- to sixfold compared with indinavir 800 mg along in steady-state pharmacokinetic interaction study, in healthy volunteers treated, for 14 days with ritonavir (4). Ritonavir increased the indinavir C_{max} up to twofold, and increased the indinavir concentration 8 h after dosing by 11- to 33-fold (Fig. 2). The estimated K_i for inhibition of indinavir metabolism in vivo—0.10 μg/mL—was very close to the in vitro K_i in human hepatic microsomes

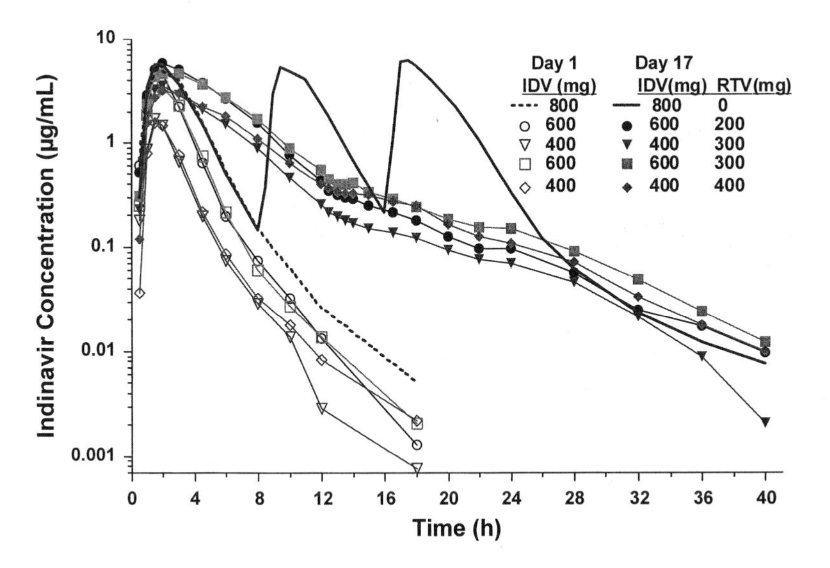

Figure 2 Influence of ritonavir on the pharmacokinetics of indinavir: mean plasma concentration–time profiles in human subjects at day 1 for oral indinavir alone at a dose of 800 mg (dashed line), 600 mg (open circles and open squares), and 400 mg (open triangles and open diamonds); and at day 17 for indinavir 800 mg alone (solid line), indinavir 600 mg plus ritonavir 200 mg (closed circles), indinavir 400 mg plus ritonavir 300 mg (closed triangles), indinavir 600 mg plus ritonavir 300 mg (closed squares), and indinavir 400 mg plus ritonavir 400 mg (closed diamonds). (From Ref. 4.)

(4). Because ritonavir is a CYP450 inducer, and baseline indinavir pharmacokinetics were assessed under noninduced conditions, the actual magnitude of metabolic inhibition would be underestimated under these circumstances. Indinavir did not appear to have a significant effect on ritonavir pharmacokinetics, when compared with historical controls in this study (4). Ritonavir coadministration significantly reduced the pharmacokinetic variability of indinavir. The coefficient of variation (CV) for indinavir AUC fell from 30 to 16%, and for C_{min} (concentration after 8 h) from 50 to 39% (4).

Indinavir's oral bioavailability, unlike that of saquinavir, is at least 60%. Estimated contribution of intestinal CYP450 3A4 to indinavir metabolism is less than 4% (4). Therefore, the pharmacokinetic benefit of ritonavir should be mainly due to decreased systemic clearance, rather than increased bioavailability. Ritonavir decreased the postabsorptive clearance of indinavir by at least two- to threefold, compared with baseline, noninduced kinetics (4). This suggests that inhibition of hepatic CYP3A4 is the main source for pharmacokinetic enhancement of indinavir by ritonavir, with reduced first-pass metabolism making a minor contribution.

For a fixed indinavir dose, increasing ritonavir from 200 to 400 mg produced relatively little increase in the indinavir AUC (4). This could be due to the increasing importance of clearance mechanisms other than CYP3A4 for indinavir as ritonavir dose increases. As the ritonavir AUC increased, indinavir clearance asymptotically approached the non-CYP3A4 clearance, which was thought to represent the combined contributions of renal clearance, glucuronidation, and CYP isoforms other than 3A4 (4). Of note, ritonavir might induce glucuronidation more effectively at higher doses.

Decreased ritonavir dose and increased indinavir dose were examined in a separate study (21). In healthy volunteers administered ritonavir for 14 days, the 24-h indinavir AUC with a 100 mg b.i.d. ritonavir–800 mg b.i.d. indinavir regimen was fourfold higher than with 800 mg q8h indinavir alone (Fig. 3). In the

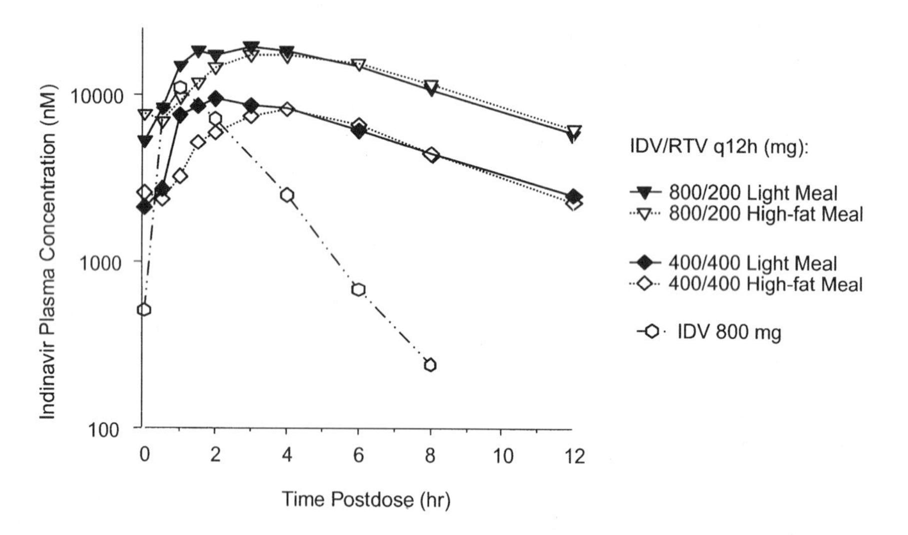

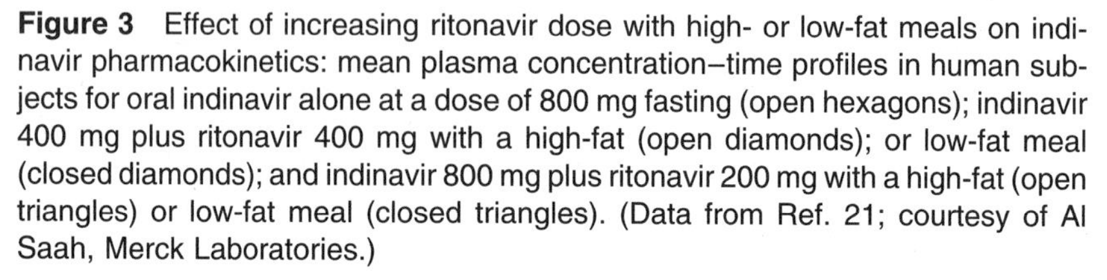
Figure 3 Effect of increasing ritonavir dose with high- or low-fat meals on indinavir pharmacokinetics: mean plasma concentration–time profiles in human subjects for oral indinavir alone at a dose of 800 mg fasting (open hexagons); indinavir 400 mg plus ritonavir 400 mg with a high-fat (open diamonds); or low-fat meal (closed diamonds); and indinavir 800 mg plus ritonavir 200 mg with a high-fat (open triangles) or low-fat meal (closed triangles). (Data from Ref. 21; courtesy of Al Saah, Merck Laboratories.)

same study, the 24-h AUC of indinavir with a 400–400 b.i.d. ritonavir–indinavir regimen was 40% lower than with the 100–800 regimen and 55% lower than with a 200–800 regimen (21). However, the mean 12-h trough concentrations of the 400–400 regimen and the 100–800 regimen were nearly the same (see Fig. 3).

In two studies, coadministration of ritonavir and indinavir abolished the effect of food on indinavir bioavailability. A high-fat meal reduced the bioavailability of oral indinavir by up to 85% (4). Doses of 100, 200, or 400 mg of ritonavir b.i.d. reversed the effect of a high- or low-fat meal on indinavir pharmacokinetics, as compared with 800 mg of indinavir given in the fasted state (21,22). Ritonavir should enhance indinavir oral bioavailability and pharmacokinetics through inhibition of CYP450 or drug transporters such as P-glycoprotein. This finding suggests that the deleterious effects of food on indinavir may be mediated by interaction with intestinal epithelial drug transporters or P450 complexes, processes potentially blocked by ritonavir.

Initial clinical trials of ritonavir–indinavir combinations have produced encouraging results. In one trial, 67 antiretroviral-naive patients taking ritonavir and indinavir plus two nucleosides lowered mean plasma viral load by 3.4 log after 24 weeks of therapy, and 67% of these subjects had viral loads of fewer than 80 copies/mL (23).

Combining 400 mg of ritonavir with 400 mg of indinavir b.i.d. produces a lower peak concentration (C_{max}) without affecting indinavir's renal clearance. Both of these factors could theoretically contribute to reduction of the risk for indinavir nephrolithiasis, which is thought to be both pH- and concentration-dependent (24). Reduced indinavir peak concentrations might reduce the risk of forming indinavir crystals in the urine, which presumably reduces the nidus for indinavir renal stones. Of 79 patients treated for a mean of 34 weeks with the 400–400 ritonavir–indinavir combination, none were reported to develop nephrolithiasis (25).

Another potential clinical benefit of this combination is elimination of the deleterious effect of food on indinavir bioavailability (21), allowing the drug to be taken without concern for meals. It is also possible that ritonavir could reduce or eliminate the need for extra hydration with indinavir, for the C_{max} is substantially reduced (4).

A theoretical disadvantage of combining ritonavir and indinavir is that the primary resistance mutations for these drugs (Val $\rightarrow$ Phe at position 82 of the HIV protease) are shared (7). This could make these two agents more prone than other dual protease inhibitor combinations to select resistant mutants. Initial clinical studies in treatment-experienced patients have reported few early-treatment failures (25), suggesting that this complication may not occur frequently in patients receiving these two drugs.

IV. RITONAVIR–NELFINAVIR COMBINATIONS

Combining ritonavir and nelfinavir may permit a reduced dosage and dosing frequency of nelfinavir, which was originally marketed as a 750 mg t.i.d. regimen. In a single-dose drug interaction study in healthy volunteers, ritonavir increased the nelfinavir AUC by 152%, whereas nelfinavir increased ritonavir's AUC by only 9% (1). A steady-state pharmacokinetic interaction study in HIV-infected volunteers evaluated the combination of 400-mg ritonavir b.i.d. with 500- or 750-mg nelfinavir b.i.d. After 5 weeks of dosing, ritonavir use was associated with a 162% increase in the 500-mg nelfinavir 24-h AUC (dose-normalized), and a 62% increase in the 750-mg AUC, as compared with historical controls taking only nelfinavir 750 mg t.i.d. (26). At the same time, the change in ritonavir's dose-normalized 24-h AUC was +13% with nelfinavir 500 mg, and −13% with the 750-mg regimen (Fig. 4).

This pharmacokinetic interaction is complicated because both drugs are CYP450 inducers as well as inhibitors. That nelfinavir's AUC did not increase significantly when the dose was increased from 500 to 750 mg b.i.d. may reflect increased autoinduction with the higher dose. In addition, there was a trend for nelfinavir to reduce ritonavir's trough concentrations at the higher 750-mg–nelfinavir dose (26). This may have decreased the magnitude of ritonavir's beneficial effect on nelfinavir pharmacokinetics.

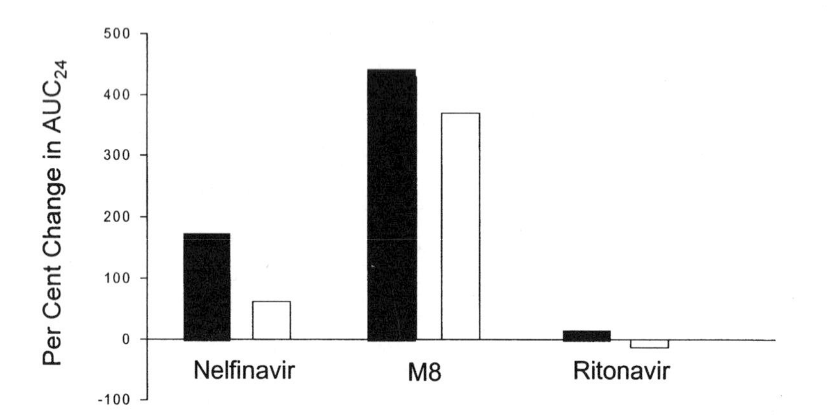

Figure 4 Pharmacokinetic interaction between nelfinavir and ritonavir: mean percentage change in the 24-h AUC of nelfinavir, the nelfinavir hydroxybutylamide metabolite M8 (AG1402), and ritonavir, normalized for drug dose (mg), after 5 weeks of ritonavir 400 mg b.i.d. plus nelfinavir 500 mg b.i.d. (closed bars), or 750 mg b.i.d. (open bars), as compared with historical controls taking nelfinavir 750 mg t.i.d. or ritonavir 400 mg b.i.d. (From Ref. 41.)

Nelfinavir is the only HIV protease inhibitor known to produce an active metabolite, the hydroxybutylamide M8 (AG1402), which is the major metabolite of nelfinavir in humans and has equipotent anti-HIV activity in vitro (27). Ritonavir had a more significant effect on the pharmacokinetics of M8 than on nelfinavir itself. After 5 weeks of dosing, ritonavir use was associated with a 430% increase in the 500-mg nelfinavir 24-h AUC, and a 370% increase in the 750-mg AUC, compared with historical controls taking nelfinavir 750 mg t.i.d. alone (26) (see Fig. 4).

M8 formation appears to be mediated mainly by CYP2C19 (28). Thus, nelfinavir is the only currently approved HIV protease inhibitor for which its major metabolite is not formed predominately by CYP3A4. M8 clearance, however, is mediated mainly by CYP3A4 (28). The discrepancy between ritonavir's effect on the pharmacokinetics of M8 and nelfinavir parent drug may be the result of induction of CYP2C19, thus increasing M8 formation, combined with inhibition of CYP3A4, thereby decreasing M8 clearance.

Ritonavir is known to be an inducer of CYP2C19 activity and a potent inhibitor of CYP3A4, but is a weak inhibitor of CYP2C19 in vitro (1,29). Therefore, it is unlikely that the increase in nelfinavir's AUC produced by ritonavir is a consequence of inhibition of systemic clearance; but it may reflect improved oral bioavailability, perhaps through inhibition of P-glycoprotein, plus the inhibition of minor metabolic pathways. Alternatively, M8 could act as an inhibitor of CYP2C19, and block the clearance of parent drug; if this is true, ritonavir's pharmacokinetic effect on nelfinavir concentrations would be indirect.

In a study conducted in 20 HIV-infected subjects, the combination of ritonavir, 400 mg b.i.d., with nelfinavir, 500 or 750 mg b.i.d., lowered viral load by a mean of 2.8 to 2.2 log, respectively, and increased CD4 cells counts by a mean of 236 and 120/mm^3 after 48 weeks (30). However, 5 of 20 patients experienced virological failure in this study, and all but 1 subject added nucleoside analogues to this regimen after 12 weeks. This regimen also produced moderate or severe diarrhea in 9 of 20 subjects (30).

V. NELFINAVIR–SAQUINAVIR COMBINATIONS

Nelfinavir increased the saquinavir AUC by up to fivefold in single-dose pharmacokinetic interaction studies, without affecting nelfinavir concentrations (1). Nelfinavir is an inducer of CYP4503A and, at steady state, the magnitude of this interaction was significantly lower. Combining nelfinavir 750 mg t.i.d. with 800 mg t.i.d. of the soft-gel formulation of saquinavir, produced a saquinavir AUC equivalent to 1200-mg t.i.d. at steady state (31). This combination was well-tolerated and was highly active against HIV in patients who were also taking two nucleoside analogues (32). However, this combination lacks many of the pharmacological and clinical benefits of other dual protease inhibitor combinations. With a trend to convert all protease inhibitor regimens to b.i.d., the t.i.d.-dosing regimens of nelfinavir and saquinavir become less relevant to clinical practice.

VI. NELFINAVIR–INDINAVIR COMBINATIONS

When nelfinavir is combined with indinavir, the indinavir AUC is increased 50% and the nelfinavir AUC is increased 80% in single-dose studies in healthy volunteers (1). However, when these two drugs were administered to patients in a b.i.d. regimen, there was little pharmacokinetic enhancement and a disappointing anti-HIV effect, with only 10 of 21 patients suppressing their plasma HIV RNA to fewer than 400 copies/mL (the lower limit of quantification) after 32 weeks (33). Presumably hepatic enzyme induction by nelfinavir resulted in reduced concentrations of both drugs, and no real pharmacokinetic benefits.

VII. INDINAVIR–SAQUINAVIR COMBINATIONS

The combination of indinavir and saquinavir was reported to be antagonistic when studied in vitro (34). Although the clinical relevance of this finding is unknown, this combination has not been pursued further in vivo, even though indinavir increased saquinavir concentrations by fivefold in single-dose studies (1). Theoretical disadvantages of dual protease inhibitor therapy (see Table 2) may discourage clinical development of some combinations.

Table 2 Potential Clinical Disadvantages of Dual Protease Inhibitor Therapy

Increased number of agents in the regimen
Increased number of potential toxicities
Increased potential for pharmacokinetic drug interactions
Increased formation of active metabolites
Decreased clearance of active metabolites
Overlapping toxicities
Same viral target for both drugs
Cross-resistance between drugs
Pharmacological antagonism between drugs

Source: Ref. 41.

VIII. AMPRENAVIR–RITONAVIR COMBINATIONS

Amprenavir has the longest elimination half-life of approved protease inhibitors (7–10 h), but has poor oral bioavailability and must be dosed as eight large 150-mg capsules twice daily (35). In single-dose pharmacokinetic interaction studies, indinavir increased the amprenavir AUC by 33%, saquinavir decreased the AUC by 32%, and nelfinavir did not change the AUC as compared with historical controls (35,36). Amprenavir decreased the indinavir AUC by 38%, decreased the saquinavir AUC by 19%, and increased the nelfinavir AUC by 15% (35,36). Amprenavir is a modest CYP4503A4 inhibitor, but is not known to be a CYP450 inducer.

Combination studies were conducted in patients with amprenavir at a dose of 800 mg t.i.d. and indinavir 800 mg q8h, nelfinavir 750 mg t.i.d., or saquinavir 800 mg t.i.d. Although most patients achieved HIV viral loads of fewer than 400 copies/mL at week 16, only 10 of 17 achieved viral loads of fewer than 20 copies/mL (37). Gastrointestinal toxicities, such as diarrhea and nausea, were common in this study.

In a study conducted in 19 HIV-infected subjects, ritonavir at a dose of 200 or 500 mg b.i.d. increased the amprenavir 12-h AUC by 2.5-fold, and increased the $C_{\min}$ up to sevenfold (38). This should allow a significant reduction in the amprenavir dose (currently 1200 mg b.i.d.) when these two drugs are given together, but also makes possible the exploration of once-daily dosing of amprenavir in combination with ritonavir. Nelfinavir (1250 mg b.i.d.) also increased the amprenavir AUC, but only by about twofold (38). Pharmacokinetic-modeling data suggest that combining ritonavir 300 mg q.d. with amprenavir 900 mg q.d. will produce amprenavir trough concentrations after 24 h that are still well above those produced when amprenavir is given alone at a dose of 1200 mg b.i.d. (39).

When amprenavir is combined with the non-nucleoside reverse transcriptase inhibitor efavirenz, which is a CYP450 inducer, amprenavir's AUC is decreased by up to 40% (35,36). Ritonavir at a dose of 200 or 500 mg b.i.d. reversed the effect of efavirenz on amprenavir clearance (38). The amprenavir AUC in the presence of ritonavir was the same, whether or not the patient was also taking efavirenz. This creates the possibility of a once-a-day antiretroviral combination including amprenavir, ritonavir, and once-daily reverse transcriptase inhibitors, such as efavirenz or didanosine.

IX. NEW COMBINATIONS IN DEVELOPMENT

Beneficial drug interactions can improve the pharmacokinetic profile of some investigational drugs. Lopinavir (ABT-378), an investigational peptidic HIV protease inhibitor, is being developed as a combination formulation with ritonavir, to take advantage of such an interaction. Lopinavir is a peptidic analogue of ritonavir, with more potent anti-HIV activity in vitro (5). Ritonavir greatly enhanced the pharmacokinetic profile of lopinavir in drug interaction studies in hepatic microsomes and in laboratory animals, presumably by improving bioavailability and slowing systemic clearance (5). Ritonavir's effect on the lopinavir AUC was severalfold greater than ritonavir's effect on the saquinavir AUC (5,40).

In studies conducted in human volunteers, mean trough concentrations of lopinavir were approximately 50-fold higher than the in vitro IC_{50} for HIV, using combination regimens of lopinavir plus ritonavir every 12 h (40). In 101 HIV-infected patients taking lopinavir, 200 or 400 mg b.i.d., with ritonavir, 100 or 200 mg b.i.d., plus two nucleoside analogues, for 24 weeks, HIV viral load was suppressed to fewer than 400 copies/mL in 93–95% of patients and to fewer than 50 copies/mL in 89% (40). Mean CD4 cell counts increased by $160/mm^3$, a result comparable with that of other highly active antiretroviral combinations. Lopinavir–ritonavir was very well tolerated: no patients dropped out of this study because of toxicity, and mild adverse reactions were seen in only a small number of patients (40).

Lopinavir is a drug that might not have been developed on its own. The availability of ritonavir to enhance its pharmacokinetic profile significantly enhanced the clinical promise of this agent. A similar strategy could be employed for other investigational protease inhibitors.

X. CONCLUSION

The need to improve regimen convenience, tolerability, and cost have spurred evaluation of dual protease inhibitor combinations that take advantage of beneficial pharmacokinetic interactions. Several of these regimens have shown great promise, and the combinations of ritonavir with saquinavir, and ritonavir with indinavir are now widely used, especially as components of salvage therapy. Converting t.i.d. regimens into b.i.d., and possibly q.d. regimens, is a particular advantage of this innovation. The discovery and development of lopinavir (ABT-378), taking advantage of its interaction with ritonavir, exemplifies the potential effect of dual protease inhibitor approaches in future therapeutics.

ACKNOWLEDGMENT

The author wishes to thank Laura Rocco for assistance in manuscript preparation.

REFERENCES

1. Flexner C. HIV protease inhibitors. N Engl J Med 1998; 338:1281–1292.
2. Volberding PA, Deeks SG. Antiretroviral therapy for HIV infection: promises and problems. JAMA 1998; 279:1343–1344.
3. Hsu A, Granneman GR, Cao G, et al. Pharmacokinetic interactions between two human immunodeficiency virus protease inhibitors, ritonavir and saquinavir. Clin Pharmacol Ther 1998; 63:453–463.
4. Hsu A, Granneman GR, Cao G, et al. Pharmacokinetic interaction between ritonavir and indinavir in healthy volunteers. Antimicrob Agents Chemother 1998; 42:2784–2791.
5. Sham HL, Kempf DJ, Molla A, et al. ABT-378, a highly potent inhibitor of the human immunodeficiency virus protease. Antimicrob Agents Chemother 1998; 42: 3218–3224.
6. Kempf DJ, Marsh KC, Kumar G, et al. Pharmacokinetic enhancement of inhibitors of the human immunodeficiency virus protease by coadministration with ritonavir. Antimicrob Agents Chemother 1997; 41:654–660.
7. Boden D, Markowitz M. Resistance to human immunodeficiency virus type 1 protease inhibitors. Antimicrob Agents Chemother 1998; 42:2775–2783.
8. Kolars JC, Lown KS, Schmeidlin–Ren P, et al. CYP 3A gene expression in human gut epithelium. Pharmacogenetics 1994; 4:247–259.
9. Drewe J, Gutmann H, Fricker G, Torok M, Beglinger C, Huwyler J. HIV protease inhibitor ritonavir: a more potent inhibitor of P-glycoprotein than the cyclosporine analog SDZ PSC 833. Biochem Pharmacol 1999; 57:1147–1152.
10. Washington CB, Duran GE, Man MC, Sikic BI, Blaschke T. Interaction of anti-HIV protease inhibitors with the multidrug transporter P-glycoprotein (P-gp) in human cultured cells. J Acquir Immune Defic Syndr Hum Retrovirol 1998; 19:203–209.
11. Greenblatt DJ, von Moltke LL, Harmatz JS, Graf JA, Mertanzis P, Hoffman JM, Shader RI. Differential impairment of triazolam and zolpidem clearance by short-term low-dose ritonavir [abstr]. Clin Pharmacol Ther 2000; 67:114.
12. Hsu A, Granneman GR, Witt G, et al. Multiple-dose pharmacokinetics of ritonavir in human immunodeficiency virus-infected subjects. Antimicrob Agents Chemother 1997; 41:898–905.
13. Hsu A, Granneman GR, Sun E, et al. Assessment of single and multiple-dose interactions between ritonavir and saquinavir [abstr]. Program and Abstracts of the 11th World AIDS Conference. Vancouver, BC, 1996.
14. Cameron DW, Japour A, Mellors J, et al. Antiretroviral safety and durability of ritonavir (RIT)–saquinavir (SQV) in protease inhibitor-naïve patients in year two of follow-up [abstr]. Program and Abstracts of 5th Conference on Human Retroviruses and Opportunistic Infections. Chicago, IL, 1998.
15. Cassano P, Hermans P, Sommereijns B, et al. Combined quadruple therapy with ritonavir–saquinavir (RTV–SQV) + nucleosides in patients (p) who failed in triple therapy with RTV, SQV or indinavir (IDV) [abstr]. Program and Abstracts of 5th Conference on Human Retroviruses and Opportunistic Infections. Chicago, IL, 1998.

16. De Truchis P, Force G, Zucman D, et al. Effects of salvage combination therapy with ritonavir + saquinavir in HIV-infected patients previously treated with protease inhibitors (PI) [abstr]. Program and Abstracts of 5th Conference on Retroviruses and Opportunistic Infections. Chicago, IL, 1998.
17. Tebas P, Kane E, Klebert M, et al. Virologic responses to a ritonavir/saquinavir containing regimen in patients who have previously failed nelfinavir [abstr]. Program and Abstracts of 5th Conference on Retroviruses and Opportunistic Infections. Chicago, IL, 1998.
18. Michelet C, Bellissant E, Ruffault A, et al. Safety and efficacy of ritonavir and saquinavir in combination with zidovudine and lamivudine. Clin Pharmacol Ther 1999; 65:661–671.
19. Gallant JE, Hall C, Barnett S, Raines C. Ritonavir/saquinavir (RTV/SQV) as salvage therapy after failure of initial protease inhibitor (PI) regimen [abstr]. Program and Abstracts of 5th Conference on Retroviruses and Opportunistic Infections. Chicago, IL, 1998.
20. Peytavin G, Bergmann JF, Leibowtich J, et al. Invirase bioavailibility is dramatically increased by ritonavir (RTV) ''baby''-doses in healthy volunteers (HV) [abstr]. Program and Abstracts of the 12th World AIDS Conference. Geneva, Switzerland, 1998.
21. Saah A, Winchell G, Seniuk M, Mehrotra D, Deutsch P. Multiple-dose pharmacokinetics (PK) and tolerability of indinavir (IDV)-ritonavir (RTV) combinations in healthy volunteers (Merck 078) [abstr]. Program and Abstracts of 6th Conference on Retroviruses and Opportunistic Infections. Chicago, IL, 1999.
22. Hsu A, Granneman GR, Heath–Chiozzi M, et al. Indinavir can be taken with regular meals when taken with ritonavir [abstr]. Program and Abstracts of the 12th World AIDS Conference. Geneva, Switzerland, 1998.
23. Rockstroh JK, Bergmann F, Wiesel W, Rieke A, Nadler M, Knechten H. Efficacy and safety of BID firstline ritonavir/indinavir plus double nucleoside combination therapy in HIV-infected individuals [abstr]. Program and Abstracts of 6th Conference on Retroviruses and Opportunistic Infections. Chicago, IL, 1999.
24. Kopp JB, Miller KD, Mican JAM, et al. Crystalluria and urinary tract abnormalities associated with indinavir. Ann Intern Med 1997; 127:119–25.
25. Workman C, Musson R, Dyer W, Sullivan J. Novel double protease combinations combining indinavir (IDV) with ritonavir (RTV): results from the first study [abstr]. Program and Abstracts of the 12th World AIDS Conference. Geneva, Switzerland, 1998.
26. Flexner C, Hsu A, Kerr B, Gallant J, Heath–Chiozzi M, Anderson R. Steady-state pharmacokinetic interactions between ritonavir (RTV), nelfinavir (NFV), and the nelfinavir active metabolite M8 (AG1402) [abstr]. Program and Abstracts of the 12th World AIDS Conference. Geneva, Switzerland, 1998.
27. Zhang KE, Wu E, Patick A, et al. Plasma metabolites of nelfinavir, a potent HIV protease inhibitor, in HIV positive patients: quantitation by LC-MS/MS and antiviral activities [abstr]. Program and Abstracts of the 6th European ISSX Meeting, Gothenburg, Sweden, 1997.
28. Lillibridge JH, Kerr BM, Shetty BV, Lee CA. Prediction and interpretation of P450 isoform selective drug interactions for the hydroxy-*t*-butylamide metabolite of the

HIV protease inhibitor nelfinavir mesylate [abstr]. Program and Abstracts of the 5th International ISSX Meeting. Cairns, Australia, 1998.
29. Kumar GN, Rodrigues AD, Buko AM, Denissen JF. Cytochrome P450-mediated metabolism of the HIV-1 protease inhibitor ritonavir (ABT-538) in human liver microsomes. J Pharmacol Exp Ther 1996; 277:423–431.
30. Gallant JE, Raines C, Heath–Chiozzi M, et al. Phase II study of ritonavir–nelfinavir combination therapy: an update [abstr]. Program and Abstracts of 6th Conference on Retroviruses and Opportunistic Infections. Chicago, IL, 1999.
31. Kravcik S, Farnsworth A, Patick A, et al. Long term follow-up of combination protease inhibitor therapy with nelfinavir and saquinavir (soft gel) in HIV infection [abstr]. Program and Abstracts of 5th Conference on Retroviruses and Opportunistic Infections. Chicago, IL, 1998.
32. Opravil M, on behalf of the SPICE study team. Study of protease inhibitor combination in Europe (SPICE): saquinavir soft gelatin capsule and nelfinavir in HIV-infected individuals [abstr]. Program and Abstracts of 5th Conference on Retroviruses and Opportunistic Infections. Chicago, IL, 1998.
33. Havlir DV, Riddler S, Squires K, et al. Coadministration of indinavir (IDV) and nelfinavir (NFV) in a twice daily regimen: preliminary safety, pharmacokinetic and anti-viral activity results [abstr]. Program and Abstracts of 5th Conference on Retroviruses and Opportunistic Infections. Chicago, IL, 1998.
34. Merrill DP, Manion DJ, Chou TC, Hirsch MS. Antagonism between human immunodeficiency virus type 1 protease inhibitors indinavir and saquinavir in vitro. J Infect Dis 1997; 176:265–268.
35. Glaxo Wellcome, Inc. Agenerase (amprenavir) capsules product monograph. Research Triangle Park, NC: Glaxo Wellcome, Inc., 1999.
36. Sadler BM, Gillotin C, Chittick GE, Symonds WT. Pharmacokinetic drug interactions with amprenavir [abstr]. Program and Abstracts of the 12th World AIDS Conference. Geneva, Switzerland, 1998.
37. Eron J, Haubrich R, Richman D, et al. Preliminary assessment of 141W94 in combination with other protease inhibitors [abstr]. Program and Abstracts of 5th Conference on Retroviruses and Opportunistic Infections. Chicago, IL, 1998.
38. Piscitelli S, Bechtel C, Sadler B, Falloon J. The addition of a second protease inhibitor eliminates amprenavir–efavirenz drug interactions and increases plasma amprenavir concentrations [abstr]. Program and Abstracts of 7th Conference on Retroviruses and Opportunistic Infections. San Francisco, CA, 2000.
39. Sadler BM, Piliero PJ, Preston SL, Yu L, Stein DS. Pharmacokinetic (PK) drug-interaction between amprenavir (APV) and ritonavir (RTV) in HIV-seronegative subjects after multiple, oral dosing [abstr]. Program and Abstracts of 7th Conference on Retroviruses and Opportunistic Infections. San Francisco, CA, 2000.
40. Murphy R, King M, Brun S, et al. ABT-378/ritonavir therapy in antiretroviral-naive HIV-I infected patients for 24 weeks [abstr]. Program and Abstracts of 6th Conference on Retroviruses and Opportunistic Infections. Chicago, IL, 1999.
41. Flexner C. Dual protease inhibitor therapy in HIV-infected patients: pharmacologic rationale and clinical benefits. Annu Rev Pharmacol Toxicol 2000; 40:651–676.

10

Protease Inhibitors: Clinical Efficacy

Roy M. Gulick
Weill Medical College of Cornell University, New York, New York

I. INTRODUCTION

The clinical development of HIV protease inhibitors represents a major step forward in the control and treatment of HIV infection. The use of combination antiretroviral regimens with protease inhibitors has contributed directly to significant declines in both morbidity and mortality among HIV-infected patients (1) (Fig. 1). Current guidelines recommend the use of combination antiretroviral regimens with a potent protease inhibitor(s) and nucleoside analogue reverse transcriptase inhibitors for the treatment of HIV infection (2–4). There are currently five available HIV protease inhibitors approved by the U. S. Food and Drug Administration (FDA): saquinavir (both hard gelatin and soft gelatin capsule formulations), ritonavir, indinavir, nelfinavir, and amprenavir. Additional investigational protease inhibitors are in clinical development. This chapter will review available clinical trial data.

II. SAQUINAVIR

The first protease inhibitor approved for the treatment of HIV disease by the FDA was saquinavir (SQV; hard-gelatin capsule formulation, hgc; Invirase, formerly known as Ro 31-8959) in December 1995. The drug is indicated in combination with nucleoside analogue reverse transcriptase inhibitors when treatment for HIV is warranted (5). The drug is supplied in 200-mg capsules and the recommended

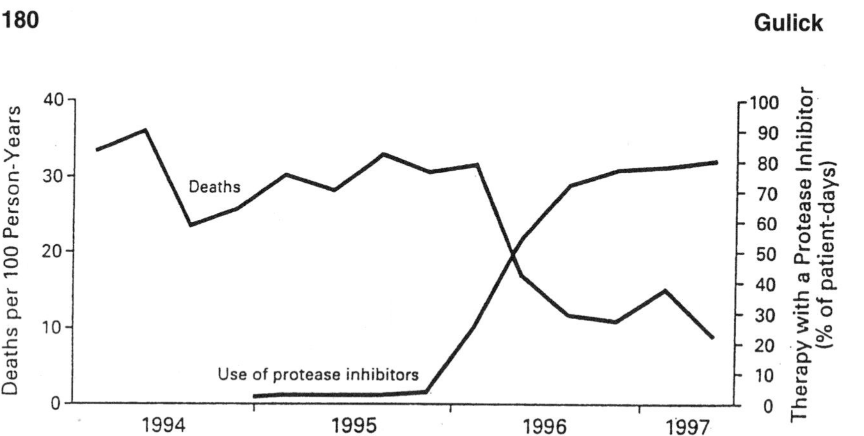

Figure 1 Contributions of combination regimens and declines of morbidity and mortality of PRIs among HIV-infected patients. (From Ref. 1, with permission. Copyright © 1998 Massachusetts Medical Society.)

dosage is 600 mg t.i.d. taken with or within 2 h after a meal. With the limited bioavailability of the hard-gelatin capsule formulation, a new formulation of saquinavir, the soft gelatin capsule (sgc; Fortovase) formulation was subsequently developed and approved for the treatment of HIV infection (6). Saquinavir sgc is also supplied in 200-mg capsules, and the recommended dosage is 1200 mg t.i.d. taken, with or within, 2 h after a meal. Current treatment guidelines recommend saquinavir sgc as one of the potent protease inhibitors (2). The use of saquinavir hgc is no longer generally recommended, unless combined with ritonavir.

A. Saquinavir hgc Monotherapy

Kitchen and colleagues (7) reported one of the first trials of saquinavir hgc: a randomized, double-blind, dose-ranging phase 1–2 study. Forty-nine zidovudine-naïve HIV-infected patients with CD4 cell counts of $500/mm^3$, or fewer, were randomized to receive saquinavir hgc at doses of 25, 75, 200, or 600 mg three times daily for 16 weeks. There were 10–12 patients within each of the dosing groups. At baseline, median CD4 cell counts were $218–391/mm^3$, and over 91% of the patients had measurable viremia. The maximum change in CD4 cell count from baseline was greatest in the 600-mg group at $+103.5/mm^3$ by week 6, with a median cell increase of $+36/mm^3$ by week 16. Median CD4 declines were observed in the other three dosing groups over 16 weeks. More patients at the 200- and 600-mg doses had 1 log10 declines in viremia than at the lower doses, although overall, there was no significant difference. The median decline in viral load was also greatest in the 600-mg group. These results supported the selection of the 600-mg–t.i.d. dose for further study.

Schapiro and colleagues (8) reported the results of the use of higher doses of saquinavir hgc monotherapy. In their study, 40 HIV-infected patients with CD4 cell counts of 200–500/mm^3 were randomized to receive saquinavir hgc at either 600 or 1200 mg, six times daily in an open-label fashion, for a planned 24-week study. At baseline, the patients had a mean plasma HIV RNA level of 24,500 (4.39 log10) copies/mL (lower dose), to 72,400 (4.86 log10) copies/mL (higher dose), and an average CD4 cell count of 346/mm^3. The majority of patients (16 of 20) in each group had not received prior antiretroviral therapy.

The mean maximal plasma HIV RNA level decline from baseline was −1.06 log10 copies/mL (week 2) in the lower-dose group and −1.34 log10 copies/mL (week 4) in the higher-dose group. The lower-dose group had an initial increase in mean CD4 cell count of +72/mm^3 by week 4, declining to +31/mm^3 by week 24. In the higher-dose group, the mean CD4 cell count increase was +121/mm^3 at week 20 and +82 at week 24. At the end of 24 weeks, change in baseline viral load was −0.48 log10 copies/mL in the lower-dose group and −0.85 log10 copies/mL in the higher-dose group. The authors noted limitations comparing their study with that of Kitchen et al. (7), but described greater improvements in the magnitude and duration of changes in viral load and CD4 cell counts with the higher doses of saquinavir hgc.

B. Double-Combination Therapy: Saquinavir hgc and Zidovudine

Vella and colleagues (9) reported the results of a study of a randomized, double-blind study of saquinavir hgc monotherapy, zidovudine monotherapy, or saquinavir in combination with zidovudine. A total of 92 HIV-infected treatment-naive patients, with CD4 cell counts of fewer than 300/mm^3 were randomized to receive one of five treatments: (1) saquinavir 600 mg t.i.d.; (2) zidovudine 200 mg t.i.d.; or saquinavir at a dose of (3) 75 mg, (4) 200 mg, or (5) 600 mg, in combination with zidovudine 200 mg t.i.d., for 16 weeks. At baseline, the groups had comparable plasma HIV RNA levels and mean CD4 cell counts from 152 to 217.

The saquinavir 600-mg–zidovudine group had the greatest median reductions in viral load of −1.6 log10 copies/mL by week 4 and −0.7 log10 copies/mL by week 16. The two monotherapy groups had maximal median reductions in viral load level of −0.5 log10, with subsequent increases toward baseline levels, with similar results seen in the lower dose saquinavir–zidovudine combination groups. CD4 cell counts increased in all five treatment groups, but the combination of saquinavir at 600 mg with zidovudine resulted in higher, more sustained increases. In the saquinavir 600-mg–zidovudine group, CD4 cell counts increased over baseline +61/mm^3 at week 16. The authors concluded that the combination of saquinavir 600 mg t.i.d. and zidovudine was superior to the other regimens in terms of viral load reduction and durability of CD4 cell count increases.

C. Saquinavir hgc: Clinical Endpoint Studies

Lalezari and colleagues (10) reported the results of a large, randomized, double-blind phase 2–3 study of saquinavir hgc and zalcitabine (Table 1). Eligible patients were HIV-infected who had taken at least 16 weeks of prior zidovudine therapy with CD4 cell counts 50–300/mm^3. A total of 940 patients were randomized to receive (1) zalcitabine monotherapy, 0.75 mg q8h; (2) saquinavir hgc monotherapy, 600 mg q8h; or (3) the combination of zalcitabine and saquinavir hgc. At baseline, the three treatment groups were well matched, with a median viral load of about 140,000 (5.1–5.2 log10) copies/mL and CD4 cell count of 160–180/mm^3.

Over a median follow-up of 73–74 weeks, the number of patients progressing to AIDS or death was 85 (27%) in the zalcitabine monotherapy group, 77 (24%) in the saquinavir hgc monotherapy group, and 46 (15%) in the combination group. The number of deaths was 28 (9%) in the zalcitabine monotherapy group, 34 (11%) in the saquinavir hgc monotherapy group, and 9 (3%) in the combination group. The authors concluded that combination therapy with zalcitabine and saquinavir hgc significantly prolonged progression to AIDS or death over zalcitabine alone. Furthermore, there was no significant difference between zalcitabine monotherapy and saquinavir hgc monotherapy.

A second clinical endpoint study was reported by Stellbrink et al. [(11); see Table 1]. In this large phase 3 study, HIV-infected patients who had taken no more than 16 weeks of prior zidovudine and had CD4 cell counts of 50–350 were randomized to one of four regimens: (1) saquinavir hgc in combination with zidovudine and zalcitabine; (2) saquinavir hgc and zidovudine; (3) zidovudine and zalcitabine; or (4) zidovudine monotherapy. During the study, patients originally randomized to zidovudine monotherapy were rerandomized to three-drug therapy. A total of 3485 patients were included, with a median baseline HIV RNA of about 113,000 (5.0–5.1 log10) copies/mL and CD4 cell count of 195–204/mm^3. Over a median duration of follow-up of 17 months, patients treated with three-drug therapy had significantly less clinical progression than those receiving the two-drug combination of zalcitabine and zidovudine (76 vs. 142 events).

D. Triple-Combination Therapy: Saquinavir hgc, Zidovudine, and Zalcitabine

Collier and colleagues (12) reported the results of AIDS Clinical Trials Group Protocol 229, a randomized, blinded phase 2 study of three treatment regimens: (1) zidovudine 200 mg t.i.d. plus zalcitabine 0.75 mg t.i.d.; (2) saquinavir hgc 600 mg t.i.d. plus zidovudine 200 mg t.i.d.; or (3) all three drugs together (see Table 1). Zidovudine was administered open-label and zalcitabine and saquinavir

Table 1 Major Efficacy Studies of Saquinavir

Study	Regimens	Population	Baseline levels	Major findings
Lalezari (10) Protocol NV 14256	1. ddC 2. SQV hgc 600 mg q8h 3. ddC + SQV	>16 wk ZDV; CD4 50–300 N = 940	HIV RNA 140K; CD4 170	ddC + SQV superior in decreasing progression to AIDS/death over 73 wk
Stellbrink (11) Protocol SV 14604	1. ZDV + ddC + SQV hgc 2. ZDV + SQV hgc 3. ZDV + ddC 4. ZDV	Antiretroviral-naive; CD4 50–350 N = 3485	HIV RNA 113K; CD4 200	ZDV + ddC + SQV hgc superior in decreasing progression to AIDS/death over 17 mo
Collier (12) ACTG Protocol 229	1. ZDV + ddC 2. ZDV + SQV hgc 3. ZDV + ddC + SQV hgc	ZDV-experienced; CD4 50 – 300 N = 302	HIV RNA 35K; CD4 154	ZDV + ddC + SQV hgc superior in decreasing HIV titers/RNA and increasing CD4 cell counts over 24 wk
Gill (14) Protocol NV 15182	1. SQV sgc 1200 mg t.i.d. added to current therapy	96% antiretroviral-experienced; 18% PI experienced; N = 444	HIV RNA 18K; CD4 201	43% of subjects decreased HIV RNA to < 400 at 24 wk
Mitsuyasu (15) Protocol NV 15355	1. SQV hgc 600 mg t.i.d. + two nucleosides 2. SQV sgc 1200 mg t.i.d. + two nucleosides	<4 weeks of prior antiretrovirals; PI naive; HIV RNA > 5K N = 171	HIV RNA 63K; CD4 428	SQV sgc regimen superior to hgc; 67% (vs. 37%) of subjects decreased HIV RNA to < 400 at 16 wk

hgc were blinded. Three hundred and two HIV-infected patients, who had previously received zidovudine therapy, with CD4 cell counts of 50–300/mm^3 were enrolled for the 24-week study, with an optional 12-week extension. At baseline, patients had taken a median of 27 months of prior zidovudine therapy, had a median HIV titer in peripheral blood mononuclear cells of 38 infectious units per million cells, and a median CD4 cell count of 145/mm^3.

The triple-drug combination suppressed HIV titers in peripheral blood mononuclear cells (PBMCs) to a greater extent and for a longer time than the two-drug combinations. The mean viral titer decreased by −0.8 log10 in the three-drug group, compared with no change in the saquinavir–zidovudine group, and less than a −0.4 log10 decrease in the zalcitabine–zidovudine group. Viral load data revealed similar trends with maximum reductions in plasma HIV RNA of −0.8 log10 in the three-drug group, −0.4 log10 in the zalcitabine–zidovudine group, and −0.1 log10 in the saquinavir hgc–zidovudine group. Viral load reductions trended back toward baseline levels over the course of the study.

The CD4-cell counts rose initially in all three treatment groups. The three-drug combination group had a significantly better CD4-cell response than either of the other two groups, with a maximum increase of about +45/mm^3 compared to about +10–30/mm^3 for the two-drug arms. At 24 weeks, 70% of the triple-drug group remained above their baseline CD4 level, compared with 63% of the saquinavir hgc–zidovudine group and 45% of the zalcitabine–zidovudine group. Trends continued through 48 total weeks of follow-up. The authors concluded that three-drug therapy produced greater antiretroviral effects than either of the two-drug arms.

E. Saquinavir sgc Monotherapy

The limited bioavailability of saquinavir hgc (4%), led to the development of an improved formulation of the drug, saquinavir soft-gelatin capsules (sgc), that increases absorption by about three- to fourfold. Lalezari and colleagues (13) reported the results of a pilot, randomized, open-label study of saquinavir hgc, 600 mg q8h, compared with three doses of saquinavir sgc, 400, 800, or 1200 mg q8h. Eligible patients were HIV-infected, had not taken any prior HIV protease inhibitor, and had CD4 cell counts of 100–500/mm^3. Eighty-eight patients were enrolled with baseline median HIV RNA copies of about 110,000 (4.8–5.2 log10)/mL and CD4 195–248/mm^3. After 8 weeks of saquinavir monotherapy, the median viral load reduction from baseline was −0.3 log10 (saquinavir hgc), −0.4 log10 (saquinavir sgc 400 mg), −0.6 log10 (saquinavir sgc 800 mg), and −0.9 log10 (saquinavir sgc 1200 mg). Median CD4 cells increased from baseline in all groups: +26/mm^3 (saquinavir hgc), +31/mm^3 (saquinavir sgc 400 mg), +95/mm^3 (saquinavir sgc 800 mg), and +74/mm^3 (saquinavir sgc 1200 mg).

From their data, the authors concluded that saquinavir sgc 1200 mg t.i.d. was the optimal dose of the drug.

Gill and colleagues (14) presented the results of a large study of saquinavir sgc 1200 mg t.i.d. in combination with other antiretrovirals (see Table 1). Four hundred forty-four HIV-infected patients were enrolled without viral load or CD4 restrictions; 96% had taken some prior antiretroviral therapy and 18% had taken prior protease inhibitors. Most patients added saquinavir sgc to their existing regimen of nuceloside analogue reverse transcriptase inhibitors, most commonly zidovudine and lamivudine. Small numbers of patients used other protease inhibitors or the nonnucleoside reverse transcriptase inhibitor, nevirapine. Median baseline viral copy load was 17,800 (4.25 log10)/mL and CD4 cell count was 201/ mm^3.

Eighty-six percent of the patients completed 24 weeks of treatment. Overall, 43% of patients had decreases in viral copy load levels to less than 400 copies/mL at week 24, including 28% of those with prior protease inhibitor experience.

F. Triple-Combination Therapy: Saquinavir hgc Versus Saquinavir sgc in Combination with Two Nucleoside Analogue Reverse Transcriptase Inhibitors

Mitsuyasu and colleagues (15) reported the results of a randomized, open-label phase 2 study of the two formulations of saquinavir, the hard-gelatin and soft-gelatin capsules in combination with two nucleoside analogues (see Table 1). Eligible patients were HIV-infected, had taken less than 4 weeks of prior antiretrovirals, and had not taken any HIV protease inhibitor, with HIV RNA copy levels at least 5000/mL. One hundred seventy-one patients were randomized to receive either saquinavir hgc 600 mg t.i.d. or saquinavir sgc 1200 mg t.i.d., both in combination with two nucleoside analogues, selected by the patient and study team and continued blinded therapy for 16 weeks. At baseline, patients had a mean HIV RNA copy number of 63,000 (4.8 log10)/mL and a mean CD4 cell count of 408–448/mm^3. The selection of concomitant nucleoside analogues were comparable in the two arms with 68% in each group using zidovudine–lamivudine, 13–21% using stavudine–lamivudine, and 7–13% using stavudine–didanosine.

At week 16, the mean change in HIV RNA copies from baseline was −2.0 log10/mL in the saquinavir sgc group, compared with −1.6 log10/mL in the saquinavir hgc group. Of 90 patients 60 (67%) had levels of less than 400/mL in the sgc group, compared with 30 of 81 (37%) in the hgc group. At the same time, 33 of 90 (37%) had copy levels of less than 50/mL in the sgc group, compared with 19 of 81 (23%) in the hgc group. CD4 cell counts showed similar mean increases of 97–115/mm^3 above baseline at 16 weeks in both groups. The authors concluded that in combination with nucleoside analogues, the soft-gelatin formulation of saquinavir provided more potent virological effects than the hard-gelatin formulation. Thompson et al. (16) reported extended study follow-up of 58 patients originally randomized to saquinavir sgc with two nucleoside analogues through 40 weeks, noting sustained virological suppression and continued mean CD4 increases (+174/mm^3) above baseline.

G. Triple-Combination Therapy: Saquinavir sgc, Zidovudine, and Lamivudine

Farthing and colleagues (17) reported the results of the SUN study, an open-label, noncomparative study of saquinavir sgc 1200 mg t.i.d., in combination with zidovudine 300 mg b.i.d. and lamivudine 150 mg b.i.d., in HIV-infected patients, who had not taken prior antiretroviral therapy, with HIV RNA levels at least 10,000/mL and CD4 cell counts at least 100/mm^3. A total of 42 patients were enrolled with a mean baseline HIV RNA copy number of 63,000 (4.8 log10)/mL and mean CD4-cell count of 419/mm^3. Preliminary data revealed that, of the 20 patients who had reached 32 weeks of follow-up, the mean viral copy load reduction was −3.4 log10/mL with 90% of these patients decreasing their viral load levels to fewer than 400, and 70% decreasing to less than 20/mL. At the same time, the mean CD4 cell count increase was +209/mm^3 above baseline. The authors concluded this was a potent triple-drug combination in antiretroviral naïve patients.

III. RITONAVIR

Ritonavir (RTV; Norvir, formerly ABT-538) is a peptidomimetic HIV protease inhibitor with potent in vitro activity and pharmacokinetic properties allowing twice daily dosing. The drug was approved by the FDA in March 1996 for the treatment of HIV infection when treatment is warranted (18). The drug is supplied in 100-mg capsules and as an oral solution (80 mg/mL). The recommended daily dose is 600 mg given twice daily with food. Current guidelines recommend ritonavir or the combination of saquinavir and ritonavir among the potent protease inhibitors (2).

A. Ritonavir Monotherapy

Danner and colleagues (19) reported the results of a phase 1–2, double-blind, randomized, placebo-controlled study of ritonavir in 84 patients with a plasma HIV p24 antigen level of at least 10 pg/mL and CD4 cell counts of more than

$50/mm^3$. Over 60% of the study patients had received prior antiretroviral therapy. At baseline, patients had a median HIV p24 antigen level of 55 pg/mL, a median HIV RNA copies of 158,000 (5.2 log10)/mL (branched-chain DNA; limit of detection 10,000 copies/mL) and a median CD4 cell count of about $122/mm^3$. After a 2-week washout period, patients received ritonavir monotherapy at doses of 300, 400, 500, or 600 mg, or placebo given twice daily for 4 weeks. After the initial 4 weeks, patients continued on their doses of ritonavir for up to 32 weeks. Patients initially randomized to placebo were randomly assigned to receive one of the four ritonavir doses after the initial 4 weeks.

During the first 4 weeks, HIV RNA and HIV p24 antigen levels decreased about 1 log10 copy/mL and CD4 cell counts increased about $100/mm^3$ over baseline levels in all of the ritonavir arms, a significant change, compared with the placebo arm. Seventy-six of the 84 patients continued into the maintenance phase of the study. On entering the maintenance phase, patients had mean HIV RNA decreases of −0.78, −0.83, −0.97, and −1.13 log10 copies/mL in the 300-, 400-, 500-, and 600-mg b.i.d. groups, respectively. Over 12 more weeks of therapy, HIV RNA levels approached baseline levels in the 300- and 400-mg groups, while treatment with 500 or 600 mg produced more sustained decreases in viral load. At 32 weeks, the 600-mg group showed a mean decrease of −0.81 log10 copies/mL that was statistically better than that seen with the 500-mg group. CD4 cell count changes paralleled the HIV RNA results. This led to the selection of ritonavir 600-mg b.i.d. as the dose for phase 2 studies.

A similar randomized, double-blinded study was reported by Markowitz and colleagues (20). Sixty-two patients, with baseline HIV of at least 25,000 RNA copies/mL (branched-chain DNA; limit of detection 10,000 copies/mL), and CD4 cell counts $50–500/mm^3$, who discontinued all antiretroviral agents for at least 14 days before study entry entered the study. Patients were randomized to receive ritonavir at doses of 200 or 300 mg, given three or four times daily, for a total daily dose of 600, 800, 900, or 1200 mg, or placebo for 4 weeks. After the initial 4 weeks, patients receiving ritonavir continued at the same dose and patients receiving placebo were randomly assigned to one of the four ritonavir doses. After 12 weeks, patients were offered open-label ritonavir.

Fifty-five patients completed 4 weeks of study, and 52 patients completed the full 12 weeks. Statistically significant decreases in HIV RNA occurred in the ritonavir groups, with a maximal decrease from −0.86 to −1.18 log10 copies/mL by day 15. At 4 weeks, the mean viral load decrease was −0.83 log10 copies/mL in the ritonavir group, compared with no change in the placebo group. By the end of 12 weeks, a persistent reduction of −0.50 log10 copies/mL was sustained at all four doses of ritonavir. In a subgroup of patients, a more sensitive HIV RNA assay (branched-chain DNA; limit of detection 400 copies/mL) revealed a maximal decrease of −1.7 log10 copies/mL by 2–3 weeks in patients taking ritonavir, with a −1.94 log10 decrease at the highest tested dose. Using this assay, patients at the lowest-tested daily dose of 600 mg appeared to have virological responses inferior to the other three dose groups. Overall, despite virological rebound in the majority of patients, 12 patients continued to demonstrate virological suppression sustained through at least 12 weeks. This study confirmed that the daily dose of ritonavir at 1200 mg was optimal for subsequent study.

B. Ritonavir: Clinical Endpoints

Cameron and colleagues (21) reported the results of a large, randomized, placebo-controlled study of ritonavir, added to concurrent antiretroviral therapy, that assessed clinical endpoints (Table 2). Eligible patients were HIV-infected adults who had taken at least 9 months of therapy with zidovudine, didanosine, zalcitabine, stavudine, singly or in combination thereof, were on a stable regimen for the 6 weeks before entering the study, and had CD4 cell counts of $100/mm^3$ or less. On entry, patients were randomized to receive either ritonavir liquid, 600 mg twice daily, or matching placebo while continuing up to two nucleoside analogue reverse transcriptase inhibitors. Patients were followed for the development of a new AIDS-defining illness or death. After 16 weeks of study, open-label ritonavir was offered to any patient who experienced an AIDS-defining event.

A total of 1090 patients were enrolled, most of whom were gay white men. Of these patients, 96% had had one or more HIV-1-associated illnesses before entering the study. Patients had a baseline median plasma HIV RNA of 251,000 (5.4 log10) copies/mL and a CD4 cell count of $20/mm^3$. Over a median follow-up of 29 weeks, an AIDS-defining illness or death occurred in 119 (22%) of the patients receiving ritonavir, compared with 205 (38%) of the patients receiving placebo, giving a hazard ratio of 0.53 (95% confidence interval [CI] 0.42–0.66) and a p value of < 0.0001 (Fig. 2). The most common study endpoint event was death, and the most common AIDS-defining illness was esophageal candidiasis. With this difference demonstrated, ritonavir was offered to all patients who continued to be followed. At a median follow-up of 51 weeks, there were 87 deaths (16%) in the ritonavir group and 126 (23%) in the original placebo group, giving a hazard ratio of 0.69 (95% CI, 0.52–0.91) and a p value of 0.0072. The addition of ritonavir as a single agent to standard antiretroviral therapy significantly reduced disease progression or death in this group of patients with advanced HIV disease.

In addition to reducing clinical progression, the addition of ritonavir was associated with a maximum reduction of HIV RNA of -1.3 log10 copies/mL (Amplicor assay; lower limit 200 copies/mL) and a mean increase from baseline in CD4 cell count of about $50–60/mm^3$. This initial profound decrease in viral load level was followed by a subsequent return toward baseline levels, reaching -0.6 log10 copies/mL below baseline levels by 4 months on the study. The

Table 2 Major Efficacy Studies of Ritonavir

Study	Regimens	Population	Baseline levels	Major findings
Cameron (21) Abbott Protocol 247	1. Current therapy 2. Current therapy + RTV 600 mg b.i.d.	> 9 mo nucleoside therapy; PI-naive; CD4 < 100 N = 1090	HIV RNA 251K; CD4 20	Adding RTV decreased clinical progression 42% over 29-wk follow-up
Connick (25) ACTG Protocol 315	1. ZDV + 3TC + RTV	> 3 mo ZDV; CD4 100–300 N = 53	HIV RNA 87K; CD4 187	58% of subjects decreased HIV RNA to < 100 at week 48; partial immunologic restoration demonstrated
Katlama (28) ALTIS Plus	1. d4T + 3TC + RTV	Enrolled on ALTIS study (d4T + 3TC), added RTV if HIV RNA > 3 K; N = 33	HIV RNA 18K; CD4 414 (ALTIS 1, N = 10) HIV RNA 44 K; CD4 262 (ALTIS 2, N = 23)	Sequential addition of RTV led to only 35% of subjects with HIV RNA < 200 by 24 wk
Saimot (29)	1. d4T + ddI + RTV	Antiretroviral-naive; CD4 50–350 N = 36	HIV RNA 43K; CD4 252	15 of 36 (42%) subjects d/c therapy early; of 16 subjects continuing at week 72, 88% had HIV RNA < 50

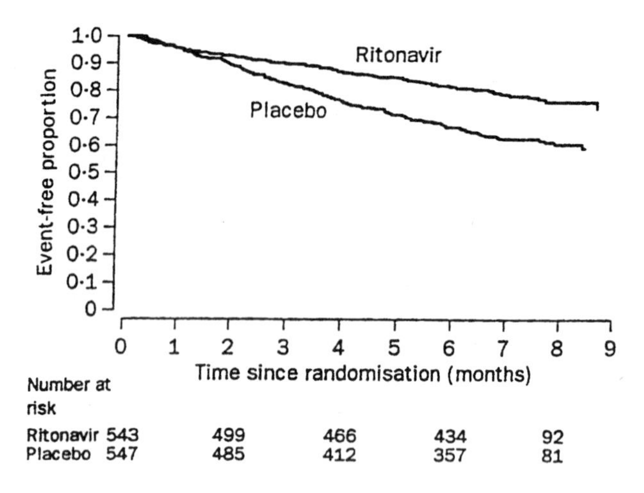

Figure 2 Disease progression or death in patients who added ritonavir or matching placebo to their antiretroviral therapy in the Abbott 247 Study. (From Ref. 21, with permission. Copyright © by the Lancet, Ltd., 1998.)

transient decrease in viral load levels was due to the addition of ritonavir as a single agent in this study, an effect similar to the use of monotherapy, that allows the rapid outgrowth of resistant virus. Despite this, the significant clinical benefit from ritonavir demonstrated in this study was unmatched previously in HIV therapeutics.

C. Double-Combination Therapy: Ritonavir and Zidovudine

One of the first studies of ritonavir given in a combination antiretroviral regimen randomized 356 antiretroviral-naive HIV-infected patients with a CD4 cell count of at least 200/mm^3 (mean baseline 364/mm^3) to receive (1) ritonavir 600 mg b.i.d., (2) zidovudine 200 mg t.i.d., or (3) the double combination of ritonavir and zidovudine at the same doses (18,22). Over the first 16 weeks of the study, the mean change from baseline in HIV RNA levels was greater in the ritonavir (−1.2 log10) and combination groups (−1.0 log10) than in the zidovudine monotherapy group (−0.5 log10). The mean CD4 cell count increases over baseline were greater in the ritonavir monotherapy group (+84/mm^3) than in the combination group (+37/mm^3) or zidovudine group (+21/mm^3). The trend toward an inferior response in the combination group compared with the ritonavir monotherapy group may well be due to increased dropout owing to gastrointestinal side effects in the combination group.

D. Triple-Combination Therapy: Ritonavir, Zidovudine, and Zalcitabine

A pilot study of ritonavir in combination with two nucleoside analogues was reported by Mathez and colleagues (23). In this open-label, uncontrolled study, 29 antiretroviral-naive patients with positive HIV cultures and CD4 cell counts fewer than 250/mm^3 were enrolled. At baseline, mean HIV RNA was 39,800 (4.6 log10) copies/mL (Amplicor assay) and mean baseline CD4 cell count was 173/mm^3. Patients received ritonavir monotherapy 600 mg b.i.d. for 2 weeks, and then added zidovudine 200 mg t.i.d. and zalcitabine 0.75 mg t.i.d. At 9 months of follow-up, mean HIV RNA decreased −2.0 log10 and CD4 cell counts increased 140 cells/mm^3 over baseline levels. This pilot study suggested that using ritonavir as part of a triple-combination antiretroviral regimen could provide sustained virological effects.

E. Triple-Combination Therapy: Ritonavir, Zidovudine, and Lamivudine

A second three-drug combination was tested in AIDS Clinical Trials Group study 315 (24) (see Table 2). In this open-label uncontrolled study, eligible patients were HIV-infected adults who had taken zidovudine for at least 3 months, but not lamivudine or any protease inhibitor, with CD4 cell counts of 100–300/mm^3. Patients discontinued all antiretroviral treatment 5 weeks before enrolling in the study. On study entry, patients received ritonavir 300 mg twice daily, increasing to 600 mg twice daily by day 7. On day 10, zidovudine 200 mg t.i.d. and lamivudine 150 mg t.i.d. were added.

Fifty-three patients enrolled in the study, with 44 receiving treatment for at least 9 of 12 weeks and 9 discontinuing the study for drug intolerance. The 44 patients who continued treatment were mostly white men who had not used intravenous drugs. These patients had a median baseline HIV RNA of 87,100 copies/mL (4.9 log10) (NASBA assay; lower limit of detection 100 copies/mL) and a median baseline CD4 cell count of 187/mm^3. On study treatment, median HIV RNA levels decreased from 87,100 copies/mL to 10,000 copies/mL by week 1; 2138 copies/mL by week 2; 1000 copies/mL by week 4; 288 copies/mL by week 8; and 166 copies/mL by week 12. The median decrease from baseline by week 12 was −2.3 log and 17 of 40 (42%) had a viral load level less than 100 copies/mL. At the same time, median CD4 cell counts increased over baseline to +82/mm^3 at week 4, and +108/mm^3 at week 12.

In addition to standard measures, additional improvements in immune function markers were noted. Of 36 patients who were anergic at baseline, 11 (31%) developed a new skin test response at week 12. Increases in lymphocyte proliferative responses to *Candida* antigen were seen in 17 of 40 patients (42%) at week

4, and 15 of 42 (36%) at week 12, although no significant increases were seen in responses to tetanus, streptokinase, or HIV antigens. Both naive and memory CD4 cells increased over 12 weeks, and the percentage of activated CD4 cells decreased from 25% at baseline, to 15% at week 4 and 12.5% at week 12. Increases in total lymphocytes, CD8 cells, and circulating B cells were also noted. In contrast, the number of circulating NK cells remained unchanged. No changes in T-cell V-beta repertoire were observed. This was one of the first studies to characterize in depth the partial immunologic restoration seen with three-drug antiretroviral therapy in a group of patients with moderately advanced HIV infection.

Forty-eight–week follow-up results for the patients in ACTG 315 were presented by Connick et al. (25). Thirty-three (62%) of the original 53 patients completed 48 weeks of therapy. By week 48, median HIV RNA had decreased −2.3 log10 copies/mL, with 19 patients (58%) decreasing to less than 100 copies/mL. At the same time, CD4 cell counts increased +175/mm^3 over baseline. Both memory and naive CD4 cells increased over the first month of therapy, after which the memory CD4 cells plateaued, while the naive CD4 cells continued to increase. DTH responses were detected in 9% of patients at baseline and 40% at week 48. No increases in lymphocyte proliferative responses to *Candida*, tetanus, streptokinase, or HIV antigens were found, suggesting incomplete immune reconstitution after 48 weeks of this potent antiretroviral therapy.

Notermans and colleagues from the Ritonavir/Lamivudine/Zidovudine Study Group (26) enrolled 33 antiretroviral naive patients to assess the effects of three-drug antiretroviral therapy on viral load in blood and tonsillar lymph tissue. Patients were randomized to receive ritonavir, lamivudine, and zidovudine started simultaneously or ritonavir monotherapy for 3 weeks, followed by the addition of lamivudine and zidovudine. Eight patients (24%) withdrew because of side effects. After 24 weeks, HIV RNA decreased −2.9 log10 copies/mL from baseline levels, with 88% of patients having viral load levels less than 200 copies/mL and CD4 cell counts of +152/mm^3 over baseline levels. Viral load burden in lymphoid tissue decreased from −8.5–9.2 log10 copies/g at baseline to −2.1 log10 copies/g of tissue by week 24 in the five patients studied. The authors concluded that significant decreases in plasma viral load are paralleled by decreases in the lymphoid tissue.

F. Ritonavir, Zidovudine, and Lamivudine for Primary HIV Infection

Hoen and colleagues (27) studied the same triple combination of ritonavir, zidovudine, and lamivudine for the treatment of symptomatic primary HIV infection in a multicenter study. Eligible patients were those with at least two symptoms of acute HIV infection, positive HIV p24 antigenemia and no more than three bands on Western blot analysis. Sixty-five patients were enrolled, with 35 followed for at least 9 months, and 24 followed for at least 12 months. Ten patients had discontinued three-drug therapy or were lost to follow-up. At baseline, patients had a median plasma HIV RNA of 158,000 (5.2 log10) copies/mL and CD4 count of 490/mm^3. In an intent-to-treat analysis, 78% of patients at 9 months and 68% at 12 months had HIV RNA less than 50 copies/mL and median CD4 increase was +278/mm^3 at 9 months. From this initial analysis, the authors concluded that treatment of primary infection with potent three-drug therapy decreased viral load levels in most patients.

G. Triple-Drug Combination: Ritonavir, Stavudine, and Lamivudine

Another three-drug combination, ritonavir, stavudine, and lamivudine, was studied in the ALTIS PLUS study by Katlama and colleagues (28) (see Table 2). The original ALTIS study evaluated stavudine and lamivudine either in naive patients (ALTIS 1) or nucleoside-experienced patients who were naive to the study drugs (ALTIS 2). After 24 weeks of the ALTIS study, patients were enrolled in ALTIS PLUS with stavudine and lamivudine if HIV RNA was fewer than 3000 copies/mL or had ritonavir added if HIV RNA was more than 3000 copies/mL. A total of 71 patients were included in ALTIS PLUS, 38 on the regimen of stavudine–lamivudine and 33 with the triple combination of stavudine–lamivudine–ritonavir. In the triple group, between day 0 and week 50, the median decrease in HIV RNA was −1.95 log10 copies/mL and the median increase in CD4 cells was +198 (ALTIS 1) and +102 (ALTIS 2). After 8 weeks of added ritonavir, 16 of 32 (50%) patients had HIV RNA levels less than 200 copies/mL; 24 weeks after adding ritonavir, this decreased to 11 of 31 (35%) of patients with levels of less than 200 copies/mL. The authors concluded that the sequential addition of ritonavir to dual nucleoside therapy does not lead to a sustained virological response in most patients.

H. Triple-Drug Combination: Ritonavir, Stavudine, and Didanosine

Saimot and colleagues (29) studied the long-term use of another three-drug combination, ritonavir, stavudine, and didanosine (see Table 2). In this pilot study, 36 treatment-naive patients who were either asymptomatic or had symptoms together with a CD4 cell count of 50–350/mm^3 were included in this uncontrolled open-label study. Of the 36 patients, 15 discontinued study drugs, mostly for treatment-related adverse events. In patients who continued treatment, plasma HIV RNA decreased below the limit of detection (Ultrasensitive assay; detection limit 50 copies/mL) in 21 of 22 patients (96%) at week 48 and in 14 of 16 (88%)

at week 72. The authors concluded that in patients fully tolerant to the therapy, stavudine–didanosine–ritonavir produced major reductions in plasma viral load.

IV. INDINAVIR

The third protease inhibitor to be approved by the FDA was indinavir (IDV; Crixivan, formerly L-735,524 or MK-0639) in March 1996. The drug is labeled currently for the treatment and HIV infection in combination with other agents (30). The drug is available in 200-mg and 400-mg capsules. The recommended dosage is 800 mg, given orally every 8 h without food, or alternatively, with a light meal. Current guidelines recommend indinavir as one of the potent protease inhibitors (2).

A. Indinavir Monotherapy

Early pilot phase 1 studies of indinavir explored the drug given at doses of 400 or 600 mg q6h showing suppression of HIV RNA and increases in CD4 cell counts in small numbers of patients for up to 24 weeks (31,32). Mellors and colleagues (33,34) reported the results of a 24-week–randomized, double-blinded phase 1–2 study in 73 p24-antigenemic, mostly zidovudine-experienced HIV-infected patients with CD4 cell counts of fewer than 500/mm^3 (median 110/mm^3). Patients were randomized to receive indinavir at either 200 or 400 mg q6h or zidovudine at 200 mg q8h for 24 weeks. Twenty-one of 23 (91%) patients in the 400-mg indinavir dose group had at least a −1 log10 decrease in HIV RNA, compared with 12 of 21 (57%) in the 200-mg indinavir dose group and only 1 of 29 (3%) in the zidovudine group. Rebounds in HIV RNA occurred in all groups, despite a subsequent increase of the indinavir dose to 600 mg q6h during the study. CD4 cell counts increased +43–65/mm^3 in the indinavir groups, compared with a loss of −11/mm^3 cells in the zidovudine group. Sixteen patients in the original zidovudine group added open-label indinavir at 600 mg q6h and experienced a median decrease in HIV RNA level of −1.98 log10/mL after 8 weeks and a median increase in CD4 cell counts of +126/mm^3 by 36 weeks of follow-up.

Steigbigel and colleagues (35) investigated higher doses of indinavir: 800 mg q8h, 1000 mg q8h, and 800 mg q6h. Seventy HIV-infected patients with a baseline median serum HIV RNA of 126,000 (5.1 log10) copies/mL and a median CD4 cell count of 250 cells/mm^3 were randomized to one of the three indinavir doses. Over 24 weeks of the study, HIV RNA decreased approximately −2 log10 copies/mL and CD4 cell counts increased approximately +100/mm^3 over baseline levels. There were no obvious differences in viral activity among the three groups. These results lead to the selection of indinavir 800 mg q8h as the optimal dose for further development.

B. Double-Combination Therapy: Indinavir and Zidovudine

Massari and colleagues (36) reported the results from Merck study 019, the first trial of indinavir in combination with other agents. In this study, 73 antiretroviral-naive patients with serum HIV RNA fewer than 20,000 copies/mL (median 79,000 (4.9 log10/mL)) and CD4 cell counts less than 500/mm^3 (median 221/mm^3) were randomized to receive (1) indinavir monotherapy at 600 mg q6h, (2) zidovudine monotherapy at 200 mg q8h, or (3) the combination of the two over 24 weeks. At 24 weeks of follow-up, HIV RNA levels decreased −1.5 log10 copies/mL in the indinavir monotherapy group, −0.3 log10 copies/mL in the zidovudine group, and −2.5 log10 copies/mL in the indinavir plus zidovudine group. CD4 cell counts increased approximately +50 cells/mm^3 over baseline more in the two indinavir groups than in the zidovudine group. The authors concluded that the combination of indinavir and zidovudine was superior to either one given as monotherapy.

Leavitt and colleagues (37) reported the preliminary results from two large phase 2 studies that compared similar treatment groups (studies 028 and 033). Four hundred ninety treatment-naive patients with CD4 50–250/mm^3 (study 028) or 50–500/mm^3 (study 033) were enrolled with mean baseline HIV RNA approximately 23,000 copies/mL and mean baseline CD4 cell count approximately 200/mm^3. Patients received indinavir, zidovudine, or the combination of the two. An interim analysis of pooled study data revealed the proportions of patients with decreases in their HIV RNA to less than 500 copies/mL at 24 weeks was 39% for indinavir, 4% for zidovudine, and 43% for the combination of the two. In further analyses, patients with higher baseline HIV RNA levels were more likely to decrease their viral loads to less than 500 copies/mL in the combination group than in the indinavir monotherapy group. At the same time, median CD4 cell counts increased +85–91/mm^3 in the indinavir groups, compared with +28/mm^3 in the zidovudine group.

C. Double-Combination Therapy: Indinavir and Stavudine

Steigbigel and colleagues (38) reported the results of an international, double-blind, randomized, placebo-controlled study of (1) indinavir 800 mg q8h, (2) stavudine 30 or 40 mg q12h (depending on weight), or (3) the combination of indinavir and stavudine. Six hundred thirteen HIV-infected patients with prior zidovudine experience and CD4 cell counts between 50 and 500/mm^3 were enrolled. At baseline, patients had a median HIV RNA of 29,000 (4.46 log10) copies/mL and a median CD4 cell count of 210/mm^3. At 52 weeks, the proportion of patients with HIV RNA of less than 500 copies/mL was 34% in the indinavir monotherapy group, 24% in the stavudine monotherapy group, and 46% in the combination group. At the same time, median CD4 cell increases were +109/mm^3 in the indinavir monotherapy group, +35/mm^3 in the stavudine group, and +144/mm^3 in the combination group. Although all treatment groups demonstrated improvements in antiretroviral activity, the combination group with indinavir and stavudine showed the greatest improvement.

D. Triple-Combination Therapy: Indinavir, Zidovudine, and Didanosine

Massari and colleagues (39) reported the first three-drug combination study with indinavir. In this study, 78 treatment-naive patients with HIV RNA of at least 20,000 copies/mL (median about 100,000 copies/mL) and CD4 cell counts of fewer than 500/mm^3 (median 150/mm^3) were randomized to receive (1) indinavir monotherapy, 600 mg q6h; (2) the two-drug combination of zidovudine 200 mg q8h and didanosine 125 or 200 mg q12h (depending on weight); or (3) all three drugs together. At 24 weeks of follow-up, viral suppression was greatest in the the triple-drug group, with approximately 60% experiencing decreases in their HIV RNA to less than 200 copies/mL. At the same time, median increases in CD4 cell counts were +90–100/mm^3 in the indinavir groups and +25/mm^3 in the zidovudine/didanosine group. Although potent, gastrointestinal intolerance ultimately limited the acceptability of this three-drug therapy.

E. Triple-Combination Therapy: Indinavir, Zidovudine, and Lamivudine

Gulick and colleagues reported the results from a second three-drug combination study (40–42) (Table 3). In this study, 97 patients who had taken zidovudine, most (81%) of whom had also taken didanosine, zalcitabine, or stavudine, but no prior lamivudine or protease inhibitor, with serum HIV RNA of at least 20,000 copies/mL and CD4 cell counts of 50–400/mm^3, were enrolled. Patients were randomly assigned to receive (1) indinavir monotherapy, 800 mg q8h; (2) the combination of zidovudine, 200 mg q8h and lamivudine 150 mg q12h; or (3) all three drugs together. At baseline, patients had taken zidovudine for an average of 30 months, had a median serum HIV RNA of 43,190 copies/mL, and a median CD4 cell count of 144/mm^3. At 24 weeks of follow-up, the proportion of patients who decreased their viral load levels to less than 500 copies/mL was 43% in the indinavir monotherapy group, 0% in the zidovudine/lamivudine group, and 90% in the three-drug combination group. The majority of patients with HIV RNA values less than 500 copies/mL also had levels less than 50 copies/mL. Median CD4 cell increases from baseline at 24 weeks were +101/mm^3 in the indinavir

Table 3 Major Efficacy Studies of Indinavir

Study	Regimens	Population	Baseline levels	Major findings
Gulick (40–42) Merck Protocol 035	1. IDV 2. ZDV + 3TC 3. ZDV + 3TC + IDV	> 6 mo ZDV therapy; 3TC, PI-naive; HIV RNA >20K; CD4 50–400 N = 97	HIV RNA 43K; CD4, 144	ZDV + 3TC + IDV superior at 24 wk; 68% of subjects with HIV RNA < 500 at 3 yr
Hirsch (43) Merck Protocol 039	1. IDV 2. ZDV + 3TC 3. ZDV + 3TC + IDV	> 6 mo ZDV therapy; 3TC, PI-naive; CD4 < 50 N = 320	HIV RNA 90K; CD4 15	ZDV + 3TC + IDV superior at 24 wk with 60% of subjects with HIV RNA < 500; sustained at 60 wk
Nguyen (46) Merck Protocol 069	1. ZDV + 3TC + IDV 1200 mg q12h 2. ZDV + 3TC + IDV 800 mg q8h	Antiretroviral-naive N = 287	Not available	Twice-daily dosing inferior to three-times daily dosing with 64% (vs. 91%) of subjects with HIV RNA < 400 at 24 wk
Hammer (53) ACTG Protocol 320	1. ZDV (or d4T) + 3TC 2. ZDV (or d4T) + 3TC + IDV	> 3 mo ZDV; 3TC, PI naive; CD4 < 200 N = 1156	HIV RNA 100K; CD4 87	Three drugs superior to two in reducing clinical progression to AIDS/death by 50% over 38 wk
Morales (56) Dupont Protocol 006	1. ZDV + 3TC + EFV 2. EFV + IDV 3. ZDV + 3TC + IDV	3TC, NNRTI, and PI-naive; HIV RNA ≥10K; CD4 ≥50 N = 450	HIV RNA 59K; CD4 345	EFV/ZDV/3TC superior in an intent-to-treat analysis at 36 wk; high drop-out rate (42%) in IDV/ZDV/3TC group

group, +46 in the zidovudine/lamivudine group, and +86 in the three-drug group. These decreases in viral load levels and increases in CD4 cell counts were maintained at approximately the same levels in the three treatment groups through 52 weeks of follow-up (40).

With the superior virological activity of the three-drug combination demonstrated at 24 weeks, the study was modified to allow all patients to receive open-label three-drug therapy after a minimum of 24 weeks of blinded therapy. The median period of blinded treatment follow-up before crossover was 41 weeks. In the group who simultaneously started three-drug therapy (the original indinavir, zidovudine, lamivudine group), 78% of patients had HIV RNA suppressed to less than 500 copies/mL at 100 weeks (41) (Fig. 3). In the two-drug group who added three drugs sequentially (the original indinavir group who added open-label zidovudine–lamivudine and the original zidovudine–lamivudine group who added open-label indinavir), only 30–45% of patients had sustained reductions in HIV RNA less than 500 copies/mL. Median CD4 cell increases in the original three-drug group (+209/mm^3) were superior to the other two groups (101–163/mm^3). The authors concluded that simultaneously initiated three-drug therapy is superior to the sequential initiation of the same drugs. Updated information from this study showed that HIV RNA suppression was durable for up to 3 years with three-drug therapy (42).

Hirsch and colleagues (43) reported a parallel study that enrolled a more immunocompromised group of patients (see Table 3). Three hundred twenty HIV-infected patients with zidovudine experience, without prior lamivudine or any protease inhibitor experience, with fewer than 50 cells/mm^3 were randomized to one of three groups: (1) indinavir monotherapy; (2) zidovudine–lamivudine; or (3) all three drugs together. At baseline, median HIV RNA was 89,510 copies/mL and CD4 cell count was 15/mm^3. At 24 weeks, the proportion of patients with HIV RNA less than 500 copies/mL was 4% in the indinavir monotherapy group, 0% in the zidovudine–lamivudine group, and 60% in the three-drug group. At the same time CD4 cell counts increased by +61/mm^3 in the indinavir group, +0/mm^3 in the zidovudine–lamivudine group, and +82/mm^3 in the three-drug group. After 24 weeks, 249 patients continued the study with open-label indinavir in combination with other approved agents. At week 60, 16% of the original indinavir group, 25% of the original zidovudine–lamivudine group, and 51% of the original three-drug group had HIV RNA suppressed to less than 500 copies/mL. The majority of those patients with less than 500 copies/mL also had HIV RNA of less than 50 copies/mL.

Leavitt and colleagues (44) presented a study of this three-drug therapy in patients with early HIV disease. In this ongoing study, treatment-naive, asymptomatic HIV-infected patients with HIV RNA of at least 1000 copies/mL and CD4 cell counts of at least 500/mm^3 received indinavir, zidovudine, and lamivudine in an open-label fashion. At baseline, HIV RNA was 7682 copies/mL. In

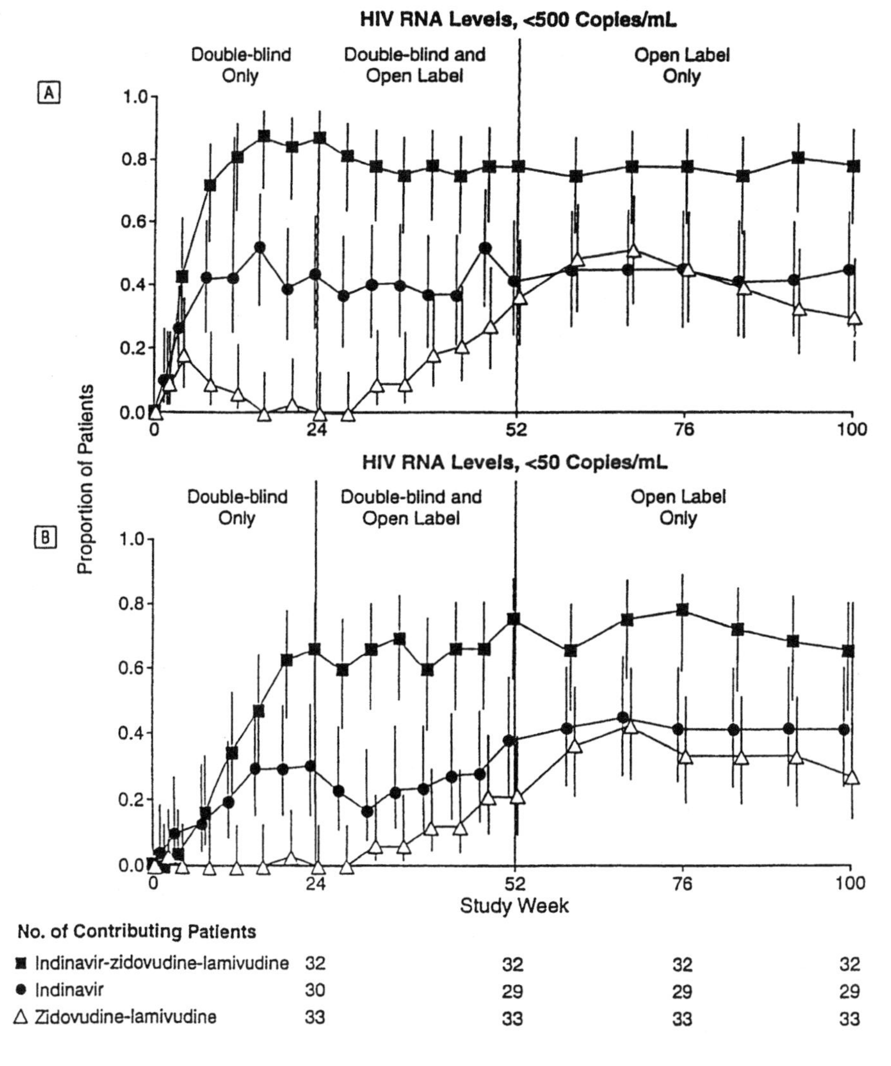

Figure 3 Proportion of patients with serum HIV RNA levels <500 copies/mL (A) and <50 copies/mL in the Merck 035 study. Bars are 95% confidence intervals. (From Ref. 41, with permission. Copyright © 1998 American Medical Association.)

a preliminary analysis, the proportion of patients with viral load levels less than 400 copies/mL was 76% (112 of 147) at week 4 and 98% (54 of 55) at week 24. At the same time the proportion with viral load levels less than 50 copies/mL was 14% (13 of 91) at week 4 and 97% (38 of 39) at week 24. The authors concluded that virological suppression was achieved in nearly all of those patients tested who had early HIV disease.

F. Triple-Combination Therapy: Twice Daily Indinavir, Zidovudine, and Lamivudine

Although a small pilot study comparing indinavir dosed 800 mg q8h versus 1000 or 1200 mg q12h in combination with zidovudine and lamivudine in 88 patients showed comparable results of the three treatment arms over 24 weeks (45), these results were not born out in a larger, more definitive study (46) (see Table 3). The larger study randomized patients to receive indinavir 1200 mg q12h or 800 mg q8h in combination with zidovudine and lamivudine at standard doses. In a preliminary analysis of 287 patients who had reached week 16, the proportion of patients with viral load levels lower than 400 copies/mL was 78% in the 800 mg q8h group and 72% in the 1200 mg q12h group. At 24 weeks in 87 patients, 91% of the q8h group compared with 64% of the q12h had viral load levels less than 400 copies/mL. Given these results, the study was stopped early. Twice-daily dosing of indinavir is no longer recommended routinely.

G. Indinavir, Zidovudine, and Lamivudine for Primary HIV Infection

Perrin and colleagues (47) reported early results from a pilot study of the use of indinavir, zidovudine, and lamivudine in patients with acute or recent HIV infection. Thirty-six patients were enrolled and followed for up to 8 months. At 3 months, 11 of 14 (79%) had HIV RNA less than 500 copies/mL and 2 of 8 (25%) had less than 50 copies/mL. At 6 months, 5 of 5 had less than 50 copies/mL. At 3 months, the median CD4 cell increase from baseline was $+91/mm^3$. These early results demonstrated the activity of three-drug therapy in acute HIV infection.

H. Triple-Combination Therapy: Induction and Maintenance with Indinavir, Zidovudine, and Lamivudine

Two studies compared the strategy of induction antiretroviral therapy with a three-drug regimen of indinavir, zidovudine, and lamivudine, followed by maintenance regimens with fewer agents. Havlir and colleagues (48) from the AIDS Clinical Trials Group Protocol 343 studied HIV-infected patients who had not taken lamivudine or any protease inhibitor, with HIV RNA of at least 1000 copies/mL and CD4 cell counts of at least 200/mm^3, who took induction therapy with indinavir 800 mg q8h, zidovudine 300 mg b.i.d, and lamivudine 150 mg b.i.d. A total of 509 patients were enrolled in the study, with 420 patients having HIV RNA levels available at weeks 16, 20, and 24. Of these 345 (82%) had viral load levels of less than 200 copies/mL at all three time points, and after 24 weeks of induction therapy, were eligible for the maintenance part of the study.

Of the 345 eligible patients, 316 patients consented to be randomized to one of the induction regimens: (1) indinavir, (2) zidovudine and lamivudine, or (3) continue three-drug therapy. At baseline before treatment, the 316 patients had a median plasma HIV RNA of 11,800 (4.1 log10) copies/mL and CD4 cell count of 448/mm^3. During the maintenance phase of the study, 23 of 104 (23%) patients receiving indinavir and 24 of 105 (23%) patients receiving zidovudine–lamivudine lost viral suppression, compared with 4 of 100 (4%) patients continuing three-drug therapy ($p < 0.001$ for each pairwise comparison). The authors concluded that viral load suppression is better sustained by continuation of three-drug therapy.

Pialoux et al. (49) reported the results of a second study exploring induction and maintenance therapy. Eligible patients were HIV-infected adults, who had not received prior antiretroviral therapy, with HIV RNA 3500–100,000 copies/mL and CD4 counts fewer than than 600/mm^3. During induction, patients received indinavir 800 mg q8h, zidovudine 300 mg q12h, and lamivudine 150 mg q12h for 3 months. Patients who decreased their viral load levels to less than 500 copies/mL were eligible to continue to the maintenance part of the study and were randomized to (1) zidovudine and lamivudine, (2) zidovudine and indinavir, or (3) continue three-drug therapy.

A total of 378 patients enrolled in the induction phase and 362 completed 3 months of induction therapy; 312 (86%) patients had reductions in HIV RNA to less than 500 copies/mL and 279 eventually were randomized to maintenance regimens. At baseline, these patients had a median HIV RNA of 39,000 (4.5 log10) copies/mL and CD4 cell count of 363/mm^3. After a median follow-up of 6 months, 29 of 93 (31%) patients receiving zidovudine–lamivudine and 21 of 94 (22%) receiving zidovudine–indinavir had viral load levels greater than 500 copies/mL, compared with 8 of 92 (9%) continuing three-drug therapy. The authors concurred that continuing three-drug therapy was superior to reducing the number of agents as maintenance therapy.

I. Triple-Combination Therapy: Indinavir, Stavudine, and Lamivudine

Gulick and colleagues (50) reported the preliminary results of an ongoing, open-label, randomized comparative study of two three-drug regimens: indinavir, zidovudine, and lamivudine; or indinavir, stavudine, and lamivudine (START I study). Two hundred HIV-infected treatment-naive patients with HIV RNA greater than 5000 copies/mL (bDNA assay) and CD4 cell counts of 200/mm^3 or more, were enrolled. A planned interim analysis was performed 24 weeks after enrollment of the first 100 patients. Preliminary results revealed that the proportion of patients decreasing their viral load levels to less than 500 copies/mL (bDNA) was 87% of the indinavir–stavudine–lamivudine group and 80% of the indinavir–zidovudine–lamivudine group. Most of the patients with viral load levels lower than 500 also had less than 50 copies/mL using an ultrasensitive assay. The preliminary conclusion is that these two regimens were comparable.

J. Triple-Combination Therapy: Indinavir, Stavudine, and Didanosine

In a parallel study, Eron and colleagues (51) reported preliminary results of an open-label, randomized, comparative study comparing indinavir combined with stavudine–didanosine or zidovudine–lamivudine (START II study.) Once again, 200 HIV-infected, treatment-naive patients with HIV RNA of more than 5000 copies/mL (bDNA assay) and CD4 cell counts at least 200/mm^3 were enrolled and a planned interim analysis was performed 24 weeks after enrollment of the first 100 patients. Preliminary results showed that at 24 weeks, 68% of the indinavir–stavudine–didanosine had viral load levels less than 500 copies/mL (bDNA), compared with 77% of the patients taking indinavir–zidovudine–lamivudine. Once again, preliminary results suggest these two regimens are comparable.

K. Indinavir: Clinical Endpoints

Leavitt and colleagues reported the results from a study that collected clinical endpoints (study 028) (52). In this study, 996 treatment-naive HIV-infected patients with CD4 cell counts between 50–250/mm^3 were enrolled. Patients were randomized to (1) indinavir 800 mg q8h; (2) zidovudine 200 mg q8h; or (3) the two-drug combination, and followed for the development of an AIDS-defining illness or death. After a median follow-up time of 40 weeks, lamivudine was added to the regimens of most patients taking zidovudine or the two-drug combination. Overall, 107 patients (11%) developed a clinical event over the course of the study. Comparing the two-drug combination to zidovudine, there was a 70% (95% confidence interval, 50–82%) reduction in the hazard of developing a clinical event. Comparing indinavir with zidovudine, there was a 61% (95% confidence interval, 38–76%) reduction in the hazard of developing a clinical event. The indinavir-containing regimens also demonstrated greater decreases in viral load levels and increases in CD4 cell counts. The authors concluded that the indinavir-containing regimens provided clinical benefit over the zidovudine monotherapy regimen.

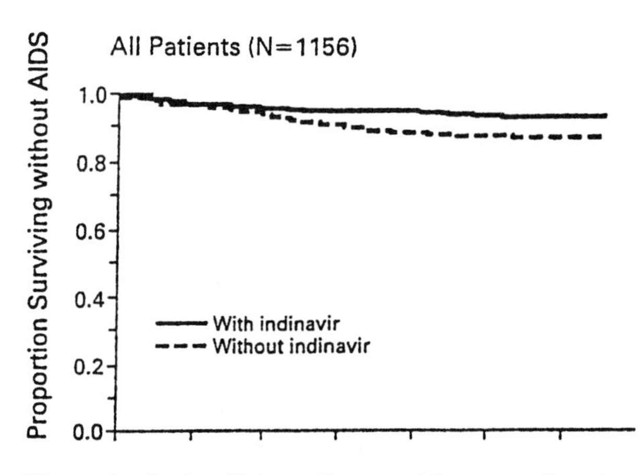

Figure 4 Kaplan-Meier estimates of the proportion of patients on zidovudine (or stavudine) and lamivudine with or without indinavir who did not progress to AIDS or death in the ACTG 320 Study. (From Ref. 53, with permission. Copyright © 1997 Massachusetts Medical Society.)

Hammer and colleagues (53) reported the results from a randomized, controlled study that collected clinical endpoints, AIDS Clinical Trials Group protocol 320 (see Table 3). A total of 1156 patients not previously treated with lamivudine or any protease inhibitor with CD4 cells less than 200/mm^3 were enrolled and randomly assigned to a two-drug regimen of zidovudine–lamivudine, with or without indinavir, and followed for the development of a new AIDS-defining event or death. Patients intolerant to zidovudine were allowed to substitute stavudine. At baseline, patients had taken an average of 21 months of prior zidovudine and had a median viral load level of 100,000 (5.0 log10) copies/mL and a median CD4 cell count of 87/mm^3. After a median duration of follow-up of 38 weeks, a total of 96 (8%) patients had AIDS-defining events or died: 63 (11%) of the patients in the two-drug group had disease progression, compared with 33 patients (6%) of the patients in the three-drug group (Fig. 4). The estimated hazard ratio was 0.50 (95% confidence interval, 0.33–0.76). Mortality was also reduced in the three-drug group (8 patients, 1.4%), compared to the two-drug group (18 patients, 3%). The estimated hazard ratio was 0.43 (95% CI, 0.19–0.99). Changes in HIV RNA and CD4 cell counts closely paralleled the clinical results. The authors noted a superior clinical benefit in the three-drug group.

L. Combination Therapy: Indinavir and Non-Nucleoside Reverse Transcriptase Inhibitors (NNRTIs)

Harris and colleagues (54) reported the results of a pilot open-label study of the use of indinavir in combination with lamivudine and the NNRTI, nevirapine. In this study, 22 HIV-infected patients with demonstrated intolerance, toxicity, or disease progression on prior nucleoside analogue-based antiretroviral regimens with CD4 cell counts less than 50/mm^3 were enrolled. Patients could have taken prior lamivudine, but not nevirapine or indinavir; concurrent use of other antiretroviral agents was not permitted. Patients received indinavir, lamivudine, and nevirapine at standard doses and were followed for 24 weeks. At baseline, patients had a median HIV RNA of 145,000 (5.16 log10) copies/mL and CD4 cell count of 30/mm^3; one subject had taken a prior protease inhibitor (ritonavir) and two had taken prior NNRTI (loviride). At 24 weeks, the median decrease in viral load was −3.1 log10 copies/mL, with 73% (11 of 15) decreasing their viral load level below 500 copies/mL and 40% (6 of 15) decreasing to less than 20 copies/mL. At the same time, the median increase in CD4 cell count was +95/mm^3. The authors concluded that this combination regimen had substantial effects that lasted up to 24 weeks.

Havlir and colleagues (55) presented the results of a study (study 266-003, cohort IV) of indinavir in combination with the NNRTI, efavirenz. In this study, 59 patients were randomized to receive the combination of indinavir and efavirenz and 42 patients were randomized to start indinavir monotherapy, and then added efavirenz and stavudine after 12 weeks. The dose of indinavir, originally 800 mg q8h, was increased to 1000 mg q8h; and the dose of efavirenz, originally 200 mg q.d., was increased to 600 mg q.d. during the study. At baseline, 71% of the patients had received prior therapy with nucleoside analogues, the mean plasma HIV RNA was 100,000 (5.0 log10) copies/mL and the mean CD4 cell count was 28/mm^3. At 72 weeks of follow-up, the proportion of patients with viral load levels less than 400 copies/mL in an intent-to-treat analysis, with last observation carried forward, was 73% in the indinavir–efavirenz group, compared with 57% in the indinavir–efavirenz–stavudine group. Mean CD4 cell counts increased by +243 in the indinavir–efavirenz group and +173 in the other group. The authors concluded that the combination of indinavir and efavirenz showed durable viral suppression.

Morales–Ramirez and colleagues (56) presented the results of a randomized, open-label comparative study (study 266-006; see Table 3). This three-armed study compared (1) efavirenz 600 mg q.d., zidovudine 300 mg b.i.d., and lamivudine 150 mg b.i.d.; (2) indinavir 1000 mg q8h and efavirenz; and (3) indinavir 800 mg q8h, zidovudine, and lamivudine. The study enrolled 450 patients who had not taken lamivudine or any protease inhibitor. At baseline, 86% had not taken prior antiretroviral therapy, mean plasma HIV RNA was 58,900 (4.77 log10) copies/mL, and mean CD4 cell count was 345/mm^3. In an intent-to-treat analysis with the last observation carried forward, preliminary results at week 24 showed the proportion of patients with viral load levels less than 400 copies/mL was 86% in the efavirenz–zidovudine–lamivudine group, 72% in the indinavir–efavirenz group, and 65% in the indinavir–zidovudine–lamivudine group.

The majority of patients with viral load levels less than 400 copies/mL were also less than 50 copies/mL. At the same time, CD4 cell count responses were not different between the three groups, with increases of $+115–134/mm^3$.

Trends in HIV RNA and CD4 responses continued through 36 weeks of follow-up. Although the viral load responses were significantly better in the efavirenz–zidovudine–lamivudine group than in the indinavir–zidovudine–lamivudine group, there was an imbalance of patient dropouts in this open-label study. Nevertheless, the triple combinations studied produced comparable viral load and CD4 responses over 36 weeks.

V. NELFINAVIR

Nelfinavir (NFV; Viracept, formerly AG1343) is a peptidomimetic HIV protease inhibitor with potent in vitro activity. The drug was approved by the FDA in March 1997 for the treatment of HIV infection when antiretroviral therapy is warranted (57). The drug is supplied in 250-mg capsules and as an oral powder (50 mg/g). The recommended daily dose for adults is 750 mg, given three times daily or 1250 mg b.i.d. with food. Current treatment guidelines recommend nelfinavir as one of the potent protease inhibitors (2).

A. Nelfinavir Monotherapy

Markowitz and colleagues (58) conducted a two-part, open-label, dose-escalating phase 1–2 study of nelfinavir in HIV-infected patients with CD4 cell counts of at least $200/mm^3$ and plasma HIV RNA greater than 20,000 copies/mL (branched-chain DNA assay; detection limit 500 copies/mL). In the first part, 35 antiretroviral-naive patients were randomized to receive nelfinavir at 500, 600, or 750 mg twice daily. After this group had received therapy for 28 days, the second group of 30 protease inhibitor-naive patients received nelfinavir at 500, 750, or 1000 mg three times daily. At baseline, the 65 patients had a mean HIV RNA of 60,300 (4.78 log10) copies/mL and a median CD4 cell count of $319/mm^3$.

At day 14 of treatment, a mean −1.0–1.6 log10 copies/mL reduction was noted in all treatment arms. However, after day 14, the viral load levels in the twice-daily–dosing arms returned toward baseline, whereas the three-times–daily dosing arms maintained viral suppression. At day 28, the nelfinavir 750-mg, three-times–daily arm demonstrated a median reduction from baseline of −1.8 log10 copies/mL, the maximal response observed among the treatment groups. At day 28, 60% of the patients taking nelfinavir 750 or 1000 mg t.i.d. decreased their viral loads below 500 copies/mL (branched-chain DNA), compared with 50% in the nelfinavir 500-mg t.i.d. group and only 26% in the b.i.d. arms. CD4 cell increases, ranging from +13 to 134/mm^3, were observed in all groups over 28 days. These data contributed to the selection of 750 mg t.i.d. as the standard dose of nelfinavir.

B. Double-Combination Therapy: Nelfinavir and Stavudine

Study 506 was a double-blind, randomized, placebo-controlled study comparing nelfinavir at a dose of either 500 or 750 mg t.i.d. in combination with stavudine 40 mg b.i.d. versus stavudine alone (57,59). Three hundred eight patients who had not taken stavudine or any protease inhibitor were enrolled with a mean baseline viral load level of 141,369 copies and a mean baseline CD4 cell count of 279/mm^3. Ninety patients were antiretroviral-naive, whereas 218 had taken prior nucleoside analogues. The maximal viral load suppression occurred at week 2 with −1.4 log10 decrease in the two double-therapy groups, compared with −0.6 log10 in the stavudine monotherapy group. Viral load levels increased after week 2 in the double-combination therapy groups, reaching about −1.0 log10 below baseline by week 24. Mean CD4 cell counts increased in all three groups, with the double-combination groups demonstrating about a +90/mm^3 rise over baseline, compared with about a +40 cell/mm^3 rise in the stavudine monotherapy group. The investigators noted the superiority of the combination regimen over stavudine alone.

C. Triple-Combination Therapy: Nelfinavir, Zidovudine, and Lamivudine

Study 511 was a double-blind, randomized, placebo-controlled study comparing nelfinavir at 500 or 750 mg t.i.d. in combination with zidovudine and lamivudine, to zidovudine–lamivudine alone (Table 4) (57,59,60). A total of 297 patients who had taken less than 1 month of prior zidovudine were enrolled, with a mean baseline plasma HIV RNA of 153,044 (5.18 log10) copies/mL and a mean baseline CD4 cell count of 288/mm^3. Using three different viral load assays—the branched-chain DNA assay, the Amplicor polymerase chain reaction (PCR) kit, and the investigational ultrasensitive PCR assay (detection limit, 50 copies/mL)—the virological response in each of the three groups was assessed.

At 6 months, the triple combination with nelfinavir 750 mg t.i.d. demonstrated the greatest virological activity, with a −2.3 log10 decrease by Amplicor (81% of patients below the detection limit) and −3.0 log10 decrease by ultrasensitive PCR (66% of patients below the detection limit). The triple combination with nelfinavir 500 mg t.i.d. showed a −2.2 log10 decrease by Amplicor (62% of patients below the limit of detection) and −2.7 log10 decrease (37% of patients below the detection limit.) Inferior results were seen in the double-nucleoside

Table 4 Major Efficacy Studies of Nelfinavir

Study	Regimens	Population	Baseline levels	Major findings
Clendeninn (60) Agouron Protocol 511	1. ZDV + 3TC + NFV 750 mg t.i.d. 2. ZDV + 3TC + NFV 500 mg t.i.d. 3. ZDV + 3TC	<1 mo prior ZDV; *N* = 297	HIV RNA 153K; CD4 288	ZDV + 3TC + NFV 750 mg was superior at 6 mo and 1 yr, with 62% of subjects with HIV RNA <50 at 52 wk
Clumeck (62) AVANTI 3	1. ZDV + 3TC + NFV 2. ZDV + 3TC	Antiretroviral-naive; CD4 150–500 *N* = 102	HIV RNA 79K CD4 283	Three drugs superior, 83% of subjects with HIV RNA < 500 at 28 wk
Petersen (64, 65)	1. d4T + 3TC + NFV 1250 mg b.i.d. 2. d4T + 3TC + NFV 750 mg t.i.d.	3TC- and PI-naive; *N* = 283	HIV RNA 200K; CD4 388	Comparable results with ~74% of subjects with HIV RNA < 400 in both groups at 32 wk
Albrecht (69) ACTG Protocol 364	Two nucleosides + 1. NFV 2. EFV 3. NFV + EFV	NNRTI- and PI-naive; HIV RNA > 500; one new nucleoside available; *N* = 196	HIV RNA 5K; CD4 389	72% of subjects with HIV RNA < 500 overall at wk 16

group with a −1.3–1.4 log10 decrease of viral load and only 5–8% decreasing below the detection limits.

At 1 year, the triple group with nelfinavir 750 mg t.i.d. showed similar decreases in viral load levels, with 76% below detection using the Amplicor and 62% below detection using the ultrasensitive assay. At the same time, the triple group with nelfinavir 500 mg t.i.d. showed 54% below detection using the Amplicor and 35% below detection using the ultrasensitive assay. Kerr et al. (61) found that baseline viral load level and plasma nelfinavir concentrations were predictive of the probability of achieving a viral load level below detection on this study. CD4 cell counts rose in all three groups with greater increases in the triple groups (more than $+120/mm^3$) than in the double-nucleoside group (more than $+80/mm^3$) at 6 months. Of 54 patients continuing therapy with zidovudine, lamivudine, and nelfinavir, 750 mg t.i.d. through 84 weeks of follow-up, more than 70% continued to have viral load levels suppressed below detection (Amplicor). The authors concluded that three-drug therapy was superior to dual-nucleoside therapy in this study.

AVANTI 3 (62) randomized 102 antiretroviral-naive patients with CD4 cell counts between 150 and $500/mm^3$ from Europe, Canada, and Australia to receive zidovudine and lamivudine, with or without nelfinavir (see Table 4). The median baseline viral load was 79,000 (4.9 log10) copies/mL and the median baseline CD4 cell count was $283/mm^3$. Preliminary analysis revealed that, at 28 weeks, the median decrease in viral load level was −0.98 log10 in the double-nucleoside group, compared with −1.85 in the triple-combination group. At the same time, 7 of 40 (18%) had viral load levels less than 500 copies/mL in the double-nucleoside group, compared with 34 of 41 (83%) of the triple-combination group. Once again, superior virological activity was shown in the zidovudine, lamivudine, and nelfinavir group compared with with the double-nucleoside group.

D. Nelfinavir, Zidovudine, and Lamivudine for Primary HIV Infection

Hecht and colleagues (63) reported the results of a pilot study of the use of nelfinavir, zidovudine, and lamivudine for the treatment of primary HIV infection. A total of 13 patients with primary HIV infection, based on the presence of HIV RNA with a negative or indeterminate HIV antibody test, or documented seroconversion within 6 months before study enrollment received open-label three-drug therapy. At baseline, the median viral load level was 77,720 copies/mL (range <500 to >1.6 million) and median CD4 cell count was $541/mm^3$. Of ten patients who had received at least 12 weeks of therapy, all ten had reductions in viral load levels to less than 500 copies/mL within 12 weeks, although two subsequently discontinued treatment. Of five patients who had completed at least 6 months of treatment, four had reductions in viral load levels to less than

50 copies/mL within 6 months, whereas the fifth patient decreased below 50 copies at 8 months. The authors concluded that this triple-drug combination was highly effective in its virological effect in acute HIV infection.

E. Triple-Combination Therapy: Twice-Daily Nelfinavir, Stavudine, and Lamivudine

Petersen and colleagues (64,65) presented the results of a multicenter, randomized study comparing twice- versus three-times-daily dosing of nelfinavir in combination with stavudine and lamivudine (see Table 4). Originally, 283 patients were randomized to receive twice daily, blinded, nelfinavir at a dose of 750, 1000, or 1250 mg b.i.d. or a standard dose of nelfinavir at 750 mg t.i.d. with standard doses of stavudine and lamivudine. When results from the 511 study demonstrated a superior virological effect with the 750 mg t.i.d dose over the 500 mg t.i.d. dose (see foregoing), this study was amended to increase the dose of all b.i.d. patients to 1250 mg nelfinavir b.i.d., and to extend the length of study follow-up to 2 years. Thus, the amended design compared nelfinavir 750 mg t.i.d. and 1250 mg b.i.d. both given with stavudine and lamivudine in an open-label fashion.

Adult patients who took less than 6 months prior nucleoside therapy, but no prior lamivudine or protease inhibitors were enrolled. Mean baseline viral load level was 200,000 (5.3 log10) copies/mL and mean baseline CD4 cell count was 388/mm^3. Over 80% of the patients had not taken prior antiretroviral agents. Pharmacokinetic profiles were comparable between the two dosing schemes. Approximately 80% of patients in both dosing groups decreased their viral load levels to less than 400 copies/mL (Amplicor). At 32 weeks, 93% of patients overall maintained their virological responses. The authors concluded that twice-daily dosing of nelfinavir appeared comparable in virological effect with three-times–daily dosing.

F. Triple-Combination Therapy: Nelfinavir, Stavudine, and Didanosine

Pednault and colleagues (66) reported the results of a small pilot study of open-label nelfinavir, stavudine, and didanosine, at standard doses, in 22 patients who had not taken prior stavudine, didanosine, or protease inhibitors, who had HIV RNA at least 10,000 copies/mL. At baseline, 11 were antiretroviral-naive and the other 11 had received prior antiretrovirals. Baseline median viral load level was 56,000 (4.75 log10) copies/mL and median CD4 cell count was 315/mm^3. Preliminary results showed viral load decreases from baseline of −1.7–2.1 log10 copies/mL in the patients tested. The authors concluded that this triple combination appeared to have potent virological activity.

G. Combination Therapy: Nelfinavir and Non-Nucleoside Reverse Transcriptase Inhibitors

Skowron and colleagues (67) presented the results of a pilot study of the pharmacokinetics and activity of triple-combination therapy with nelfinavir, stavudine, and nevirapine, all administered at standard doses. Patients had not taken prior NNRTI or protease inhibitors, had viral load levels at least 5000 copies/mL (PCR), and CD4 cell counts at least $100/mm^3$. A total of 22 of 25 patients completed the initial 5-week pharmacokinetic portion of the study, with 20 patients continuing into the extension portion of the study. Three patients discontinued therapy, and 17 patients remained on study therapy. Preliminary viral load information revealed that 15 of 20 (75%) had viral load levels less than 400 copies/mL (Amplicor) with 3 of 13 (23%) of those patients with viral load levels less than 50 copies/mL at 5 weeks. At 13 weeks, 15 of 17 (88%) had viral load levels less than 400 copies/mL, and 9 of 12 (75%) of those had levels less than 50 copies/mL. At 21 weeks, 8 of 9 (89%) had levels less than 400 copies/mL and 6 of 6 had levels less than 50 copies/mL. The authors concluded that this three-drug combination was highly effective.

Kagan and colleagues (68) presented the results of study 024, a pilot study of the double combination of nelfinavir and the NNRTI, efavirenz. In this open-label, single-arm study, 63 patients without prior NNRTI or protease inhibitor treatment, with HIV RNA at least 10,000 copies/mL and CD4 at least $50/mm^3$, were enrolled. At baseline, 30 patients were antiretroviral-naive and 30 had taken nucleoside analogues, mean HIV RNA level was 37,000 (4.57 log10) copies/mL and mean CD4 count was $370/mm^3$. Preliminary intent-to-treat results at 16 weeks of study showed a mean decrease from baseline viral load of -1.6 log10 copies/mL with 45 of 62 (73%) patients experiencing decreases in viral load levels to less than 400 copies/mL and 31 of 56 (55%) to less than 50 copies/mL. At the same time, mean CD4 increase from baseline was $+59$ cells/mm^3. The authors concluded that the double-drug combination of nelfinavir and efavirenz appeared to have virological effectiveness comparable with three-drug regimens.

Albrecht and colleagues (69) presented the results of AIDS Clinical Trials Group (ACTG) Study 364, a study of the virologic activity of nelfinavir, efavirenz, or both in combination with nucleoside analogues (see Table 4). This randomized, partially double-blind study enrolled patients who had taken prior nucleoside analogues through participation in prior ACTG studies. One hundred ninety-six patients who had viral load levels of at least 500 copies/mL were randomly assigned to receive (1) nelfinavir, (2) efavirenz, or (3) both in combination with two nucleoside analogues, at least one of which the patient had not taken previously. At baseline, median plasma HIV RNA was 5386 copies/mL and mean CD4 cell count was $389/mm^3$. Pooled preliminary results at week 16 revealed that 125 of 173 (72%) patients had viral load levels less than 500 copies/mL, with 35 of 173 (20%) experiencing virological failure with viral load levels of at least 2000 copies/mL. At the same time, mean CD4 cell counts increased $+80/\text{mm}^3$ across all the treatment arms. Additional analyses are underway.

VI. AMPRENAVIR

Amprenavir (APV; Agenerase, formerly 141W94 and VX-478) is peptidomimetic protease inhibitor with potent in vitro activity and pharmacokinetic properties, allowing twice-daily dosing. In addition, absorption characteristics allow the drug to be taken either with or without food. The drug was recently approved by the FDA. The drug is supplied as 150-mg capsules and the current recommended dose is 1200 mg twice daily.

A. Amprenavir Monotherapy

Schooley and colleagues (70) presented the first study of amprenavir in HIV-infected individuals. In this open-label, dose-escalating study, patients who had not received any protease inhibitor were randomly assigned to one of four treatment groups: amprenavir at doses of 300 mg given twice daily or three times daily; 900 mg twice daily; or 1200 mg twice daily, for a period of 4 weeks. Forty-two patients were enrolled in the study with a baseline median HIV RNA of about 63,000 (4.8 log10) copies/mL and CD4 about $283/\text{mm}^3$. Subjects were distributed among the treatment arms, with 7–12 patients in each group.

In this phase 1 study, the median maximum change in HIV RNA from baseline was −0.58 log10 copies/mL in the 300-mg b.i.d. group, −1.0 log10 in the 300-mg t.i.d. group, −1.7 log10 in the 900-mg b.i.d. group, and −2.0 log10 in the 1200-mg b.i.d. group. Corresponding median maximum changes in CD4 cell counts from baseline were $+64/\text{mm}^3$ in the 300-mg b.i.d. group, $+85/\text{mm}^3$ in the 300-mg t.i.d. group, $+35/\text{mm}^3$ in the 900-mg b.i.d. group, and $+110/\text{mm}^3$ in the 1200-mg b.i.d. group. Three patients discontinued the study prematurely for adverse events. In this pilot study, the authors noted significant dose-related antiretroviral activity of amprenavir, with the maximum effect demonstrated at a dose of 1200-mg b.i.d. This dose was selected for subsequent phase 2 studies.

B. Triple-Combination Therapy: Amprenavir, Zidovudine, and Lamivudine

The first phase 2 study of amprenavir was AIDS Clinical Trials Group (ACTG) study 347, first reported by Murphy and colleagues (Table 5) (71). In this phase 2, multicenter, randomized, double-blind, placebo-controlled study, subjects were randomized to receive amprenavir 1200 mg b.i.d. with or without zidovudine, 300 mg b.i.d., and lamivudine 150 mg b.i.d. for 24 weeks. Eligible patients were HIV-infected and had not taken prior lamivudine or any HIV protease inhibitor,

Table 5 Major Efficacy Studies of Amprenavir

Study	Regimens	Population	Baseline levels	Major findings
Murphy (71) ACTG Protocol 347	1. ZDV + 3TC + APV 1200 mg b.i.d. 2. APV 1200 mg b.i.d.	3TC- and PI-naive; HIV RNA > 5K; CD4 > 50 $N = 92$	HIV RNA 31K; CD4 305	Three-drug regimen superior with 73% (vs. 26%) of subjects with HIV RNA < 500 at 12 wk
Bart (75)	1. APV 1200 mg b.i.d. + ABC 300 mg b.i.d.	Antiretroviral-naive; HIV RNA > 5K; CD4 > 400 $N = 40$	HIV RNA 25K; CD4 747	90% of subjects with HIV RNA < 500 at 24 wk

with HIV RNA at least 5000 copies/mL (Amplicor assay) and CD4 cell counts of at least 50/mm^3. In this study, a virological endpoint was defined as (1) 1 log10 or greater increase in HIV RNA above the nadir level, (2) an increase in HIV RNA to above the baseline level, or (3) HIV RNA greater than 500 copies/mL after at least 16 weeks of therapy.

Ninety-two patients were enrolled; 48 (52%) of patients had never taken antiretroviral therapy, with 44 (48%) having some prior antiretoviral therapy. The median baseline HIV RNA was 30,707 copies/mL (range 500–758,577) and median baseline CD4 cell count was 305/mm^3 (range 30–1276).

After a median follow-up of 88 days (range 0–138), a planned interim review of the data showed that nine subjects had reached a virological endpoint. It was further revealed that all nine patients had been assigned to the amprenavir monotherapy arm ($p = 0.0009$). With that information, the blinded portion of the study was formally stopped and patients were unblinded for treatment assignment.

Patients on three drugs were encouraged to continue receiving therapy, whereas patients on the monotherapy regimen were strongly encouraged to enroll into a rollover protocol, ACTG 373, in which indinavir, nevirapine, stavudine, and lamivudine were given in an open-label fashion (72). A final data analysis showed that 15 of 42 (36%) of the amprenavir monotherapy patients and 1 of 43 (2%) of the amprenavir, zidovudine, and lamivudine patients had reached a virological endpoint ($p = 0.0001$).

In an intent-to-treat analysis, median HIV RNA levels were reduced −1.5–1.7 log10 below baseline in both groups by week 2, but diverged after that time. In the amprenavir monotherapy group, median HIV RNA levels reached a nadir of −1.5 log10 below baseline at weeks 2 and 4, then trended back toward baseline with a level of −0.9 log10 below baseline at weeks 8 and 12. In contrast, in the amprenavir, zidovudine, and lamivudine group, median HIV RNA levels were reduced to −1.7 log10 below baseline at week 2; −1.9 log10 at week 4; and −2.1–2.2 log10 at weeks 8 and 12.

In patients who had received study treatment for at least 12 weeks, 6 of 23 (26%) in the monotherapy group had HIV RNA levels less than 500 copies/mL, compared with 19 of 26 (73%) in the triple-therapy group. At 24 weeks, 27 of the original 46 patients continued to take their assigned three-drug therapy, and 17 (63%) of these had HIV RNA levels less than 500 copies/mL (Fig. 5).

There were similar increases in CD4 cell counts in both treatment arms. The median CD4 cell count increase above baseline at week 12 was +67/mm^3 in the amprenavir monotherapy group, and +48/mm^3 for the triple-therapy group. For the 27 subjects continuing three-drug therapy at 24 weeks, the median increase above baseline was +65/mm^3.

The investigators concluded that amprenavir in combination with zidovudine and lamivudine had significant antiretroviral activity that was superior to that

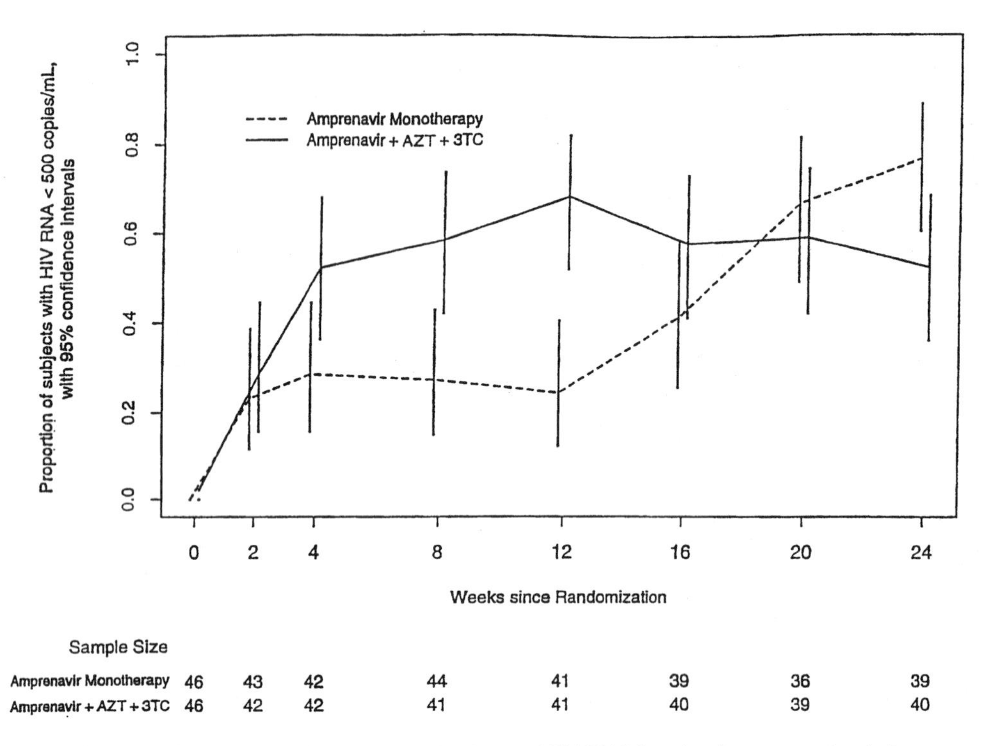

Figure 5 Proportion of subjects with plasma HIV RNA levels of <500 copies/mL in the ACTG 347 Study. Bars are 95% confidence intervals. AZT, zidovudine; 3TC, lamivudine. (From Ref. 71, with permission. © 1999 by the Infectious Diseases Society of America.)

seen when amprenavir was given alone. Possible explanations for the transient response in the amprenavir monotherapy group include preexisting viral resistance or a greater than expected degree of protein binding to amprenavir. Preliminary resistance analyses revealed that only one of four patients tested experienced viral rebound on the three-drug combination with mutations in the HIV protease (73).

Another phase 2 study of amprenavir in combination with zidovudine and lamivudine was presented by Haubrich and colleagues (74). Eligible patients were HIV-infected individuals, without prior lamivudine or protease inhibitor use, with HIV RNA more than 10,000 copies/mL and CD4 cell counts of at least 150/mm^3. Subjects were randomly assigned to amprenavir at doses of 900, 1050, or 1200 mg b.i.d., in combination with zidovudine 300 mg b.i.d. and lamivudine 150 mg b.i.d., for 48 weeks.

A total of 60 patients were enrolled, with 20 patients randomized to each treatment arm. Baseline median HIV RNA was 79,000 (4.9 log10) copies/mL and baseline median CD4 cell count was 366/mm^3. Preliminary results of an intent-to-treat analysis showed median changes of −2.0–2.4 log10 copies/mL in baseline HIV RNA, and CD4 cell counts of +99–135/mm^3 at each treatment dose at 24 weeks of follow-up. The authors concluded that amprenavir was generally well tolerated and showed excellent antiretroviral activity at the doses tested.

C. Double-Drug Combination Therapy: Amprenavir and Abacavir

Bart and colleagues (75) reported an open-label study investigating the antiretroviral activity of the two-drug combination of amprenavir, 1200 mg b.i.d., and abacavir, 300 mg b.i.d., in both peripheral blood and lymph node tissue (see Table 5). Forty HIV-infected subjects without prior antiretroviral treatment and with HIV RNA greater than 5000 copies/mL and CD4 cell counts of at least 400/mm^3 were enrolled in the 72 week study. Baseline mean HIV RNA was 25,000 (4.4 log10) copies/mL, baseline mean CD4 cell count was 747/mm^3 in peripheral blood, and baseline mean percentage of CD4 cells in lymph node tissue was 37%.

Preliminary results revealed that baseline mean HIV RNA was reduced −1.6 log10 copies/mL by week 2, and sustained at that level through week 24. At the same time, 20 of 35 (57%) patients by week 2 and 19 of 21 (90%) by week 8 had HIV RNA reduced to less than 500 copies/mL, with 90–100% of patients continuing at that level through week 24. At week 24, 9 of 11 (82%) patients had HIV RNA reduced to less than 50 copies/mL. The mean percentage of CD4 cells in lymph nodes had increased to 58% by 24 weeks, comparable with the 56% of CD4 cells seen in lymph nodes obtained from 8 HIV-negative control patients. The authors concluded that the combination amprenavir and abacavir is well tolerated, effectively suppressed HIV replication, and improved CD4 cell counts in both peripheral blood and lymph node tissue.

VII. OTHER PROTEASE INHIBITORS: ABT-378 (LOPINAVIR)

ABT-378 (lopinavir) is an investigational peptidomimetic HIV protease inhibitor developed through rational structure-based drug design to bind in the active site of the HIV protease enzyme while minimizing contact with the residue at position 82, a common site for mutation seen in patients who have experienced virological failure on ritonavir or indinavir regimens. Initial studies showed potent in vitro activity against both wild-type and ritonavir-resistant HIV. Pharmacokinetic studies showed ABT-378 levels were enhanced 50- to 100-fold by concomitant administration with ritonavir at doses of 200 mg twice daily or less, and that the drug should be dosed with food.

Japour and colleagues presented the first report of ABT-378 in HIV-infected subjects (76). In part I of this randomized phase 2 study, 32 HIV-infected, antiretroviral-naive subjects, with baseline HIV RNA of at least 5000 copies/mL (median 100,000 copies/mL) and baseline median CD4 counts of 424/mm^3, were enrolled. Patients were randomized to receive either ABT-378 at 200 or 400 mg twice daily in combination with ritonavir 100 mg twice daily. After 3 weeks, all patients added open-label stavudine and lamivudine at standard doses to their regimens.

After the initial 3 weeks of ABT-378–ritonavir therapy, HIV RNA decreased −2.0 logs below baseline levels. Sixteen of 17 patients (94%) at week 16 and 10 of 11 (91%) patients at week 24 taking ABT-378–ritonavir–stavudine–lamivudine had HIV RNA reduced to less than 400 copies/mL. CD4 counts increased +120/mm^3 over baseline at week 12 and +150/mm^3 at week 24. An additional 69 patients were enrolled on the two regimens in part II of the study.

In vitro, ABT-378 has shown activity against HIV with substitutions characteristic of ritonavir resistance. The first study of ABT-378 in protease inhibitor-experienced patients was recently enrolled. In this phase 1–2, blinded, randomized study, 70 patients with at least 12 weeks of protease inhibitor use and HIV RNA between 1000 and 100,000 copies/mL were eligible for enrollment. Subjects discontinued their current antiretroviral regimens and then were randomized to receive ABT-378, 400 mg twice daily, together with ritonavir at either 100 or 200 mg twice daily, for 2 weeks, then added two nucleoside analogue reverse transcriptase inhibitors, with at least one that the patient had not previously taken, and nevirapine 200 mg b.i.d.

IX. OTHER PROTEASE INHIBITORS: TIPRANAVIR

Tipranavir (TPV, formerly PNU-140690) is an investigational nonpeptidic protease inhibitor that is structurally unrelated to the other available peptidomimetic protease inhibitors. Having completed initial dose-ranging pharmacokinetic studies in healthy volunteers, tipranavir recently entered trials in HIV-infected individuals.

Wang and colleagues (77) reported the results of the first phase 1–2 study of tipranavir in HIV-infected patients, assessing initial antiretroviral activity of the drug. A total of 24 HIV+ patients receiving stable dual-combinations of HIV reverse transcriptase inhibitors for at least 2 months (median 18 months) without prior use of HIV protease inhibitors who had HIV RNA of at least 4000 copies/mL (median 26,000 or 4.42 log10 copies/mL) and CD4 cell counts of at least

$50/mm^3$ (median $407/mm^3$) entered the study. Patients added tipranavir at doses of 900, 1200, or 1500 mg three times daily to their nucleoside regimens and were followed through 12 weeks.

At day 11, mean HIV reductions were −1.0, −1.3, and −1.2 log10 copies/mL below baseline for the 900-, 1200-, and 1500-mg t.i.d. groups, respectively. At 4 weeks, mean HIV reductions below baseline were −0.5, −1.1, and −0.9 log10 for the three groups. One, two, and three patients in the 900-, 1200-, and 1500-mg t.i.d. groups, respectively, had HIV RNA reduced to less than 400 copies/mL. By week 12, HIV RNA had trended back toward baseline levels in each of the groups. Clinical records suggested that compliance was associated with reductions in HIV RNA. Interestingly, no consistent pattern of protease gene substitutions was shown in genotypic analysis of week 12 viral isolates.

The study population consisted of patients on a stable double-nucleoside regimen with an HIV RNA of at least 4000 copies, implying the presence of some degree of viral resistance to the baseline drugs. Adding a single agent to these regimens would be expected to blunt both the initial viral load reduction and the chance for a durable antiretroviral response. Also in this study, tipranavir was provided in 150-mg capsules, thus requiring patients to take six to ten capsules three-times daily, which may adversely affect adherence. A newer formulation of the drug in 300-mg capsules is in development.

Laboratory studies have suggested that tipranavir has virological activity against HIV resistant to the currently available protease inhibitors. Currently in progress is the first study of tipranavir in HIV-infected subjects who have experienced virological failure with a protease inhibitor-containing regimen. Subjects will replace their current protease inhibitor with tipranavir and be followed for virological activity. Other laboratory studies have demonstrated synergistic antiretroviral activity with combinations of tipranavir and ritonavir and provide a basis for clinical studies.

X. COMBINATION PROTEASE INHIBITOR THERAPY

Use of more than one protease inhibitor in an antiretroviral regimen may provide both advantages and disadvantages. The potential advantages of using two protease inhibitors together are to improve pharmacokinetic properties to allow twice-daily dosing or to decreased pill counts; to improve antiretroviral activity; and to avoid the development of resistance. The potential disadvantages of this approach are targeting one enzyme; antiretroviral antagonism; the development of resistance and cross-resistance; and increased adverse events. Favorable protease inhibitor combinations will maximize antiretroviral activity while minimizing drug intervals, pill counts, and adverse events. Several double protease inhibitor regimens have undergone evaluation.

A. Saquinavir and Ritonavir

The observation that ritonavir causes substantial increases in plasma drug levels of saquinavir hgc when the drugs are coadministered led to testing of the first dual protease regimen: saquinavir hgc and ritonavir (77a). Mellors and colleagues (78) reported the follow-up results of the first study of the combination of saquinavir and ritonavir, study 462 (Table 6). This prospective, open-label, randomized study examined four dosing regimens of saquinavir and ritonavir; after 12 weeks, patients who had not decreased HIV RNA to less than 200 copies/mL added reverse transcriptase inhibitors to their regimens. Eligible patients were HIV-infected, had not taken prior protease inhibitors, had CD4 cell counts of 100–500/mm^3, and were randomized to receive: (1) saquinavir 400 mg b.i.d. and ritonavir 400 mg b.i.d.; (2) saquinavir 400 mg b.i.d. and ritonavir 600 mg b.i.d.; (3) saquinavir 400 mg t.i.d. and ritonavir 400 mg t.i.d.; or (4) saquinavir 600 mg b.i.d. and ritonavir 600 mg b.i.d.

A total of 141 patients were enrolled with baseline median HIV RNA of 43,000 (4.63 log10) copies/mL and CD4 cell count of 273/mm^3. More than half of the patients reduced the doses of saquinavir–ritonavir to 400 mg b.i.d. by week 24, due to drug intolerance. Based on preliminary results, all patients were allowed to switch to saquinavir–ritonavir 400 mg b.i.d. after 48 weeks (Fig. 6). Seventy-three percent (103 of 141) remain on study at 72 weeks; 33 patients had reverse transcriptase inhibitors added to their regimens, 27 for HIV RNA more than 200 copies/mL. Of 96 patients with available data at week 72 of follow-up, 86 (90%) had HIV RNA less than 200 copies/mL; 23 of 27 patients (85%) who had added reverse transcriptase inhibitors for HIV RNA greater than 200 copies/mL had HIV RNA less than 200 at week 72. Median CD4 cell count increase from baseline was +188/mm^3.

Angel and colleagues (79) reported the immunological results from a subset of 42 patients on the 462 study followed at a single study site. Peripheral blood mononuclear cells from 41 patients were analyzed for proliferative responses to HIV p24 antigen and phytohemagglutinin (PHA) at baseline, week 4, and week 24. At baseline, peripheral blood monocyte cells (PBMC) from 2 patients proliferated in response to p24 antigen, compared with 9 after 4 weeks, and 10 after 24 weeks. In response to PHA, PBMC from 7 patients proliferated at baseline, compared with 26 after 4 weeks and 25 after 24 weeks. The authors concluded that this potent antiretroviral therapy is paralleled by a rapid, sustained improvement in functional measures of cell-mediated immunity.

Lorenzi and colleagues (80) reported the results of a pilot study of saquinavir–ritonavir in advanced HIV infection. The study enrolled 18 HIV-infected patients who experienced virological failure or toxicity with reverse transcriptase inhibitors and had not taken prior protease inhibitors, with CD4 less than 50/mm^3, who received saquinavir and ritonavir, both given at 600 mg b.i.d. At baseline, median HIV RNA level was 178,000 (5.25 log10) copies/mL, CD4 cell

Table 6 Major Efficacy Studies of Dual Protease Inhibitors

Study	Regimens	Population	Baseline levels	Major findings
Cameron (77a) and Mellors (78) Abbott Protocol 462	1. Saquinavir hgc + ritonavir (four-dosing regimens)	PI-naive; CD4 100–500 $N = 141$	HIV RNA 43K; CD4 273	400 mg b.i.d. of both drugs optimal dose with > 90% of subjects with HIV RNA < 200 at 72 wk
Gisolf (82) Prometheus Study	1. d4T + SQV hgc 400 mg b.i.d. + RTV 400 mg b.i.d. 2. SQV hgc + RTV	d4T- and PI-naive; $N = 111$	HIV RNA 20K; CD4 260	Three-drug combination superior with 87% (vs. 64%) of subjects with HIV RNA < 400 at 24 wk
Reijers (86) ADAM Study	1. d4T + 3TC + SQV hgc 600 mg t.i.d. + NFV 750 mg t.i.d.	Antiretroviral-naive; HIV RNA >1K; CD4 >200 $N = 43$	HIV RNA 34K; CD4 400	91% of subjects with HIV RNA < 50 at 24 wks; two-drug maintenance inferior to four drugs

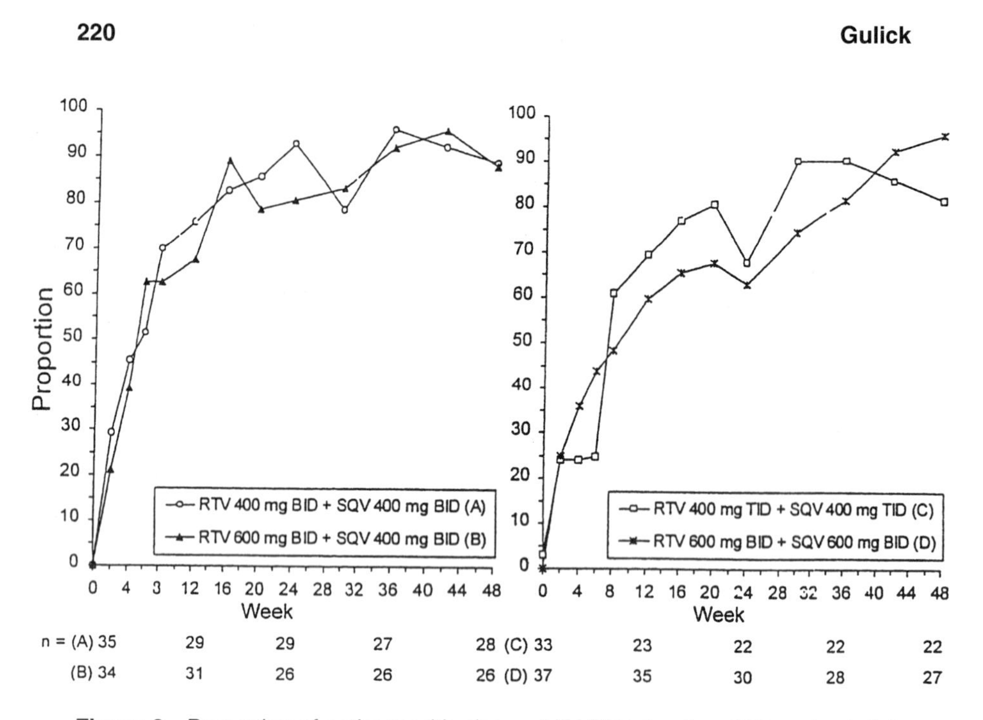

Figure 6 Proportion of patients with plasma HIV RNA levels <200 copies/mL in the Abbott 462 Study. The sample size is presented below the time points. BID, twice daily; TID, three times daily; RTV, ritonavir; SQV, saquinavir hgc. (From Ref. 77a, with permission.)

count was 12/mm^3, and 17 antiretroviral-experienced patients had a median of 35 months of treatment. Among the 18 patients, only 9 completed 13 weeks of study therapy. Those 9 patients had a median decrease in HIV RNA of −2.0 log copies/mL, with 2 of 9 decreasing to less than 100 copies/mL, and increased CD4 cell count +19/mm^3. The authors noted that only a minority of these advanced patients appeared to respond to the dual-protease inhibitor combination.

In another study of moderately advanced patients, Battegay et al. (81) reported the results using a combination of saquinavir 600 mg b.i.d., ritonavir 400–600 mg b.i.d., and stavudine 30–40 mg b.i.d. Eligible patients were HIV-infected, who had not taken prior stavudine or any protease inhibitor, with HIV RNA greater than 10,000 copies/mL and CD4 counts less than 250/mm^3. At baseline the 64 enrolled patients had a median HIV RNA of more than 100,000 copies/mL and a CD4 cell count of 80/mm^3. After 9 weeks of treatment, in an intent-to-treat analysis, 79% had decreased their viral load levels more than 2 log10 copies/mL, and 62% had levels less than 200 copies/mL. At the same time CD4 cell counts increased over 100/mm^3. The authors concluded this antiretroviral combination to be highly potent in moderately advanced patients.

Gisolf and colleagues (82) reported the results from the Prometheus study, an open-label, multicenter, randomized study of saquinavir and ritonavir, both given at 400 mg b.i.d., with or without stavudine (see Table 6). Patients who did not achieve a viral load level of less than 400 copies/mL by week 18 were allowed to add nucleoside analogues. Eligible patients were HIV-infected who had not taken prior stavudine or any protease inhibitor. Two hundred eight patients were enrolled, 50% of whom had received prior nucleoside analogues, with a baseline HIV RNA of 20,000 (4.3 log10) copies/mL and CD4 cell count of 260/mm^3.

Of 111 patients who had completed 24 weeks, serum HIV RNA decreased to less than 400 copies/mL in 64% of patients taking saquinavir–ritonavir and 87% taking three drugs. At the same time, mean CD4 cell counts increased +145/mm^3 over baseline. Six patients in the dual protease inhibitor arm were eligible to add nucleoside analogues and, after adding stavudine and lamivudine, had HIV RNA decreased to less than 400 copies/mL by week 36. From these preliminary results, the authors concluded it was uncertain whether there were differences in long-term virological responses between the regimens.

Markowitz and colleagues (83) reported the results of an open-label four-drug combination of saquinavir, ritonavir, zidovudine or stavudine, and lamivudine in a cohort of patients with documented recent HIV infection. Fourteen patients who acquired HIV infection within 90 days of study entry, began the four-drug antiretroviral combination. In follow-up, 12 of 14 remained on therapy from 44 to 72 weeks. Using monthly viral load determinations, 11 of 12 had HIV RNA suppressed to less than 25 copies/mL at least 50% of the time, and 6 of 12 at least 75% of the time. No patient had increasing HIV RNA documented. At 1 year, 8 patients tested failed to have evidence of multiply spliced mRNA isolated from their PBMC. The authors concluded that this four-drug regimen provides durable suppression of viral load levels in recently infected patients.

B. Saquinavir and Nelfinavir

Levels of saquinavir sgc are increased by concomitant administration of nelfinavir by about a factor of five. Kravcik and colleagues (84) presented the results of long-term follow-up of a small pilot study exploring the pharmacokinetics and virological activity of the combination of saquinavir and nelfinavir. Fourteen patients, with baseline HIV RNA of 39,900 copies/mL and CD4 327/mm^3, received saquinavir sgc 800 mg t.i.d. and nelfinavir 750 mg t.i.d. together for pharmacokinetic testing, and then continued to be followed for antiviral effects. At 11 months of follow-up, the median decrease in HIV RNA was −2.2 log10 copies/mL, with 9 of 10 decreasing their HIV RNA to less than 500 copies/mL, the limit of detection of the assay. At the same time, the median increase in CD4 cell counts was +172/mm^3 over baseline. The authors concluded that this combination provides potent, durable suppression of HIV RNA.

Farnsworth and colleagues (85) reported preliminary results of the Women

First study, using a combination of saquinavir hgc, nelfinavir, stavudine, and lamivudine. Eligible patients were HIV-infected women, who had not taken prior protease inhibitors, with plasma HIV RNA higher than 10,000 copies/mL; who were randomized to receive either (1) saquinavir hgc 1000 mg b.i.d. and nelfinavir 1250 mg b.i.d. or (2) saquinavir hgc 600 mg t.i.d. and nelfinavir 750 mg t.i.d., all in combination with stavudine and lamivudine. Preliminary results on 40 enrolled patients, with 10 patients treated for at least 20 weeks, revealed comparable pharmacokinetic parameters in the two groups. At 20 weeks, both groups showed median decreases in HIV RNA of −2.4–2.5 log10 copies/mL, with 80% of patients decreasing their HIV RNA to less than 400 copies/mL and median CD4 increases of +225–256/mm^3. The authors noted comparable results between twice- and three-times-daily dosing of saquinavir and nelfinavir in the preliminary results from this study.

Reijers and colleagues (86) reported the results of the Amsterdam Duration of Antiretroviral Medication (ADAM) study, a trial that evaluated induction therapy with a combination antiretroviral regimen consisting of saquinavir hgc, nelfinavir, stavudine, and lamivudine, followed by maintenance therapy with a reduced number of drugs (see Table 6). Eligible patients were HIV-infected, who had not taken prior antiretrovirals, with HIV RNA at least 1000 copies/mL and CD4 cell counts at least 200/mm^3. After 26 weeks of induction therapy with saquinavir hgc 600 mg t.i.d., nelfinavir 750 mg t.i.d., stavudine 40 mg b.i.d., and lamivudine 150 mg b.i.d., patients were randomized to receive maintenance therapy with (1) saquinavir hgc and nelfinavir, (2) stavudine and nelfinavir, or (3) continue four-drug therapy. When saquinavir sgc became available during the course of the study, all patients were switched to saquinavir sgc 800 mg t.i.d.

Sixty-two patients were enrolled, of whom 43 patients had been followed at least 26 weeks. At baseline, the 43 patients had median HIV RNA 34,000 (4.53 log10) copies/mL and CD4 cell count 400/mm^3. At the conclusion of induction therapy, plasma HIV RNA had decreased by −2.8 log10, to the limit of detection, with 39 of 43 patients (91%) decreasing their HIV RNA to less than 50 copies/mL. At the same time, median CD4 counts increased +200/mm^3 over baseline.

At week 26, 31 of 39 patients taking induction therapy were randomized to the two maintenance regimens or to continue four-drug therapy. Of these, 25 patients were followed for at least 36 weeks. At week 36, 9 of 14 (64%) patients receiving maintenance therapy had HIV RNA levels more than 50 copies/mL, compared with 1 of 11 (9%) receiving continued induction therapy ($p = 0.01$). The authors concluded that although the four-drug induction regimen provided rapid suppression of HIV RNA to fewer than 50 copies/mL, suppression was not sustained in a considerable number of patients receiving maintenance therapy, making this an inferior treatment strategy.

C. Ritonavir and Indinavir

The combination of ritonavir and indinavir has the potential to allow twice-daily dosing, reduced drug doses, and eliminates food restrictions. Workman and colleagues (87) reported a pilot study of 24 patients in two groups who received a four-drug combination of ritonavir 400 mg b.i.d., indinavir 400 mg b.i.d., stavudine, and lamivudine. Group A consisted of 12 patients taking ritonavir, saquinavir, stavudine, and lamivudine for at least 6 months, with HIV RNA less than 400 copies/mL for at least 6 months and stable CD4 cell counts. Group B consisted of 12 patients who had not taken prior antiretrovirals, with a baseline mean HIV RNA of 65,000 copies/mL. In group A, all patients maintained HIV RNA levels less than 400 copies/mL for up to 36 weeks. In group B, in 10 patients who had reached week 12, all patients had HIV RNA levels less than 400 copies/mL.

D. Ritonavir and Nelfinavir

The combination of ritonavir and nelfinavir has the potential to allow twice-daily dosing and reduced pill counts. Gallant and colleagues (88) presented the results of a single-site, open-label, multidose comparison pilot study of ritonavir and nelfinavir. Eligible patients were HIV-infected, who had not taken prior protease inhibitors. Twenty patients were enrolled, 10 in each cohort, with median baseline HIV RNA of 32,000 copies/mL and CD4 cell count of $325/mm^3$. In cohort I, patients received ritonavir 400 mg q12h in combination with nelfinavir 500 mg q12h; cohort II patients received ritonavir 400 mg q12h in combination with nelfinavir 750 mg q12h.

At 16 weeks, in cohort I, in 7 patients, mean reduction in HIV RNA was −2.1 log10, with 5 to 7 reducing HIV RNA to less than 400 copies/mL and mean CD4 increase was $+202/mm^3$. In cohort II, mean reduction in HIV RNA was −2.7 log10 (8 patients), with 7 of 8 reducing HIV RNA to less than 400 copies/mL and CD4 increase was $+37/mm^3$ (7 patients). Four patients in cohort I and 1 patient in cohort II had confirmed HIV RNA increases of at least 1 log10. Of the remaining 15 patients, 12 patients added reverse transcriptase inhibitors at 16–20 weeks of follow-up. The authors concluded that this was a promising dual protease inhibitor regimen that may require other antiretroviral agents for optimal antiretroviral activity.

E. Indinavir and Nelfinavir

Combining indinavir and nelfinavir may allow twice-daily dosing of both drugs. Saah and colleagues (89) presented the results of a pilot study of the combination of indinavir and nelfinavir. Eligible patients were HIV-infected who had not taken prior protease inhibitors, with serum HIV RNA of at least 30,000 copies/mL and CD4 cell counts of at least $100/mm^3$. Patients initially received indinavir 1000 mg q12h and nelfinavir 750 mg q12h. Noting an enhancement of indinavir levels by nelfinavir, but a lack of demonstrated pharmacokinetic effect of indinavir on nelfinavir levels, the dose of nelfinavir was later increased to 1000 mg q12h.

Twenty-one patients entered the study. After a median of 24 weeks, 11 of 18 (61%) patients receiving study treatment had HIV RNA decreased to fewer than 400 copies/mL, with 6 of those decreasing HIV RNA to less than 50 copies/mL. Median CD4 cell count increased $133/mm^3$ from baseline. Further assessment of indinavir–nelfinavir is planned, including a further dose increase of nelfinavir to 1250 mg q12h.

F. Amprenavir and Other Protease Inhibitors (Saquinavir sgc, Indinavir, or Nelfinavir)

Eron and colleagues (90) reported the results of a pilot study of amprenavir in combination with a second protease inhibitor. Patients were HIV-infected, who had not taken prior protease inhibitors, with HIV RNA levels greater than 10,000 copies/mL and CD4 cell counts more than $200/mm^3$. Patients were randomized to receive amprenavir 800 mg t.i.d. as monotherapy or combined with either (1) saquinavir sgc 800 mg t.i.d.; (2) indinavir 800 mg t.i.d.; or (3) nelfinavir 750 mg t.i.d. Patients receiving amprenavir monotherapy added zidovudine and lamivudine at week 3.

Preliminary results from a total of 16 patients at week 4 revealed median HIV RNA decreases of −2.5 log (amprenavir + saquinavir); −2.6 log10 (amprenavir + indinavir); −3.2 log10 (amprenavir + nelfinavir); and −1.7 (amprenavir, zidovudine, and lamivudine). In total, 13 of 16 patients had decreases in HIV RNA to less than 400 copies/mL by week 4. As amprenavir is now routinely dosed at 1200 mg q12h, future studies will explore twice-daily dual protease inhibitor regimens.

XI. COMPARATIVE STUDIES OF PROTEASE INHIBITORS

A. Cohort Studies

1. Saquinavir Versus Ritonavir Versus Indinavir

Kirk et al. (91) reported preliminary results from a subset of patients from the EuroSIDA study, a cohort consisting of 4491 unselected patients with a baseline CD4 less than $500/mm^3$ from 50 centers in 17 European countries. Of 1749 patients starting protease inhibitor therapy, 539 started saquinavir, 384 started ritonavir, and 826 started indinavir therapy. In this cohort of patients, at baseline before starting the protease inhibitor, 45% had AIDS, 95% previously had taken nucleoside analogues, and the median CD4 cell count was 125/mm^3. In 586 patients, viral load levels were available and the median was 7900 (3.9 log10) copies/mL. After a median 8 months of follow-up, 214 patients had clinical progression, including 105 deaths, without demonstrable differences in outcome among the three protease inhibitor groups. Further follow-up is planned.

Casado et al. (92) reported the results of a cohort of 400 HIV-infected patients who began therapy with a protease inhibitor. In this cohort, 57% had AIDS, 91% had taken nucleoside analogues, median HIV RNA was 29,000 (4.46 log10) copies/mL, and median CD4 cell count was 86/mm^3. One hundred eleven (28%) started saquinavir, 105 (26%) started ritonavir, and 184 (46%) started indinavir. At least one new nucleoside was added in 66% of the 365 pretreated patients. A total of 335 completed 12 months of therapy, among whom 149 (45%) had a reduction in HIV RNA below 200 copies/mL and a CD4 cell increase of at least 100/mm^3 over baseline levels. This was true of 38% of the saquinavir group, 40% of the ritonavir group, and 51% of the indinavir group. In a logistic regression analysis, the use of saquinavir was independently associated with virological failure (RR, 1.55; 95% CI, 1.03–2.27; $p = 0.03$).

Egger and colleagues (93) followed patients from the Swiss HIV Cohort Study who started treatment with a protease inhibitor-containing regimen for the development of a new AIDS event or death. Six hundred seventy-five patients began treatment with saquinavir hgc (54 patients, 8%), indinavir (399 patients, 59%), or ritonavir (221 patients, 33%) in combination with zidovudine and lamivudine. At baseline, 25–37% of the patients had AIDS, mean HIV RNA was 20,000–32,000 (4.3–4.5 log10) copies/mL, and median CD4 cell count was 112–201/mm^3, without significant differences among the groups. At the time of report, there were 29 new AIDS events and 12 deaths in the group as a whole. The adjusted hazard ratios, using indinavir as the reference (1.0) were ritonavir (0.87 for AIDS, 0.48 for death) and saquinavir hgc (2.8 for AIDS, and 3.5 for death). The authors concluded that compared with indinavir and ritonavir, patients treated with saquinavir hgc had more disease progression and mortality.

Suter et al. (94) reported the results from a cohort of patients who started protease inhibitor therapy with either saquinavir or indinavir as part of a triple-combination regimen. Three hundred thirteen patients were followed, 169 (54%) received saquinavir and 144 (46%) received indinavir. At baseline, the cohorts were comparable relative to CDC stage III (73–81%), nucleoside experience (62–71%), mean HIV RNA (158,000–200,000 or 5.2–5.3 log10 copies/mL), and median CD4 cell count (220/mm^3). Over 8 months of follow-up, 36% of patients on saquinavir regimens reduced their HIV RNA to less than 200 copies/mL, compared with 49% on indinavir regimens. Over the same time, there were no differences in clinical events (21 for saquinavir, 18 for indinavir).

B. Randomized Studies

1. Ritonavir Versus Indinavir Versus Saquinavir–Ritonavir

Kirk and colleagues (95) reported the results of a randomized, open-label, controlled trial of patients starting protease inhibitors: (1) ritonavir 600 mg b.i.d., (2) indinavir 800 mg q8h, or (3) the combination of saquinavir hgc and ritonavir, both given 400 mg b.i.d. All protease inhibitors were given with two nucleoside analogues (Table 7). Eligible patients had not taken prior protease inhibitors and had a clinical indication for protease inhibitor therapy, as determined by their physician. A total of 284 patients were enrolled, 165 (58%) who had taken prior nucleoside analogues, and 42% who had not taken prior antiretrovirals. Patients had a baseline median HIV RNA 50,000 (4.7 log10) copies/mL and CD4 cell count of 176/mm^3. Intent-to-treat analyses were performed, and patients were stratified for prior antiretroviral experience.

Overall, at week 24, the proportion of patients who decreased their HIV RNA to less than 200 copies/mL was no different among the three groups (67% for ritonavir, 71% for indinavir, 82% for saquinavir–ritonavir). However, there were significant differences in the proportion of patients who decreased viral load to less than 20 copies/mL (45% for ritonavir, 40% for indinavir, 57% for saquinavir–ritonavir; $p = 0.05$). The difference among the groups is accounted for by significant differences in the treatment-naive patient stratum; there were no differences among the treatment-experienced patients. CD4 cell count increases were similar among the three treatment groups (+110–132/mm^3 in treatment-naive patients, +66–75/mm^3 among treatment-experienced patients). The authors concluded that the combination of saquinavir and ritonavir had better virological activity than ritonavir or indinavir over 24 weeks in patients without prior antiretroviral experience.

2. Saquinavir sgc Versus Nelfinavir Versus Saquinavir–Nelfinavir

Moyle and colleagues (96) presented the results of the SPICE study (study of protease inhibitor combinations in Europe), a comparison of regimens including (1) saquinavir sgc, (2) nelfinavir, and (3) the combination of the two (see Table 7). Eligible patients were HIV-infected, had not taken prior protease inhibitors, and were able to start one new nucleoside analogue, with HIV RNA level of at least 10,000 copies/mL. Patients were randomized to receive (1) saquinavir sgc 1200 mg t.i.d. plus two nucleoside analogues; (2) nelfinavir 750 mg t.i.d. plus

Table 7 Major Efficacy Studies Comparing Protease Inhibitors

Study	Regimens	Population	Baseline levels	Major findings
Kirk (95)	Two nucleosides + 1. RTV 600 mg b.i.d. 2. IDV 800 mg q8h 3. SQV hgc 400 mg b.i.d. + RTV 400 mg b.i.d.	Clinical indication for PI; PI-naive; *N* = 284	HIV RNA 50K; CD4 176	SQV + RTV regimen superior with 57% (vs. 40–45%) of subjects with HIV RNA < 20 at 24 wk
Moyle (96) SPICE Study	Two nucleosides + 1. SQV sgc 1200 mg t.i.d. 2. NFV 750 mg t.i.d. 3. SQV sgc + NFV 4. SQV sgc + NFV without nucleosides	PI-naïve; one new nucleoside available; HIV RNA >10K *N* = 157	HIV RNA 63K; CD4 301	SQV + NFV + two nucleosides superior with 83% (vs. 55–70%) of subjects with HIV RNA <400 at 32 wk
Boucher (97) CHEESE Study	ZDV + 3TC + 1. SQV sgc 1200 mg t.i.d. 2. IDV 800 mg q8h	<12 mo of ZDV; naive to 3TC and PI; HIV RNA >10K; CD4 <500 *N* = 62	HIV RNA 79–100K; CD4 301–337	Overall, 50% of subjects with HIV RNA < 20 at 24 wk in each group; CD4 + 177 (SQV) and +93 (IDV)

two nucleosides; (3) saquinavir sgc plus nelfinavir plus two nucleosides; or (4) saquinavir sgc plus nelfinavir. A total of 157 patients, of whom 85 (54%) were treatment-naive, were enrolled, with a mean baseline HIV RNA of 63,000 (4.8 log10) copies/mL and CD4 cell count of 301/mm^3.

At 32-weeks follow-up, preliminary data indicate that the proportion of patients who decreased HIV RNA to less than 400 copies/mL was 70% (saquinavir sgc + nucleosides), 55% (nelfinavir + nucleosides), 83% (saquinavir sgc + nelfinavir + nucleosides), and 69% (saquinavir sgc + nelfinavir without nucleosides). Similar trends were seen among the proportion of patients who reduced HIV RNA to less than 50 copies/mL. CD4 cell counts increased from 73 to 161/mm^3. The authors concluded that saquinavir sgc, nelfinavir, and two nucleoside analogues produced potent virological suppression. Further follow-up is planned.

3. Saquinavir sgc Versus Indinavir

Boucher et al. reported the results (97) from a comparative study of saquinavir sgc and indinavir, each in combination with zidovudine and lamivudine (see Table 7). Eligible patients were HIV-infected, who had taken less than 12 months of zidovudine and no other antiretrovirals, with HIV RNA levels greater than 10,000 copies/mL, and CD4 counts of less than 500/mm^3 or HIV-related symptoms. Patients were randomized to receive either saquinavir sgc 1200 mg t.i.d. or indinavir 800 mg q8h in combination with zidovudine 200 mg t.i.d. and lamivudine 150 mg b.i.d.

A total of 62 patients were enrolled, with a median baseline HIV RNA of 79,000–100,000 (4.9–5.0 log10) copies/mL and CD4 cell counts of 301–337/mm^3. After 24 weeks of follow-up, about 50% of patients had reduced their viral load levels to less than 20 copies/mL in each of the two treatment groups. Median increases in CD4 cell counts were +163/mm^3 in the saquinavir group and +12/mm^3 in the indinavir group. The authors concluded that the regimens had similar potent virological activity and noted a possible immunological advantage in the saquinavir group. The minimal CD4 cell response in the indinavir group has not been explained. Patients continue in follow-up.

REFERENCES

1. Palella FJ, Delaney KM, Moorman AC, et al. Declining morbidity and mortality among patients with advanced human immunodeficiency virus infection. N Engl J Med 1998; 338:853–860.
2. Panel on Clinical Practices for Treatment of HIV Infection. Guidelines for the use of antiretroviral agents in HIV-infected adults and adolescents. Ann Intern Med 1998; 128:1079–1100.
3. Carpenter CJ, Fischl MA, Hammer SM, et al. Antiretroviral therapy for HIV infec-

tion in 1998: updated recommendations of the International AIDS Society–USA Panel. JAMA 1998; 280:78–86.
4. Gazzard B, Moyle G, BHIVA Guidelines Writing Committee. 1998 revision to the British HIV Association guidelines for antiretroviral treatment of HIV seropositive individuals. Lancet 1998; 352:314–316.
5. Package insert, Invirase, Roche Pharmaceuticals, Hoffmann–La Roche, Nutley, NJ, Jan 1997.
6. Package insert, Fortovase, Roche Pharmaceuticals, Hoffmann–La Roche, Nutley, NJ, Nov 1997.
7. Kitchen VS, Skinner C, Ariyoshi K, et al. Safety and activity of saquinavir in HIV infection. Lancet 1995; 345:952–955.
8. Schapiro JM, Winters MA, Stewart F, et al. The effect of high-dose saquinavir on viral load and CD4+ T-cell counts in HIV-infected patients. Ann Intern Med 1996; 124:1039–1050.
9. Vella S, Lazzarin A, Carosi G, et al. A randomized controlled trial of a protease inhibitor (saquinavir) in combination with zidovudine in previously untreated patients with advanced HIV infection. Antiviral Ther 1996; 1:129–140.
10. Lalezari J, Haubrich R, Burger HU, et al. Improved survival and decreased progression of HIV in patients treated with saquinavir (Invirase, SQV) plus Hivid (zalcitabine, ddC). 11th International Conference on AIDS, Vancouver, Canada, July 7–12, 1996: abstr LB.B.6033.
11. Stellbrink H–J. Clinical and survival benefit of saquinavir (SQV) in combination with zalcitabine (ddC) and zidovudine (ZDV) in untreated/minimally treated HIV-infected patients. 6th European Conference of Clinical Aspects and Treatment of HIV-Infection, Hamburg, Germany, Oct 11–15, 1997:abstr 212.
12. Collier AC, Coombs RW, Schoenfeld DA, et al. Treatment of human immunodeficiency virus infection with saquinavir, zidovudine, and zalcitabine. N Engl J Med 1996; 334:1011–1017.
13. Lalezari J. Selecting the optimum dose for a new soft gelatin capsule formulation for saquinavir. J Acquir Defic Syndr Hum Retrovirol 1998; 19:195–196.
14. Gill MJ, Beall G, Beattie D, et al. Safety of saquinavir soft gelatin capsule (SQV–SGC) in combination with other antiretroviral agents: multicenter study NV15182: 24 week analysis. 37th International Conference on Antimicrobial Agents and Chemotherapy, Toronto, Ontario, Canada, Sept 28–Oct 1, 1997:abstr I-90.
15. Mitsuyasu RT, Skolnik PR, Cohen SR, et al. Activity of the soft gelatin formulation of saquinavir in combination therapy in antiretroviral-naive patients. AIDS 1998; 12:F103–F109.
16. Thompson M. Activity of soft gelatin capsule formulation of saquinavir in combination with two nucleosides in treatment-naive HIV-1–seropositive persons. 12th International Conference on AIDS, Geneva, Switzerland, June 28–July 3, 1998:abstr 12145.
17. Farthing C, Sension M, Pilson R, et al. Fortovase (saquinavir, SQV) soft gel capsule (SGC) in combination with AZT and 3TC in antiretroviral-naive HIV-1 infected patients. 12th International Conference on AIDS, Geneva, Switzerland, June 28–July 3, 1998:abstr 12301.
18. Package insert, Ritonavir, Abbott Laboratories, Inc., March 1997.

19. Danner SA, Carr A, Leonard JM, et al. A short-term study of the safety, pharmacokinetics, and efficacy of ritonavir, an inhibitor of HIV-1 protease. N Engl J Med 1995; 333:1528–1533.
20. Markowitz M, Saag M, Powderly M, et al. A preliminary study of ritonavir, an inhibitor of HIV-1 protease, to treat HIV-1 infection. N Engl J Med 1995; 333: 1534–1539.
21. Cameron DW, Health–Chiozzi M, Danner S, et al. Randomised placebo-controlled trial of ritonavir in advanced HIV-1 disease. Lancet 1998; 351:543–549.
22. Kempf DJ, Rode RA, Xu Y, et al. The duration of viral suppression during protease inhibitor therapy for HIV-1 infection is predicted by plasma HIV-1 RNA at the nadir. AIDS 1998; 12:F9–F14.
23. Mathez D, Bagnarelli P, DeTruchis P, et al. A triple combination of ritonavir + AZT + ddC as a first line treatment of patients with AIDS: update. 11th International Conference on AIDS, Vancouver, Canada, July 7–12, 1996:abstr Mo.B.175.
24. Lederman MM, Connick E, Landay A, et al. Immunologic responses associated with 12 weeks of combination antiretroviral therapy consisting of zidovudine, lamivudine, and ritonavir: results of AIDS Clinical Trials Group Protocol 315. J Infect Dis 1998; 178:70–79.
25. Connick E, Lederman M, Kotzin B, et al. Immunologic effects of 48 weeks of AZT/3TC/ritonavir. 5th Conference of Retroviruses and Opportunistic Infections, Chicago, IL, Feb 1–5, 1998:abstr LB14.
26. Notermans DW, Jurriaans S, de Wolf F, et al. Decrease of HIV-1 RNA levels in lymphoid tissue and peripheral blood during treatment with ritonavir, lamivudine, and zidovudine: ritonavir/3TC/ZDV Study Group. AIDS 1998; 12:167–173.
27. Hoen B, Harzic M, Dumon B, et al. Efficacy of zidovudine, lamivudine, and ritonavir combination in patients with symptomatic primary HIV-1 infection: the ANRS 053/53b trial. Can eradication be obtained? 5th Conference of Retroviruses and Opportunistic Infections, Chicago, IL, February 1–5, 1998:abstr 524.
28. Katlama C, Valantin MA, Calvez V, et al. ALTIS PLUS: long-term d4T–3TC with and without ritonavir. 5th Conference of Retroviruses and Opportunistic Infections, Chicago, IL, February 1–5, 1998:abstr 376.
29. Saimot AG, Landman R, Damond F, et al. Stavudine (d4T), didanosine (ddI), and ritonavir as a triple therapy in antiretroviral-naive patients: results at 72 weeks. 12th International Conference on AIDS, Geneva, Switzerland, June 28–July 3, 1998:abstr 22401.
30. Package insert, Crixivan, Merck and Company, Inc., Feb 1998.
31. Deutsch P, Teppler H, Squires K, et al. Antiviral activity of L-735,524, an HIV protease inhibitor, in infected patients. 34th International Conference on Antimicrobial Agents and Chemotherapy, Orlando, FL, Oct 4–7, 1994:abstr I-59.
32. Stein DS, Fish DG, Bilello JA, et al. A 24-week open-label phase I/II evaluation of the HIV protease inhibitor MK-639 (indinavir). AIDS 1996; 10:485–492.
33. Mellors J, Steigbigel R, Gulick R, et al. Antiretroviral activity of the oral protease inhibitor, MK-639, in p24 antigenemic, HIV-1 infected patients with <500 CD4/mm^3. 35th International Conference on Antimicrobial Agents and Chemotherapy, San Francisco, CA, Sept 17–20, 1995:abstr I-172.
34. Mellors J, Steigbigel R, Gulick R, et al. A randomized, double blind study of the

oral HIV protease inhibitor, L-735,524 vs. zidovudine (ZDV) in p24 antigenemic, HIV-1 infected patients with <500 CD4 cells/mm^3. 2nd National Conference on Human Retroviruses and Related Infections, Washington DC, Jan 29–Feb 2, 1995: abstr 183.

35. Steigbigel RT, Berry P, Mellors J, et al. Efficacy and safety of the HIV protease inhibitor indinavir sulfate (MK 639) at escalating dose. 3rd Conference on Retroviruses and Opportunistic Infections, Washington, DC, Jan 28–Feb 1, 1996:abstr 146.
36. Massari F, Staszewski S, Berry P, et al. A double-blind, randomized trial of indinavir (MK-639) alone or with zidovudine vs. zidovudine alone in zidovudine naive patients. 35th International Conference on Antimicrobial Agents and Chemotherapy, San Francisco, CA, Sept 17–20, 1995:abstr LB-6.
37. Leavitt R, Massari F, Nessly M, et al. Antiviral activity of indinavir (IDV) plus zidovudine (ZDV) compared to IDV or ZDV alone in antiretroviral naive patients. 36th International Conference on Antimicrobial Agents and Chemotherapy, New Orleans, LA, Sept 15–18, 1996:abstr I-109.
38. Steigbigel RT, Copper D, Clumeck N, et al. Indinavir with stavudine vs. indinavir alone vs. stavudine alone in zidovudine experienced, HIV-infected patients. Merck protocol 037 study group. 12th International Conference on AIDS, Geneva, Switzerland, June 28–July 3, 1998:abstr 12335.
39. Massari F, Conant M, Mellors J, et al. A phase II open-label, randomized study of the triple combination of indinavir, zidovudine (ZDV) and didanosine (ddI) versus indinavir alone and zidovudine/didanosine in antiretroviral naive patients. 3rd Conference on Retroviruses and Opportunistic Infections, Washington, DC, Jan 28–Feb 1, 1996:abstr 200.
40. Gulick RM, Mellors JW, Havlir D, et al. Treatment with indinavir, zidovudine, and lamivudine in adults with human immunodeficiency virus infection and prior antiretroviral therapy. N Engl J Med 1997; 337:734–739.
41. Gulick RM, Mellors JW, Havlir D, et al. Simultaneous vs. sequential initiation of therapy with indinavir, zidovudine, and lamivudine for HIV-1 infection: 100 week follow-up. JAMA 1998; 280:35–41.
42. Gulick R, Mellors J, Havlir D, et al. Treatment with indinavir (IDV), zidovudine (ZDV) and lamivudine (3TC): three-year follow-up. 6th Conference on Retroviruses and Opportunistic Infections, Chicago, IL, Jan 31–Feb 4, 1999:abstr 388.
43. Hirsch MS, Protocol 039 Study Group. Indinavir (IDV) in combination with zidovudine (ZDV) and lamivudine (3TC) in ZDV-experienced patients with CD4 cell counts $\leq$ 50 cells/mm^3. 4th Conference on Retroviruses and Opportunistic Infections, Washington, DC, Jan 22–26, 1997:abstr LB-7.
44. Leavitt R, McMahon D, Meibohm A, et al. Indinavir (IDV) and lamivudine (3TC) in treatment-naive patients with $\geq$ 500 CD4 cells/mm^3. 12th International Conference on AIDS, Geneva, Switzerland, June 28–July 3, 1998:abstr 12311.
45. Nguyen B–W, Haas DW, Ramirez–Ronda C, et al. A pilot, multicenter, open-label, randomized study to compare the safety and efficacy of indinavir sulfate (IDV) administered every 8 hours versus every 12 hours in combination with zidovudine (ZDV) and lamivudine (3TC). 37th International Conference on Antimicrobial Agents and Chemotherapy, Toronto, Ontario, Canada, Sept 28–Oct 1, 1997:abstr I-91.

46. Nguyen BY, Haas DW, Ramirez-Ronda C, et al. Thirty-two-week follow-up of indinavir sulfate (IDV) administered q8 hours (h) versus q12 h in combination with zidovudine (ZDV) and lamivudine (3TC). 5th Conference on Retroviruses and Opportunistic Infections, Chicago, IL, Feb 1–5, 1998: abstr 374.
47. Perrin L, Markowitz M, Calandra G, et al. An open treatment study of acute HIV infection with zidovudine, lamivudine, and indinavir sulfate. 4th Conference on Retroviruses and Opportunistic Infections, Washington DC, Jan 22–26, 1997:abstr 238.
48. Havlir DV, Marschner IC, Hirsch MS, et al. Maintenance antiretroviral therapies in HIV-infected subjects with undetectable plasma HIV RNA after triple-drug therapy. N Engl J Med 1998; 339:1261–1268.
49. Pialoux G, Raffi F, Brun–Vezinet F, et al. A randomized trial of three maintenance regimens given after three months of induction therapy with zidovudine, lamivudine, and indinavir in previously untreated HIV-1-infected patients. N Engl J Med 1998; 339:1269–1276.
50. Gulick RM, Squires K, Powderly W, et al. An open-label, randomized, comparative study of d4T + 3TC + IDV vs ZDV + 3TC + IDV in treatment-naive HIV-infected patients (START I). 12th International Conference on AIDS, Geneva, Switzerland, June 28–July 3, 1998:abstr 12223.
51. Eron J, Peterson D, Murphy R, et al. An open-label, randomized, comparative study of d4T + ddI + IDV vs ZDV + 3TC + IDV in treatment-naive HIV-infected patients (START II). 12th International Conference on AIDS, Geneva, Switzerland, June 28–July 3, 1998:abstr 12225.
52. Leavitt R, Lewi DS, Uip D, et al. Subgroup analysis of clinical and surrogate marker efficacy by gender in a study of indinavir (IDV) alone and with zidovudine (ZDV) in treatment-naive HIV-1 patients with 50–250 CD4 cells/mm^3. 12th International Conference on AIDS, Geneva, Switzerland, June 28–July 3, 1998:abstr 60358.
53. Hammer SM, Squires KE, Hughes MD, et al. A controlled trial of two nucleoside analogues plus indinavir in persons with human immunodeficiency virus infection and CD4 cell counts of 200 per cubic millimeter or less. N Engl J Med 1997; 337: 725–739.
54. Harris M, Durakovic C, Rae S, et al. A pilot study of nevirapine, indinavir, and lamivudine among patients with advanced human immunodeficiency virus disease who have had failure of combination nucleoside therapy. J Infect Dis 1998; 177: 1514–1520.
55. Havlir D, Hicks C, Kahn J, et al. Durability of antiviral activity of efavirenz (EFV, DMP-266) in combination with indinavir. 38th International Conference on Antimicrobial Agents and Chemotherapy, San Diego, CA, Sept 24–27, 1998:abstr I-104.
56. Morales–Ramirez J, Tashima K, Hardy D, et al. A phase II, multi-center, randomized, open label study to compare the antiretroviral activity and tolerability of efavirenz (EFV) + indinavir (IDV), versus EFV + zidovudine (ZDV) + lamivudine (3TC), versus IDV + 3TC at > 36 weeks [DMP-266-006]. 38th International Conference on Antimicrobial Agents and Chemotherapy, San Diego, CA, Sept 24–27, 1998:abstr I-103.
57. Package insert, Viracept, Agouron Pharmaceuticals, Inc., Dec 1997.

58. Markowitz M, Conant M, Hurley A, et al. A preliminary evaluation of nelfinavir mesylate, an inhibitor of human immunodeficiency virus (HIV)-1 protease, to treat HIV infection. J Infect Dis 1998; 177:1533–1540.
59. Powderly W, Sension M, Conant M, Stein A, Clendeninn N. The efficacy of Viracept (nelfinavir mesylate, NFV) in pivotal phase II/III double-blind randomized controlled trials as monotherapy and in combination with d4T or AZT/3TC. 4th Conference on Retroviruses and Opportunistic Infections, Washington, DC, Jan 22–26, 1997:abstr 370.
60. Clendeninn N, Quart B, Anderson R, et al. Analysis of long-term virologic data from the Viracept (nelfinavir, NFV) 511 protocol using 3 HIV-RNA assays. 5th Conference on Retroviruses and Opportunistic Infections, Chicago, IL, Feb 1–5, 1998:abstr 372.
61. Kerr B, Pithavala Y, Zhang M, et al. Virologic response–plasma drug concentration relationship in phase III study of nelfinavir mesylate (VIRACEPT). 12th International Conference on AIDS, Geneva, Switzerland, June 28–July 3, 1998:abstr 12304.
62. Clumeck N, AVANTI 3 Study Group. A randomized, double blind comparative trial to evaluate the safety, efficacy, and tolerance of AZT/3TC vs AZT/3TC/nelfinavir in antiretroviral naive patients. 5th Conference on Retroviruses and Opportunistic Infections, Chicago, IL, Feb 1–5, 1998:abstr 8.
63. Hecht FM, Smith D, Cooper D, et al. Treatment of primary HIV infection with nelfinavir, zidovudine, and lamivudine. 12th International Conference on AIDS, Geneva, Switzerland, June 28–July 3, 1998:abstr 12197.
64. Petersen A, Johnson M, Clendeninn N, et al. Nelfinavir study 542: comparison of BID and TID dosing of nelfinavir (NFV) in combination with stavudine (d4T) and lamivudine (3TC): an interim look. 5th Conference on Retroviruses and Opportunistic Infections, Chicago, IL, Feb 1–5, 1998:abstr 270.
65. Petersen A, Johnson M. Long-term comparison of BID and TID-dosing Viracept (nelfinavir) in combination with stavudine (d4T) and lamivudine (3TC) in HIV patients. 12th International Conference on AIDS, Geneva, Switzerland, June 28–July 3, 1998:abstr 12224.
66. Pedneault L, Elion R, Adler M, et al. Stavudine (d4T), didanosine (ddI) and nelfinavir combination therapy in HIV-infected subjects: antiviral effect and safety in an ongoing pilot study. 4th Conference on Retroviruses and Opportunistic Infections, Washington, DC, Jan 22–26, 1997:abstr 241.
67. Skowron G, Leoung G, Berman BY, et al. Stavudine (d4T), nelfinavir (NFV), and nevirapine (NVP): suppression of HIV-1 RNA to fewer than 50 copies/mL during 5 months of therapy. 12th International Conference on AIDS, Geneva, Switzerland, June 28–July 3, 1998:abstr 12275.
68. Kagan S, Jemsek J, Martin DG, et al. Initial effectiveness and tolerability of nelfinavir (NFV) in combination with efavirenz (EFV, SUSTIVA, DMP-266) in antiretroviral therapy naïve or nucleoside analogue experienced HIV-1 infected patients: characterization in a phase II, open-label, multicenter study at >36 weeks (Study DMP-266-024). 38th Interscience Conference on Antimicrobial Agents and Chemotherapy, San Diego, CA, Sept 24–27, 1998:abstr I-102.
69. Albrecht M, Katzenstein D, Bosch RJ, Liou SH, Hammer SM. ACTG 364: virologic

efficacy of nelfinavir (NFV) and/or efavirenz (EFV) in combination with new nucleoside analogs in nucleoside experienced subjects. 12th International Conference on AIDS, Geneva, Switzerland, June 28–July 3, 1998:abstr 125/12203.

70. Schooley RT. Preliminary data on the safety and antiviral activity of the novel protease inhibitor 141W94 in HIV-infected patients with 140 to 400 CD4+ cells/mm^3. 36th International Conference on Antimicrobial Agents and Chemotherapy, New Orleans, LA, Sept 15–18, 1996:abstr LB7a.
71. Murphy RL, Gulick RM, DeGruttola V, et al. Treatment with amprenavir alone or amprenavir with zidovudine and lamivudine in adults with human immunodeficiency virus infection. J Infect Dis 1999; 179:808–816.
72. Murphy RL, Gulick R, Smeaton L, et al. Treatment with indinavir, nevirapine, stavudine, and 3TC following therapy with an amprenavir-containing regimen [abstr]. AIDS 1998; 12:S9.
73. De Pasquale MP, Murphy R, Gulick R, et al. Mutations selected in HIV plasma RNA during 141W94 therapy. 4th Conference on Retroviruses and Opportunistic Infections, Washington, DC, Jan 22–26, 1997:abstr 406A.
74. Haubrich R. Phase 2 study of amprenavir, a novel protease inhibitor, in combination with zidovudine/3TC. 12th International Conference on AIDS, Geneva, Switzerland, June 28–July 3, 1998:abstr 12321.
75. Bart PA, Rizzardi GP, Gallant S, et al. Combination 1592/141W94 therapy in HIV-1 infected antiretroviral naive subjects with CD4+ counts greater than 400 cells/microliter and viral load greater than 500 copies/mL. 12th International Conference on AIDS, Geneva, Switzerland, June 28–July 3, 1998:abstr 365.
76. Japour A, Murphy R, Hicks C, et al. Safety and efficacy of ABT-378/ritonavir in antiretroviral naive HIV patients: preliminary phase II results. 12th International Conference on AIDS, Geneva, Switzerland, June 28–July 3, 1998:abstr 12460.
77. Wang Y, Tutton CM, Borin MT, et al. The safety, pharmacokinetics, and efficacy of PNU-140690, a new non-peptidic HIV protease inhibitor, in a phase I/II study. 12th International Conference on AIDS, Geneva, Switzerland, June 28–July 3, 1998: abstr 41176.

77a. Cameron DW, Japour AJ, Xu Y, et al. Ritonavir and saquinavir combination therapy for the treatment of HIV infection. AIDS 1999; 13:213–224.

78. Mellors J, Japour AJ, Leonard J, et al. Ritonavir (RTV)–saquinavir (SQV) in protease inhibitor-naive patients after 72 weeks. 12th International Conference on AIDS, Geneva, Switzerland, June 28–July 3, 1998:abstr 12295.
79. Angel JB, Kumar A, Parato K, et al. Improvement in cell-mediated immune function during potent anti-human immunodeficiency virus therapy with ritonavir plus saquinavir. J Infect Dis 1998; 177:898–904.
80. Lorenzi P, Yerly S, Abderrakim K, et al. Toxicity, efficacy, plasma drug concentrations and protease mutations in patients with advanced HIV infection treated with ritonavir plus saquinavir. AIDS 1997; 11:F95–F99.
81. Battegay M, Bernasconi E, Flepp M, et al. Saquinavir (SQV) in combination with ritonavir (RTV) and d4T in patients with advanced HIV disease. 12th International Conference on AIDS, Geneva, Switzerland, June 28–July 3, 1998:abstr 60714.
82. Gisolf EH, de Wolf F, Pelgrom J, et al. Treatment of ritonavir (RTV)/saquinavir (SQV) versus RTV/SQV/stavudine (d4T) (the Prometheus study): preliminary re-

sults. 12th International Conference on AIDS, Geneva, Switzerland, June 28–July 3, 1998:abstr 12274.
83. Markowitz M, Talal A, Vesanen MS, et al. Durable suppression of virus replication with saquinavir in combination with ritonavir and double nucleoside inhibitor therapy. 12th International Conference on AIDS, Geneva, Switzerland, June 28–July 3, 1998:abstr 22395.
84. Kravcik S, Farnsworth A, Patick A, et al. Long term follow-up of combination protease inhibitor therapy with nelfinavir and saquinavir (soft gel) in HIV infection. 4th Conference on Retroviruses and Opportunistic Infections, Washington, DC, Jan 22–26, 1997:abstr 394c.
85. Farnsworth A, Squires K, Currier JS, et al. Women First: a study of the effects of treatment in women + HIV-infected with combination nelfinavir, saquinavir, stavudine, and lamivudine. 12th International Conference on AIDS, Geneva, Switzerland, June 28–July 3, 1998:abstr 12305.
86. Reijers MHE, Weverling GJ, Jurriaans S, et al. Maintenance therapy after quadruple induction therapy in HIV-1 infected individuals: Amsterdam Duration of Antiretroviral Medication (ADAM) study. Lancet 1998; 352:185–190.
87. Workman C, Musson R, Dyer W, Sullivan J. Novel double protease combinations—combining indinavir (IDV) with ritonavir (RTV): results from first study. 12th International Conference on AIDS, Geneva, Switzerland, June 28–July 3, 1998:abstr 22372.
88. Gallant J, Heath–Chiozzi M, Anderson R, Fields C, Flexner C. Phase II study of ritonavir–nelfinavir combination therapy: an update. 12th International Conference on AIDS, Geneva, Switzerland, June 28–July 3, 1998:abstr 12207.
89. Saah A, Riddler S, Havlir DV, et al. Co-administration of indinavir and nelfinavir: pharmacokinetics, tolerability, anti-viral activity, and preliminary viral resistance. 12th International Conference on AIDS, Geneva, Switzerland, June 28–July 3, 1998: abstr 22352.
90. Eron J, Haubrich R, Richman D, et al. Preliminary assessment of 141W94 in combination with other protease inhibitors. 4th Conference on Retroviruses and Opportunistic Infections, Washington, DC, Jan 22–26, 1997:abstr 6.
91. Kirk O. Preliminary comparison of outcome in patients starting either ritonavir (R), indinavir (I) or saquinavir (S) in 1749 patients from the euroSIDA study. 12th International Conference on AIDS, Geneva, Switzerland, June 28–July 3, 1998:abstr 12273.
92. Casado JL, Perez-Elias MJ, Antela A, et al. Predictors of long-term response to protease inhibitor therapy in a cohort of HIV-infected patients. AIDS 1998; 12: F131–F135.
93. Egger M, Telenti A, Wirz M, Battegay M. Disease progression and survival on triple therapy: indinavir, ritonavir or saquinavir combined with 3TC and AZT. 12th International Conference on AIDS, Geneva, Switzerland, June 28–July 3, 1998:abstr 12317.
94. Suter F, Maggiolo F, Bottura P, et al. Saquinavir vs. indinavir in triple drug therapy. 12th International Conference on AIDS, Geneva, Switzerland, June 28–July 3, 1998: abstr 12337.
95. Kirk O, Katzenstein TL, Gerstoft J, et al. Combination therapy containing ritonavir

plus saquinavir has superior short-term antiretroviral efficacy: a randomized trial. AIDS 1999; 13:F9–F16.
96. Moyle G, The SPICE study team. Study of protease inhibitor combination in Europe (SPICE): saquinavir soft gelatin capsule (SQV-sgc) plus nelfinavir (NFV) in HIV-infected individuals. 12th International Conference on AIDS, Geneva, Switzerland, June 28–July 3, 1998:abstr 12222.
97. Boucher CA, Borleffs J. Long-term evaluation of saquinavir soft-gel-capsule or indinavir as part of combination triple therapy (CHEESE Study). 12th International Conference on AIDS, Geneva, Switzerland, June 28–July 3, 1998:abstr 12267.

11
Toxicities and Adverse Effects of Protease Inhibitors

Marshall J. Glesby
Weill Medical College of Cornell University, New York, New York

The toxicity profiles of protease inhibitors are of major concern from both the standpoints of patient safety and quality of life. Furthermore, the tolerability of these drugs significantly affects a patient's acceptance of treatment and adherence to combination antiretroviral therapy; the latter is critically important to reduce the likelihood of the development of resistance to particular drugs or classes of antiretroviral agents.

The assessment of the toxicities and adverse effects of protease inhibitors is complicated by several issues. The frequent occurrence of adverse events owing to the underlying HIV disease process and concurrent use of multiple medications makes it difficult to attribute causality to these events in the contexts of clinical trials and clinical practice. Not only can concurrent medications have their own adverse effects, but also their interaction with protease inhibitors can lead to additional toxicities. The task of attributing causality to adverse events is simplified in the cases of the early placebo-controlled studies of protease inhibitors, in which differences in the incidence of events can be compared between groups receiving active drug and placebo. Clinical trials, in general, however, are useful only to delineate relatively common toxicities of short-term drug therapy, because relatively few patients are exposed to drug for short periods of time.

With the prolongation of survival of patients with advanced HIV disease attributable to combination antiretroviral therapy, long-term toxicities of protease inhibitors are of increasing importance. Preliminary information on some of these toxicities has come from case reports and case series, the interpretation of which is complicated by the lack of a defined denominator of a population exposed to

Table 1 Major Side Effects of Protease Inhibitors[a]

Side effect	Indinavir	Nelfinavir	Ritonavir	Saquinavir	Amprenavir
Nausea	++	+	++	++	+
Vomiting	+	NR	++	+	−
Diarrhea	+	++	++	++	−
Asthenia or fatigue	−	−	++	−	−
Nephrolithiasis or flank pain	+	NR	NR	NR	NR
Hyperbilirubinemia	+	NR	−	−	−
High serum aminotransferase concentration	+	+	+	+	+
Hypertriglyceridemia	NR	NR	+	NR	−
Paresthesias	NR	NR	+	−	+
Rash	NR	+	+	−	+

[a] One plus sign indicates toxicity of moderate or severe intensity reported in less than 10% of treated patients but occurring at least twice as often as in concurrently treated patients not taking the protease inhibitor. Two plus signs indicate toxicity in at least 10% of treated patients and occurring at least twice as often as in control patients. A minus sign indicates toxicity occurring less than twice as often in treated patients as in control patients, or in less than 3% of treated patients. NR denotes not reported. Data are for the standard doses of the drugs and are from the product package inserts (5,10–13) for all drugs except for amprenavir which was taken from a phase 3 clinical trial (110). Preliminary data suggest that hyperglycemia and fat redistribution may occur with all protease inhibitors, but no controlled incidence data are available.
Source: Adapted from Ref. 109.

the drug. More definitive data on the incidence of long-term toxicities of protease inhibitors will come from observational postmarketing studies.

Overall, the short-term safety profile of protease inhibitors is favorable. Although there is considerable overlap in the toxicities of individual protease inhibitors, each drug has a distinct profile that is important to consider in selecting a drug regimen for a particular patient. The relative incidences of the most frequent adverse effects of the marketed protease inhibitors are summarized in Table 1. The remainder of this chapter is organized by organ system in which the major toxicities manifest.

I. HEMATOLOGICAL

There is a possible association between protease inhibitor use and spontaneous bleeding in HIV-infected hemophiliacs. In July 1996, the U.S. Food and Drug Administration (FDA) informed health care providers that 11 cases of spontaneous hematomas and 5 hemarthroses had been reported in 15 European patients with advanced HIV disease receiving protease inhibitor therapy (1). Increased bleeding has been reported in both hemophilia A (factor VIII deficiency) and B (factor IX deficiency). A preliminary report of 17 hemophiliac patients who added indinavir to a regimen of zidovudine and lamivudine noted bleeding episodes in 5 (29.4%) patients (2). The mechanism for increased bleeding and whether protease inhibitors are, in fact, responsible are still unknown. Investigators have described prolongation of bleeding time and a defect in platelet aggregation in one of four hemophiliac patients studied just before and 3 h after ritonavir administration (3).

Other hematological abnormalities have not been commonly attributed to protease inhibitor therapy, with the exception of hemolytic anemia in association with indinavir use (4,5).

II. GASTROINTESTINAL

Gastrointestinal symptoms are among the most common adverse effects reported with all of the available protease inhibitors. Symptoms include anorexia, nausea, vomiting, abdominal pain, acid reflux, flatulence, and diarrhea. The mechanism of these toxicities is not understood, and clinical management involves symptomatic treatment, including antiemetics and antimotility agents. The promotility agent cisapride should not be coadministered with protease inhibitors owing to the potential for life-threatening drug interaction.

III. DERMATOLOGICAL

Rashes have been reported with all available protease inhibitors, albeit relatively infrequently. In particular, the incidence of rash appears to be greatest with amprenavir. Although a typical morbilliform eruption is most common, a few cases of exfoliative dermatitis and Stevens–Johnson syndrome have been reported with amprenavir (6). Photosensitivity has been described with saquinavir (7).

Alopecia has been reported in association with indinavir use. In a report of five cases of alopecia (1.5% of patients taking indinavir at the institution), two patients had alopecia areata, rather than the typical telogen effluvium pattern seen with most drugs associated with hair loss (8). The investigators raised the unresolved question of whether autoimmunity caused by immune reconstitution may have been responsible for the hair loss (8).

An association between indinavir use and the development of paronychia and pyogenic granuloma-like lesions of the great toes has been reported (9). The investigators hypothesized that inhibition of endogenous proteases might lead to hypertrophy of the nail fold and the subsequent development of pathology (9).

IV. HEPATIC

Asymptomatic liver enzyme elevations are relatively common in patients receiving protease inhibitor therapy, with only a minority (<10%) of patients having transaminase levels greater than five times the upper limit of normal in pivotal trials of the approved drugs (5,10–13). A unique finding in approximately 10% of patients taking indinavir is asymptomatic, indirect hyperbilirubinemia, resembling Gilbert's syndrome. Less than 1% of patients with hyperbilirubinemia have coexisting transaminase elevations (5). It is not known whether indinavir administration to the mother during the peripartum period may exacerbate physiological jaundice of the newborn.

Cases of severe hepatitis with decreased hepatic synthetic function and fatal hepatic failure have been reported in association with protease inhibitor use (14–16). In some of these patients, coexisting hepatitis A, B, or C, and the concomitant use of other potentially hepatotoxic medications, confounds the attribution of causality. Some cases of flares of hepatitis B after protease inhibitor initiation may be due to an improved immune response to the hepatitis B virus (17). In contrast, flares of hepatitis C may be due to increased hepatitis C virus replication, perhaps related to an interaction with HIV viremia that is waning in response to the protease inhibitor (16).

Clinicians should carefully monitor hepatic function in patients coinfected with hepatitis B or C who are receiving protease inhibitor therapy. In some patients, treatment of the viral hepatitis may be indicated before initiating protease inhibitor therapy.

V. NEUROLOGICAL

Neurological side effects are not common with protease inhibitors, with the exception of circumoral paresthesias, seen frequently with ritonavir and occasionally with amprenavir. Headache and dizziness have been reported infrequently in clinical trials of protease inhibitors.

VI. URINARY TRACT

Nephrolithiasis is an important toxicity of indinavir therapy and has been reported in 4% of patients receiving the drug in clinical trials (5). The cores of the renal calculi are composed of indinavir-based monohydrate, implicating the drug itself as the promoter of calculus formation (18). Presumably because of pH-dependent solubility of indinavir, nephrolithiasis has been associated with a higher urine pH (19); however, the clinical usefulness of the statistical association is uncertain.

Patients receiving indinavir should consume at least 1.5 L of liquids each day in an attempt to prevent nephrolithiasis (5), and an even greater quantity may be necessary in warmer climates (20). The management of nephrolithiasis usually involves hydration and pain control. Temporary interruption of indinavir therapy, ideally along with the other antiretrovirals used in combination, may also be considered in some patients.

Kopp et al. described a spectrum of urinary tract abnormalities in 240 patients receiving indinavir therapy (21). Eleven percent of their cohort developed symptomatic urinary tract disease, which manifested as one of three syndromes: renal colic and nephrolithiasis (3.8%); flank or back pain, with crystalluria (5%); or dysuria or urgency, with crystalluria (2.5%). Asymptomatic crystalluria was present in 20% of patients, but did not predict the onset of symptoms.

Indinavir has also been associated with renal parenchymal injury in the absence of nephrolithiasis in a case report of a patient whose renal biopsy demonstrated interstitial fibrosis, tubular atrophy with chronic inflammation, and crystal formation within collecting ducts (22). Reversible nephrotoxicity without nephrolithiasis in association with indinavir use has also been reported elsewhere (23), and asymptomatic, mild creatinine elevations may, in fact, be relatively common, with a prevalence of 19% in one study (24). In the latter study, the mean creatinine elevation was 42% above baseline, and risk factors included low body weight, duration of treatment with indinavir, concomitant use of trimethoprim–sulfamethoxazole, and low-baseline CD4 cell count (24).

Cases of renal failure in association with use of ritonavir and combined use of ritonavir and saquinavir have also been reported (25–27). Interpretation of these cases, however, is confounded by the existence of underlying chronic renal insufficiency in one case (25) and the concomitant use of foscarnet, a nephrotoxic drug, in three of the remaining six cases (26,27). Nonetheless, ritonavir appears to occasionally cause acute renal insufficiency.

VII. ENDOCRINE AND METABOLIC

Reports of endocrinological and metabolic derangements in association with protease inhibitor use became prevalent about 18 months after the licensure of the first protease inhibitor in the United States. Investigators have reported dramatic aberrations in glucose and lipid metabolism and body fat distribution. In some patients, several abnormalities have coexisted, leading to the use of the term *syndrome*, whereas other patients have manifested only isolated metabolic abnormalities. The major categories of metabolic abnormalities will be considered separately.

A. Hyperglycemia

In June 1997, the U.S. FDA issued a Public Health Advisory summarizing 83 cases of diabetes mellitus or hyperglycemia in HIV-infected patients receiving protease inhibitor therapy (28). Fourteen of these patients had preexisting diabetes, with a loss of glucose control; 27 patients required hospitalization; and 5 patients had ketoacidosis. The average onset was 76 days after initiation of protease inhibitor therapy, but 1 case occurred as early as 4 days after initiation. Cases were reported with all four available protease inhibitors, and updated data from November 1997 consisted of 141 cases with indinavir, 31 with saquinavir, 23 with ritonavir, 24 with the combination of ritonavir and saquinavir, and 15 with nelfinavir (29).

Given the available data, new-onset diabetes in patients receiving protease inhibitors appears to be uncommon, although most reports have been from cross-sectional, rather than longitudinal, studies; however, asymptomatic impaired glucose tolerance does appear to be more common than frank diabetes (30,31). A longitudinal analysis from an observational database at the Johns Hopkins Hospital HIV Clinic reported an incidence of severe hyperglycemia, defined as random glucose level higher than 200 mg/dL, of 0.35/100 person-months [95% confidence interval (CI), 0.01–0.90] (32). However, follow-up was of relatively limited duration in this retrospective cohort, and it is not known if the incidence rate is related to cumulative exposure to drug. Some have suggested that the prevalence of diabetes mellitus among patients receiving protease inhibitors may not even exceed that of the general age-adjusted population (33). Others have reported a higher than expected prevalence of impaired glucose tolerance among HIV-infected patients who were treatment-naive or receiving nucleoside analogue therapy alone (30,34), thereby suggesting that factors other than protease inhibitors may play a role. Hadigan et al. found that elevated fasting insulin and insulin/glucose ratios were associated with increased truncal adiposity, but not protease inhibitor use in a study of 75 HIV-infected women (35).

A causal relation between protease inhibitor use and hyperglycemia has not yet been proved, but there are increasing amounts of data to support such a link. There are anecodotal reports of reversibility of hyperglycemia in some patients after withdrawal of protease inhibitor therapy (36) or change to an alternative protease inhibitor (37). One reported patient had correction of hyperglycemia after withdrawal of protease inhibitor therapy and a recurrence on subsequent rechallenge (38). In a controlled study, Mulligan et al. examined patients initiating treatment with a protease inhibitor or lamivudine and reported increases in glucose and insulin levels in the protease inhibitor group, but not the lamivudine controls (39).

Although a few cases of ketoacidosis have been reported (28,32,40), many patients respond to oral hypoglycemic therapy, suggesting that insulin resistance, rather than deficiency is the underlying problem. Further evidence for insulin resistance comes from reports of elevated fasting insulin and C-peptide levels (39,41) and impaired insulin sensitivity by intravenous insulin tolerance test (42). Insulin-sensitizing agents, such as troglitazone (43) and metformin (44), have been explored as potential therapies for hyperinsulinemia or frank diabetes, but safety data are still limited. Potential safety concerns of these drugs include overlapping toxicities with antiretroviral agents, including hepatotoxicity and lactic acidosis.

B. Hyperlipidemia

Hypertriglyceridemia has long been described as a feature of advanced HIV infection and has been associated with elevated circulating levels of interferon-alpha (45,46). More recently, protease inhibitors as a class, and ritonavir in particular, have been associated with significant hypertriglyceridemia (30,47–49). The clinical significance of hypertriglyceridemia in the setting of protease inhibitor use is not yet clear, although sequelae such as pancreatitis (48,50) and cholelithiasis in the short term, and promotion of atherosclerosis in the longer term, are of concern.

Hypercholesterolemia has also been associated with protease inhibitor use, but the data are still limited owing to the retrospective nature of many of the studies and the general lack of fasting lipid profiles. Increases in cholesterol have been reported after initiation of therapy with all marketed protease inhibitors, and the greatest increases have been seen with ritonavir and the combination of ritonavir and saquinavir (50–52). The prevalence of hyperlipidemia was estimated to be 33% in a retrospective review of 124 patients attending an HIV clinic (53). A high prevalence of lipid abnormalities was also described in a series of women on a protease inhibitor therapeutic regimen, who reported changes in body habitus (54). In the 17 women with available lipid profiles, 76% had low high-density lipoprotein cholesterol (HDL), 47% elevated low-density lipoprotein cholesterol (LDL), 41% elevated total cholesterol, and 35% elevated triglycerides (54).

There are limited data available on the treatment of hyperlipidemia in association with protease inhibitors. Dietary management and exercise were of little benefit in the largest intervention study reported to data (55). Pharmacological treatment has thus far focused on the hydroxymethylglutaryl-coenzyme A (HMG CoA) reductase inhibitors (statin drugs) and fibrates, with modest improvements in lipid levels (55,56). There is theoretical potential for drug interactions between most of the statin drugs and protease inhibitors. No recommendations can be made about coadministration of these drugs until formal pharmacokinetic studies have been completed.

Considerable attention has been given to reports of accelerated atherosclerosis and vascular complications associated with hyperlipidemia in the setting of protease inhibitor use. Henry et al. reported two cases of angina pectoris and angiographically documented coronary artery thromboses in patients 26 and 37 years of age who were receiving protease inhibitor therapy (53). Other cases of myocardial infarction, transient ischemic attacks, and peripheral vascular disease have been reported, although many of the patients had other concomitant risk factors for atherosclerosis (57–61). The temporal association between initiation of protease inhibitor therapy, subsequent increases in lipid levels, and the occurrence of vascular events suggests a potential contributing role of protease inhibitor therapy. Retrospective reviews of clinical and clinical trials databases have yielded conflicting results on whether the incidence of ischemic cardiac events is increased in patients receiving protease inhibitors. One study reported a fivefold increased risk of myocardial infarction in association with protease inhibitor use (60), whereas others found no significant association (62,63).

C. Altered Fat Distribution

1. Clinical Manifestations and Epidemiology

Changes in body fat distribution are among the most striking abnormalities associated with, but not conclusively proved to be caused by, protease inhibitor therapy. Patients appear to have either peripheral fat wasting (lipodystrophy or lipoatrophy), central fat accumulation, or a mixed picture of fat wasting and accumulation. The fat-wasting component consists of thinning of the face, buttocks, arms, or legs with prominence of veins owing to subcutaneous fat loss, whereas the fat accumulation component consists of increased abdominal girth, breast enlargement, or dorsocervical and submandibular fat pad enlargement (41,54,64–74) (Fig. 1).

Researchers have used sophisticated measures of body composition to further characterize the fat redistribution being observed clinically. Studies using dual energy x-ray absorptiometry (DEXA) have confirmed the central accumulation and peripheral depletion of fat (31,41). Cross-sections obtained by computed tomography (CT) scanning at the L4 vertebral level have demonstrated increased ratios of visceral adipose tissue to total adipose tissue (66), and whole-body magnetic resonance imaging (MRI) has confirmed the visceral nature of the abdominal fat accumulation (75) (Fig. 2).

Incidence data for these alterations in fat distribution are still lacking owing to the cross-sectional nature of the initial studies, and estimates of prevalence range widely from 7 to 83% (31,54,67). A major limitation of studies performed

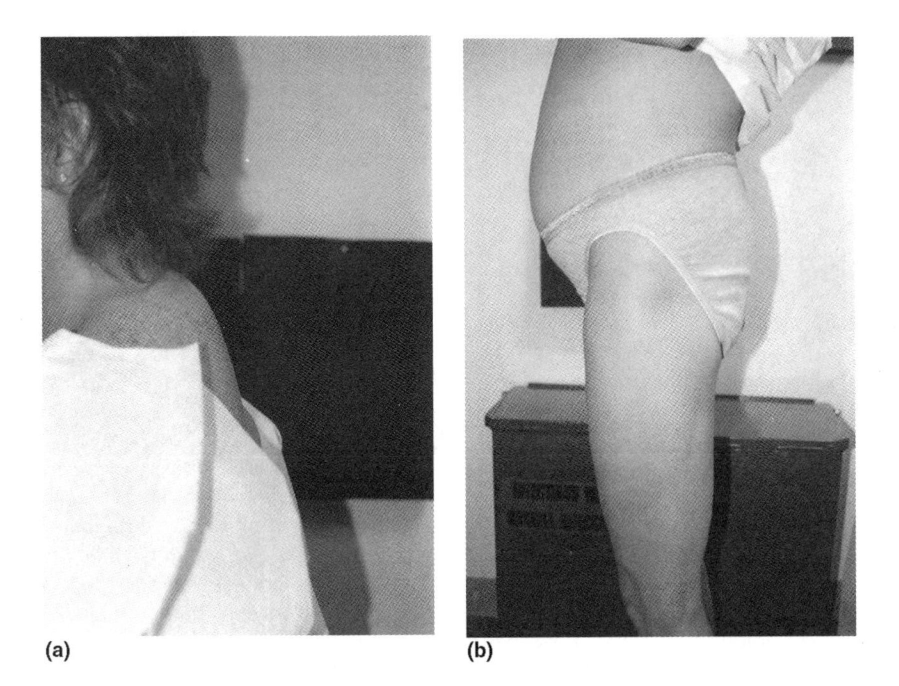

(a) (b)

Figure 1 A 36-year-old HIV-infected woman who developed (a) a prominent dorsocervical fat pad and (b) truncal obesity and flattening of the buttocks while receiving a protease inhibitor–containing antiretroviral regimen.

to date is the lack of a consistent case definition for the syndrome or syndromes. Most studies rely on patient self-report or clinician assessment for diagnosis. Similarly, no consensus has been reached on nomenclature, and terms such as peripheral lipodystrophy, fat redistribution, pseudo-Cushing's syndrome, pseudo-cachexia, truncal obesity or adiposity, multiple symmetrical lipomatosis, and, colloquially, ''protease paunch'' and ''Crix belly'' have been widely used to describe these patients. The terms lipodystrophy syndrome and fat redistribution syndrome have been used to describe the constellation of peripheral fat wasting, central fat accumulation, hyperlipidemia, and insulin resistance or hyperglycemia.

2. Pathogenesis

The pathogenesis of the metabolic derangements described in association with protease inhibitor use is unknown, and whether or not protease inhibitors are the cause of these derangements is unproved. In fact, many case series and cross-

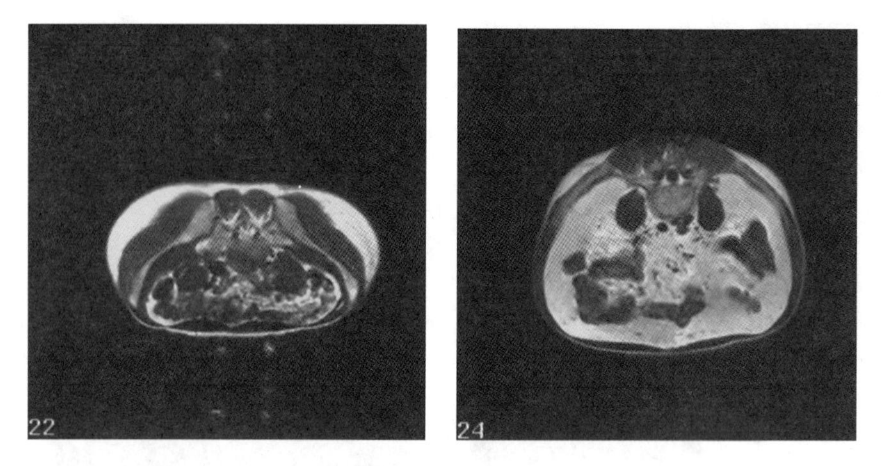

Figure 2 Magnetic resonance imaging cross-sections of the lower abdomen in a 37-year-old HIV-infected man with truncal obesity (right) and an HIV-negative control matched for age, race, sex, height, and weight (left). Fat appears white on these T1-weighted images. Note the thinner band of subcutaneous fat and greatly increased quantity of visceral fat surrounding loops of bowel (which appear black) in the HIV-infected patient compared with the control. (Courtesy of Ellen Engelson, St. Luke's–Roosevelt Hospital Center, New York, New York.)

sectional studies have included patients who developed abnormal fat distribution in the absence of protease inhibitor therapy (41,64,65,76–84). Researchers have put forward a number of hypotheses to explain the pathogenesis of fat redistribution, most of which have focused on the role of protease inhibitors, and all of which are still unproved.

Because some patients have features resembling Cushing's syndrome, researchers have studied the hypothalamic–pituitary–adrenal axis in patients with fat redistribution. No consistent abnormalities have been demonstrated, and hypercortisolism has effectively been ruled out as a cause (85). Hypersensitivity of the hypothalamic–pituitary–adrenal axis has been proposed as an underlying etiology (86), whereas others have postulated that local cortisol metabolism may be affected, noting that the *vpr* gene product of HIV stimulates glucocorticoid receptors on some cells (87).

Carr et al. have published an elegant unifying hypothesis to try to explain the pathogenesis of many of the metabolic abnormalities (88). By searching human protein and genomic databases, they determined that significant homologies exist between the catalytic site of HIV-1 protease and two human proteins involved in lipid metabolism: cytoplasmic retinoic acid-binding protein type 1 (CRABP-1) and low-density lipoprotein–receptor-related protein (LRP). The authors postulate that protease inhibitors bind to these proteins, inhibit their function, and cause adipocyte apoptosis, lipid release from adipocytes, and reduced triglyceride clearance. Protease inhibitor inhibition of cytochrome P450 may also contribute to the altered fat metabolism. The resultant hypertriglyceridemia leads to increased central adiposity, insulin resistance, and diabetes in susceptible persons and, through the influence of estrogen, deposition of breast fat in women. Impaired peripheral fat storage results in peripheral lipodystrophy. Other toxicities attributed to protease inhibitor therapy are potentially explained by the hypothesis, including dry skin and lips, alopecia, and nail pathology, owing to altered retinoic acid metabolism, and increased bleeding in hemophiliacs, owing to inhibition of LRP's uptake of tissue plasminogen activator. In vitro and animal experiments to test aspects of the Carr hypothesis have yielded conflicting results (89–91), and the hypothesis fails to explain the increasing numbers of cases of fat redistribution being described in patients naive to protease inhibitors.

An alternative hypothesis for the pathogenesis of some of the metabolic abnormalities has been proposed by Kotler, based on retrospective analyses of comprehensive body composition assessments done in New York City (86). In these studies, body fat distribution (e.g., visceral versus subcutaneous adiposity) was similar in groups of HIV-infected patients studied before and after the availability of protease inhibitors, but both groups differed significantly from HIV-negative controls. A multivariate analysis of patients in the protease inhibitor era, classified by type of antiretroviral treatment received, found no association between protease inhibitor use and abnormal fat distribution. However, the log viral load near the time of body composition studies was inversely correlated with abnormal fat distribution, suggesting that the degree of viral suppression was the important factor, rather than the class of antiretroviral drugs used (86). The suggestion that the metabolic abnormalities are somehow the body's (mal)adaptation to suppression of chronic HIV replication is a robust theory because it may account for patients who manifest these abnormalities on regimens that do not include a protease inhibitor.

Investigators have also considered the possibility that mitochondrial toxicity from nucleoside analogues may play a role in the development of fat redistribution. Mitochondrial dysfunction has been implicated in the pathogenesis of multiple symmetric lipomatosis, a condition that superficially resembles HIV-associated fat redistribution (92,93). Furthermore, some patients who have developed fat redistribution while receiving nucleoside analogue therapy alone have had mild increases in serum lactate levels suggestive of mitochondrial toxicity (78).

Other unproved hypotheses that directly or indirectly implicate protease inhibitors have been discussed. Cytochrome P450 inhibition by protease inhibitors could theoretically potentiate the effects of glucocorticoids (87). Alternatively, the weight gain seen in some patients achieving viral suppression with protease inhibitor therapy may be primarily in the form of fat, akin to what may occur with refeeding of patients with starvation or anorexia nervosa. Or perhaps protease inhibitors inhibit human enzymes that degrade insulin, thereby resulting in the primary abnormality of hyperinsulinemia, the secondary effects of which result in fat redistribution (94).

From the available data, the pathogenesis of the fascinating metabolic derangements that have been described is likely multifactorial. Ongoing research, including prospective observational studies and clinical trials, should shed light on this important clinical problem.

3. Management of Fat Redistribution

There are several reasons why fat redistribution may be more than just a cosmetic concern for HIV-infected patients. The coexistence of truncal obesity, insulin resistance, and hyperlipidemia resembles a previously described disorder unrelated to HIV infection, termed metabolic syndrome X or visceral fat syndrome (95,96). Because this syndrome is associated with type II diabetes mellitus and accelerated atherosclerosis, there is potential for increased cardiovascular morbidity in afflicted HIV-infected patients. Anecdotally, some patients have suffered functional problems from fat accumulation, such as limited neck motion and sleep apnea. Furthermore, the psychological effect of body shape changes and the implications for adherence to antiretrovirals are not trivial issues.

At this time, no evidence-based recommendations can be made for the management of patients with altered body fat distribution, with or without associated metabolic abnormalities. Periodic monitoring of fasting blood sugars and lipid profiles, with treatment of diabetes and hyperlipidemia if indicated, should probably be routine. Attempts should be made to assess and modify other cardiovascular risk factors, such as smoking and hypertension. Dietary counseling and exercise should also be encouraged.

Investigators have reported reductions in truncal obesity and dorsocervical fat pad size in a small number of patients treated with recombinant human growth hormone (97–99), and systematic study of this drug as treatment for truncal obesity is ongoing. Others have explored the use of insulin-sensitizing agents to treat the insulin resistance aspect of the syndrome. Both metformin (44) and troglitazone (43) have been studied in small numbers of patients, some of whom have appeared to have had beneficial changes in body composition. There is also a report of modest reduction in dorsocervical fat pad size in a patient treated with ketoconazole, a drug that inhibits adrenal steroid production (65). This patient had elevated basal urine free cortisol, but a dexamethasone suppression test that ruled out Cushing's syndrome. Liposuction helped one patient with dorsocervical fat pad enlargement with short-term follow-up (65), but would not be expected to help with truncal obesity, given that the fat accumulation is visceral rather than subcutaneous.

Whether regression of fat redistribution can occur with changing to an alternative protease inhibitor or to a potent protease inhibitor-sparing antiretroviral regimen remains to be seen. Several small studies have been reported in which patients with fat redistribution and HIV viral loads below the level of detection substituted non-nucleoside reverse transcriptase inhibitors (nevirapine or efavirenz) for the protease inhibtor. In general, these studies have shown continued viral suppression in the short-term and variable effects on fat redistribution, with some patients reporting subjective improvements (100–105). Clearly, the risks versus benefits of stopping protease inhibitor therapy must be carefully assessed on a case-by-case basis, because the natural history of fat accumulation is unknown as are the long-term effects of stopping protease inhibitor therapy (106).

D. Hyperuricemia

Hyperuricemia has been reported to be more prevalent in patients on protease inhibitor-containing regimens compared with those receiving dual nucleoside therapy in one study (107). In this report, two patients had episodes of gout while receiving protease inhibitor therapy (107). The relation, if any, between hyperuricemia and other metabolic abnormalities has not yet been delineated, although hyperuricemia may be a component of the metabolic syndrome X in the absence of HIV infection (108).

VIII. CONCLUSIONS

Protease inhibitors are well tolerated by most patients, and many of the toxicities that occur after short-term administration can be managed with symptomatic treatment. Of more concern are the potential longer-term adverse events that have been described in patients taking protease inhibitors. In particular, the occurrence of vascular events in patients with hyperlipidemia in the setting of protease inhibitor use has raised questions about the risk/benefit ratio of these drugs in patients with less-advanced HIV disease who may need to be treated for many years. It is important to note that investigators have not yet conclusively established a causal relation between protease inhibitor use and fat redistribution, and the potential risk of accelerated atherosclerosis with protease inhibitor use remains to be defined. Further research into the pathogenesis of these adverse events will shed light on their etiology. Given that many of the toxicities associated with protease inhibitors, as a class, so far appear to occur in a relatively small proportion of patients, the identification of risk factors through careful epidemiological investigations may help with the weighing of the risks and benefits of protease inhibitor use for individual patients.

REFERENCES

1. Fiegal DW Jr. FDA Public Health Advisory: HIV protease inhibitors and patients with hemophilia. 1996.
2. Ruiz I, Altisent C, Ocana I, Tusell JM, Puig LI, Pahissa A. Indinavir in therapy-experienced HIV haemophiliac patients [abstr]. Sixth European Conference on Clinical Aspects and Treatment of HIV Infection, Hamburg, Germany, Oct 11–15, 1997.
3. Ginsburg C, Salmon–Ceron D, Vassilief D, et al. Unusual occurrence of spontaneous haematomas in three asymptomatic HIV-infected haemophilia patients a few days after the onset of ritonavir treatment. AIDS 1997; 11:388–389.
4. Prazuck T, Semaille C, Roques S. Fatal acute haemolysis in an AIDS patient treated with indinavir [letter]. AIDS 1998; 12:531–533.
5. Crixivan (indinavir sulfate) capsules product monograph. West Point, PA: Merck & Co., 1998 (package insert).
6. Haubrich R. Phase 2 study of amprenavir, a novel protease inhibitor, in combination with zidovudine/3TC [abstr]. 12th World AIDS Conference, Geneva, Switzerland, June 28–July 3, 1998.
7. Winter AJ, Pywell JM, Ilchyshyn A, Fearn J, Natin D. Photosensitivity due to saquinavir [letter]. Genitourin Med 1997; 73:323.
8. d'Arminio Monforte A, Testa L, Gianotto M, et al. Indinavir-related alopecia [letter]. AIDS 1998; 12:328.
9. Bouscarat F, Bouchard C, Bouhour D. Paronychia and pyogenic granuloma of the great toes in patients treated with indinavir [letter]. N Engl J Med 1998; 338:1776–1777.
10. Viracept (nelfinavir mesylate) tablets and oral powder product monograph. La Jolla, CA: Agouron Pharmaceuticals, 1997 (package insert).
11. Norvir (ritonavir) capsules and oral solution product monograph. North Chicago, IL: Abbott Laboratories, 1997 (package insert).
12. Invirase (saquinavir mesylate) capsules product monograph. Nutley, NJ: Roche Laboratories, Hoffmann–La Roche, Inc. 1997 (package insert).
13. Fortovase (saquinavir) soft gelatin capsules product mongraph. Nutley, NJ: Roche Laboratories, Hoffmann–La Roche, Inc., 1997 (package insert).
14. Bräu N, Leaf HL, Wieczorek RL, Margolis DM. Severe hepatitis in three AIDS patients treated with indinavir. Lancet 1997; 349:924–925.
15. Jeurissen FJF, Schneider MME, Borleffs JCC. Is the combination of hepatitis and indinavir potentially dangerous? [letter]. AIDS 1998; 12:441–442.
16. Vento S, Garofano T, Renzini C, Casali F, Ferraro T, Concia E. Enhancement of hepatitis C virus replication and liver damage in HIV-coinfected patients on antiretroviral therapy [letter]. AIDS 1998; 12:116–117.

17. Carr A, Cooper DA. Restoration of immunity to chronic hepatitis B infection in HIV-infected patient on protease inhibitor. Lancet 1997; 349:995–996.
18. Daudon M, Estepa L, Viard JP, Joly D, Jungers P. Urinary stones in HIV-1 positive patients treated with indinavir. Lancet 1997; 349:1294–1295.
19. Gerber J, Johnson S. Evaluation of urinary pH and specific gravity in the development of indinavir induced renal stones [abstr]. 12th World AIDS Conference, Geneva, Switzerland, June 28–July 3, 1998.
20. Bach MC, Godofsky EW. Indinavir nephrolithiasis in warm climates [letter]. J Acquir Immune Defic Syndr 1997; 14:296–297.
21. Kopp JB, Miller KD, Mican JAM, et al. Crystalluria and urinary tract abnormalities associated with indinavir. Ann Intern Med 1997; 127:119–125.
22. Tashima KT, Horowitz JD, Rosen S. Indinavir nephropathy [letter]. N Eng J Med 1997; 336:138–140.
23. Chen SCA, Nankivell BJ, Dwyer DE. Indinavir-induced renal failure [letter]. AIDS 1998; 12:440–441.
24. Boubaker K, Bally F, Vogel G, Meuwly JY, Glauser MP, Telenti A. Change in renal function associated with indinavir [abstr]. 12th World AIDS Conference, Geneva, Switzerland, June 28–July 3, 1998.
25. Chugh S, Bird R, Alexander EA. Ritonavir and renal failure [letter]. N Engl J Med 1997; 336:138.
26. Duong M, Sgro C, Grappin M, Biron F, Boibieux A. Renal failure after treatment with ritonavir [letter]. Lancet 1996; 348:693–694.
27. Witzke O, Plentz A, Schäfers RF, Reinhardt W, Heemann U, Philipp T. Side-effects of ritonavir and its combination with saquinavir with special regard to renal function [letter]. AIDS 1997; 11:836–838.
28. Lumpkin MM. FDA Public Health Advisory: Reports of diabetes and hyperglycemia in patients receiving protease inhibitors for the treatment of human immunodeficiency virus (HIV). 1997.
29. Dubé MP, Sattler FR. Metabolic complications of antiretroviral therapies. AIDS Clin Care 1998; 10:41–44.
30. Behrens G, Dejam A, Schmidt H, et al. Impaired glucose tolerance, beta cell function and lipid metabolism in HIV patients under treatment with protease inhibitors. AIDS 1999; 13:F63–F70.
31. Carr A, Samaras K, Thorisdottir A, Kaufmann GR, Chisholm DJ, Cooper DA. Diagnosis, prediction, and natural course of HIV-1 protease-inhibitor-associated lipodystrophy, hyperlipidaemia, and diabetes mellitus: a cohort study. Lancet 1999; 353:2093–2099.
32. Keruly JC, Chaisson RE, Moore RD. Diabetes and hyperglycemia in patients receiving protease inhibitors [abstr]. 5th Conference on Retroviruses and Opportunistic Infections, Chicago, IL, Feb 1–5, 1998.
33. Fisher A, Stenzel M. Increased prevalence of diabetes mellitus in patients with HIV infection [abstr]. 12th World AIDS Conference, Geneva, Switzerland, June 28–July 3, 1998.
34. Glesby M, Desai A, Erbelding E, Kostman J. High prevalence of impaired glucose tolerance prior to initiating protease inhibitor therapy [abstr]. 6th Conference on Retroviruses and Opportunistic Infections, Chicago, IL, Jan 31–Feb 4, 1999.

35. Hadigan C, Miller K, Corcoran C, Anderson E, Basgoz N, Grinspoon S. Fasting hyperinsulinemia and changes in regional body composition in human immunodeficiency virus-infected women. J Clin Endocrinol Metab 1999; 84:1932–1937.
36. Dubé MP, Johnson DL, Currier JS, Leedom JM. Protease inhibitor-associated hyperglycemia. Lancet 1997; 350:713–714.
37. Dubé MP, Johnson DLJ, Currier JSC, Leedom JML. Protease inhibitor-associated hyperglycemia: results of switching from indinavir to nelfinavir [abstr]. 12th World AIDS Conference, Geneva, Switzerland, June 28–July 3, 1998.
38. Eastone JA, Decker CF. New-onset diabetes mellitus associated with use of protease inhibitor [letter]. Ann Intern Med 1997; 127:948.
39. Mulligan K, Tai VW, Algren H, Chernoff DN, Lo JC, Schambelan M. Evidence of unique metabolic effects of protease inhibitors [abstr]. 5th Conference on Retroviruses and Opportunistic Infections, Chicago, IL, Feb 1–5, 1998.
40. Besson C, Jubault V, Viard JP, Pialoux G. Ketoacidosis associated with protease inhibitor therapy [letter]. AIDS 1998; 12:1400–1401.
41. Carr A, Samaras K, Burton S, et al. A syndrome of peripheral lipodystrophy, hyperlipidaemia and insulin resistance in patients receiving HIV protease inhibitors. Lancet 1998; 12:F51–F58.
42. Walli R, Herfort O, Michl GM, et al. Treatment with protease inhibitors associated with peripheral insulin resistance and impaired oral glucose tolerance in HIV-1-infected patients. AIDS 1998; 12:F167–F173.
43. Walli RK, Michl GM, Bogner JR, Goebel FD. Effects of the PPAR-activator troglitazone on protease inhibitor associated peripheral insulin resistance [abstr]. 6th Conference on Retroviruses and Opportunistic Infections, Chicago, IL, Jan 31–Feb 4, 1999.
44. Saint–Marc T, Touraine JL. Effects of metformin on insulin resistance and central adiposity in patients receiving effective protease inhibitor (PI) therapy [abstr]. 6th Conference on Retroviruses and Opportunistic Infections, Chicago, IL, Jan 31–Feb 4, 1999.
45. Grunfeld C, Kotler DP, Hamadeh R, Tierney A, Wang J, Pierson RN Jr. Hypertriglyceridemia in the acquired immunodeficiency syndrome. Am J Med 1989; 86: 27–31.
46. Grunfeld C, Kotler DP, Shigenaga JK, et al. Circulating interferon-a levels and hypertriglyceridemia in the acquired immunodeficiency syndrome. Am J Med 1991; 90:154–162.
47. Cameron DW, Heath–Chiozzi M, Danner S, et al. Randomised placebo-controlled trial of ritonavir in advanced HIV-1 disease. Lancet 1998; 351:543–549.
48. Sullivan AK, Feher MD, Nelson MR, Gazzard BG. Marked hypertriglyceridemia associated with ritonavir therapy [letter]. AIDS 1998; 12:1393–1394.
49. Sullivan AK, Nelson MR. Marked hyperlipidaemia on ritonavir therapy [letter]. AIDS 1997; 11:938–939.
50. Henry K. Lipid abnormalities associated with use of protease inhibitors: prevalence, clinical sequelae and treatment [abstr]. 12th World AIDS Conference, Geneva, Switzerland, June 28–July 3, 1998.
51. Chang E, Deleo M, Liu YT, Tetreault D, Beall G. The effects of antiretroviral

protease inhibitors (PIs) on serum lipids and glucose in HIV-infected patients [abstr]. 12th World AIDS Conference, Geneva, Switzerland, June 28–July 3, 1998.
52. Pollner J, Aronson NE, McHugh S, Nielson R, Hawkes C. Significant increases in serum cholesterol are seen among HIV patients taking protease inhibitors [abstr]. 12th World AIDS Conference, Geneva, Switzerland, June 28–July 3, 1998.
53. Henry K, Melroe H, Huebsch J, et al. Severe premature coronary artery disease with protease inhibitors. Lancet 1998; 351:1328–1328.
54. Dong K, Flynn MM, Dickinson BP, et al. Changes in body habitus in HIV+ women after initiation of protease inhibitor therapy [abstr]. 12th World AIDS Conference, Geneva, Switzerland, June 28–July 3, 1998.
55. Henry K, Melroe H, Huebsch J, Hermundson J, Simpson J. Atorvastatin and gemfibrozil for protease-inhibitor-related lipid abnormalities. Lancet 1998; 352:1031–1032.
56. Hewitt RG, Shelton MJ, Esch LD. Gemfibrozil effectively lowers protease inhibitor-associated hypertriglyceridemia in HIV-1-positive patients [letter]. AIDS 1999; 13:868–869.
57. Behrens G, Schmidt H, Meyer D, Stoll M, Schmidt RE. Vascular complications associated with use of HIV protease inhibitors [letter]. Lancet 1998; 352:1958.
58. Gallet B, Pulik M, Genet P, Chedin P, Hiltgen M. Vascular complications associated with use of HIV protease inhibitors [letter]. Lancet 1998; 351:1958–1959.
59. Vittecoq D, Escault L, Monsuez JJ. Vascular complications associated with use of HIV protease inhibitors [letter]. Lancet 1998; 351:1959.
60. Jütte A, Schwenk A, Franzen C, et al. Increasing morbidity from myocardial infarction during HIV protease inhibitor treatment? [letter]. AIDS 1999; 13:1796–1797.
61. Sullivan AK, Nelson MR, Moyle GJ, Newell AM, Feher MD, Gazzard BG. Coronary artery disease occurring with protease inhibitor therapy. Int J STD AIDS 1998; 9:711–712.
62. Klein D, Sidney S, Hurley L, Iribarren C. Do protease inhibitors increase the risk for coronary heart disease among HIV positive patients? [abstr]. 6th Conference on Retroviruses and Opportunistic Infections, Chicago, IL, Jan 31–Feb 4, 1999.
63. Coplan P, Nikas A, Saah A, et al. No association observed between indinavir therapy for HIV/AIDS and myocardial infarction in 4 clinical trials with 2,825 subjects [abstr]. 6th Conference on Retroviruses and Opportunistic Infections, Chicago, IL, Jan 31–Feb 4, 1999.
64. Lo JC, Mulligan K, Tai VW, Algren H, Schambelan M. "Buffalo hump" in men with HIV-1 infection. Lancet 1998; 351:867–870.
65. Miller KK, Daly PA, Sentochnik D, et al. Pseudo-Cushings's syndrome in human immunodeficiency virus-infected patients. Clin Infect Dis 1998; 27:68–72.
66. Miller KD, Jones E, Yanovski JA, Shankar R, Feuerstein I, Falloon J. Visceral abdominal-fat accumulation associated with use of indinavir. Lancet 1998; 351: 871–875.
67. Rosenberg HE, Mulder J, Sepkowitz KA, Giordano MF. "Protease-paunch" in HIV+ persons receiving protease inhibitor therapy: incidence, risks, and endocrinologic evaluation [abstr]. 5th Conference on Retroviruses and Opportunistic Infections, Chicago, IL, Feb 1–5, 1998.

68. Striker R, Conlin D, Marx M, Wiviott L. Localized adipose tissue hypertrophy in patients receiving human immunodeficiency virus protease inhibitors. Clin Infect Dis 1998; 27:218–220.
69. Viraben R, Aquilina C. Indinavir-associated lipodystrophy. AIDS 1998; 12:F37–F39.
70. Herry I, Bernard L, de Truchis P, Perronne C. Hypertrophy of the breasts in a patient treated with indinavir. Clin Infect Dis 1997; 25:937–938.
71. Roth VR, Kravcik S, Angel JB. Development of cervical fat pads following therapy with human immunodeficiency virus type 1 protease inhibitors. Clin Infect Dis 1998; 27:65–67.
72. Hengel RL, Geary JAM, Vuchetich MA, et al. Multiple symmetrical lipomatosis associated with protease inhibitor therapy [abstr]. 5th Conference on Retroviruses and Opportunistic Infections, Chicago, IL, Feb 1–5, 1998.
73. Mann M, Piazza–Hepp T, Koller E, Gibert C. Abnormal fat distribution in AIDS patients following protease inhibitor therapy: FDA summary [abstr]. 5th Conference on Retroviruses and Opportunistic Infections, Chicago, IL, Feb 1–5, 1998.
74. Ruane PJ. Atypical accumulations of fatty tissue [abstr]. 37th Interscience Conference on Antimicrobial Agents and Chemotherapy, Toronto, Ontario, Sept 28–Oct 1, 1997.
75. Engelson ES, Kotler DP, Tan Y, et al. Fat distribution in HIV-infected patients reporting truncal enlargement quantified by whole-body magnetic resonance imaging. Am J Clin Nutr 1999; 69:162–169.
76. Gervasconi C, Ridolfo AL, Trifirò G, et al. Redistribution of body fat in HIV-infected women undergoing combined antiretroviral therapy. AIDS 1999;13:465–471.
77. Madge S, Kinlock-de-Loes S, Mercey D, Johnson MA, Weller IVD. Lipodystrophy in patients naive to HIV protease inhibitors [letter]. AIDS 1999; 13:735–737.
78. Carr A, Miller J, Law M, Cooper DA. A syndrome of lipodsytrophy (LD), lactic acidaemia and liver dysfunction associated with HIV nucleoside analogue reverse transcriptase therapy: contribution to PI-related LD syndrome [abstr]. Antiviral Ther 1999; 4(suppl 2):19.
79. Mercié P, Malvy D, Daucourt V, et al. Case report of lipodystrophy observations in patients naive of protease inhibitor treatment, Aquitaine cohort, 1999 [abstr]. Antiviral Ther 1999; 4(suppl 2):27.
80. Mallal S, John M, Moore C, James I, McKinnon E. Protease inhibitors and nucleoside analogue reverse transcriptase inhibitors interact to cause subcutaneous fat wasting in patients with HIV infection [abstr]. Antiviral Ther 1999; 4(suppl 2):28–29.
81. Galli M, Ridolfo AL, Gervasconi C, Ravasio L, Adorni F, Moroni M. Incidence of fat tissue abnormalities in protease inhibitor-naive patients treated with NRTI combinations [abstr]. Antiviral Ther 1999; 4(suppl 2):29.
82. Saint–Marc T, Touraine JL. Reversibility of peripheral fat wasting (lipoatrophy) on stopping stavudine therapy [abstr]. Antiviral Ther 1999; 4(suppl 2):31.
83. Moyle G, Dent N, Baldwin C. Body fat redistribution in persons on non-PI-containing regimens [abstr]. Antiviral Ther 1999; 4(suppl 2):47–48.
84. Saint–Marc T, Partisani M, Poizot–Martin I, et al. A syndrome of peripheral fat

wasting (lipodystrophy) in patients receiving long-term nucleoside analogue therapy. AIDS 1999; 13:1659–1667.
85. Yanovski JA, Miller KD, Kino T, et al. Endocrine and metabolic evaluation of human immunodeficiency virus-infected patients with evidence of protease inhibitor-associated lipodystrophy. J Clin Endocrinol Metab 1999; 84:1925–1931.
86. Kotler DP, Rosenbaum K, Wang J, Pierson RN. Studies of body composition and fat distribution in HIV-infected and control subjects. J Acquir Immune Defic Syndr 1999; 20:228–237.
87. Hirsch MS, Klibanski A. [Editorial response] What price progress? Pseudo-Cushing's syndrome associated with antiretroviral therapy in patients with human immunodeficiency virus infection. Clin Infect Dis 1998; 27:73–75.
88. Carr A, Samaras K, Chisholm DJ, Cooper DA: Pathogenesis of HIV-1-protease inhibitor-associated peripheral lipodystrophy, hyperlipidaemia, and insulin resistance. Lancet 1998; 352:1881–1883.
89. Lenhard J, Furfine E, Paulik M, Croom D, Spaltenstein A, Weiel J. Effect of HIV protease inhibitors on in vitro adipogenesis and in vivo fat deposition [abstr]. Antiviral Ther 1999; 4(suppl 2):3.
90. Stevens GJ, Chen M, Grecko R, et al. Investigations into the proposed mechanisms of HIV-associated peripheral lipodystrophy, hyperlipidaemia and insulin resistance [abstr]. Antiviral Ther 1999; 4(suppl 2):34.
91. Gagon A, Angel JB, Sorisky A. Protease inhibitors and adipocyte differentiation in cell culture. Lancet 1998; 352:1032.
92. Brinkman K, Smeitink JA, Romijn JA, Reiss P. Mitochondrial toxicity induced by nucleoside-analogue reverse-transcriptase inhibitors is a key factor in the pathogenesis of antiretroviral-therapy-related lipodystrophy. Lancet 1999; 354:1112–1115.
93. Kakuda TN, Brundage RC, Anderson PL, Fletcher CV. NRTI-induced mitochondrial toxicity as an aetiology for fat redistribution syndrome [abstr]. Antiviral Ther 1999; 4(suppl 2):41.
94. Martinez E, Casamitjana R, Conget I, Gatell JM. Protease inhibitor-associated hyperinsulinemia [letter]. AIDS 1998; 12:2077–2079.
95. Björntorp P. Abdominal obesity and the development of noninsulin-dependent diabetes mellitus. Diabetes Metab Rev 1998; 4:615–622.
96. Matsuzawa Y, Shimomura I, Nakamura T, Keno Y, Tokunaga K. Pathophysiology and pathogenesis of visceral fat obesity. Ann NY Acad Sci 1995; 748:399–406.
97. Torres R, Unger K. Treatment of dorsocervical fat pads and truncal adiposity with Serostim (recombinant human growth hormone) in patients with AIDS maintained on HAART [abstr]. 12th World AIDS Conference, Geneva, Switzerland, June 28–July 3, 1998.
98. Mauss S, Wolf E, Jaeger H. Reversal of protease inhibitor-related visceral abdominal fat accumulation with recombinant human growth hormone [letter]. Ann Intern Med 1999; 131:313–314.
99. Engelson ES, Glesby M, Sheikhan J, et al. Effect of recombinant human growth hormone in the treatment of visceral fat accumulation in HIV infection: interim analysis [abstr]. Antiviral Ther 1999; 4(suppl 2):11.
100. Martinez E, Conget I, Lozano L, Casamitjana R, Gatell JM. Reversion of metabolic

abnormalities after switching from HIV-1 protease inhibitors to nevirapine. AIDS 1999; 13:805–810.

101. Ruiz L, Negredo E, Bonjoch A, et al. A multicentre, randomized open-label, comparative trial of the clinical, immunological and virological benefit of switching the PI by nevirapine in HAART-experienced patients suffering lipodystrophy [abstr]. Antiviral Ther 1999; 4(suppl 2):35–36.
102. Harris M, Larsen G, Bell M, Montaner JSG. Replacing a PI with efavirenz for management of PI toxicity [abstr]. Antiviral Ther 1999; 4(suppl 2):46–47.
103. Moyle GJ, Baldwin C, Comitis S, Dent N, Gazzard BG. Changes in visceral adipose tissue and blood lipids in persons reporting fat redistribution syndrome switched from PI therapy to efavirenz [abstr]. Antiviral Ther 1999; 4(suppl 2):48.
104. Rozenbaum W, Adda N, Nguyen T, et al. Prospective follow-up of a PI substitution for efavirenz in patients with HIV-related lipodystrophy syndrome [abstr]. Antiviral Ther 1999; 4(suppl 2):55.
105. Carr A, Thorisdottir A, Samaras K, Kaufmann GR, Chisholm DJ, Cooper DA. Reversibility of protease inhibitor (PI) lipodystrophy syndrome on stopping PIs or switching to nelfinavir [abstr]. 6th Conference on Retroviruses and Opportunistic Infections, Chicago, IL, Jan 31–Feb 4, 1999.
106. Lipsky JJ. Abnormal fat accumulation in patients with HIV-1 infection [editorial]. Lancet 1998; 351:847–848.
107. Bernasconi E, Carota A, Magenta M, Pons M, Russotti M, Moccetti T. Metabolic changes in HIV-infected patients treated with protease inhibitors [abstr]. 12th World AIDS Conference, Geneva, Switzerland, June 28–July 3, 1998.
108. Zavaroni I, Mazza S, Fantuzzi M, et al. Changes in insulin and lipid metabolism in males with asymptomatic hyperuricaemia. J Intern Med 1993; 234:25–30.
109. Flexner C: HIV-protease inhibitors. N Engl J Med 1998; 338:1281–1292.
110. Goodgame J, Hanson C, Vafidis I, Stein A, Jablonowski H. Amprenavir (141W94, APV)/3TC/ZDV exerts durable antiviral activity in HIV-1-infected antiretroviral therapy-naive subjects through 48 weeks of therapy [abstr]. 39th Interscience Conference on Antimicrobial Agents and Chemotherapy, San Francisco, CA, Sept 26–29, 1999.

12

Resistance to HIV-1 Protease Inhibitors

Benjamin Young
Rose Medical Center and University of Colorado Health Sciences Center, Denver, Colorado

Daniel R. Kuritzkes
University of Colorado Health Sciences Center, Denver, Colorado

I. INTRODUCTION

Human immunodeficiency virus (HIV)-1 protease (PR) is a viral-encoded enzyme that is essential for the maturation of viral particles to yield infectious virions. The enzyme functions to cleave the Gag–Pol polyprotein precursor to its constituent proteins, including reverse transcriptase and integrase. The Gag–Pol substrates are bound to the protease in the substrate-binding site, and are thought to be held in position by flexible domains, called ''flaps.'' There are multiple cleavage sites in the Gag and Gag–Pol polyprotein precursor; each site differs somewhat in amino acid sequence and rates of chemical reaction with the protease. Given protease's central role in the life cycle of HIV-1, compounds that inhibit the enzyme became the subject of intensive structure-based research efforts. These efforts yielded the several protease inhibitors that are now approved for clinical use in HIV-1-infected individuals. These compounds are among the most potent antiretroviral agents available, and their use in combination therapy has produced dramatic virological and clinical results (1–5).

II. RESISTANCE TO PROTEASE INHIBITORS: GENERAL PRINCIPLES

Rapid HIV-1 viral replication and high levels of mutation permit shifts in the viral population in response to selective pressures, such as immune surveillance or drug therapy. If viral replication is not completely suppressed, drug-resistant variants are likely to emerge.

Analysis of the PR gene sequences from protease inhibitor-resistant viral isolates reveals a set of amino acid substitutions scattered throughout the length of the protein. Of the 99 amino acids that constitute HIV-1 protease, 42 substitutions, occurring at 27 sites, have been associated with protease inhibitor resistance (Fig. 1) (6). Many mutations map to the substrate-binding domain of the enzyme (e.g., codons 82 and 84) (7,8). Indeed, substitution at residues surrounding the binding site were postulated to play a role in protease inhibitor resistance even before the identification of protease–inhibitor-resistant virus (9). Mutations outside of the substrate-binding site, such as those in the flap region (e.g., codons 45 through 56), can also contribute to drug resistance (10–12), as do mutations in the Gag–Pol substrate (13).

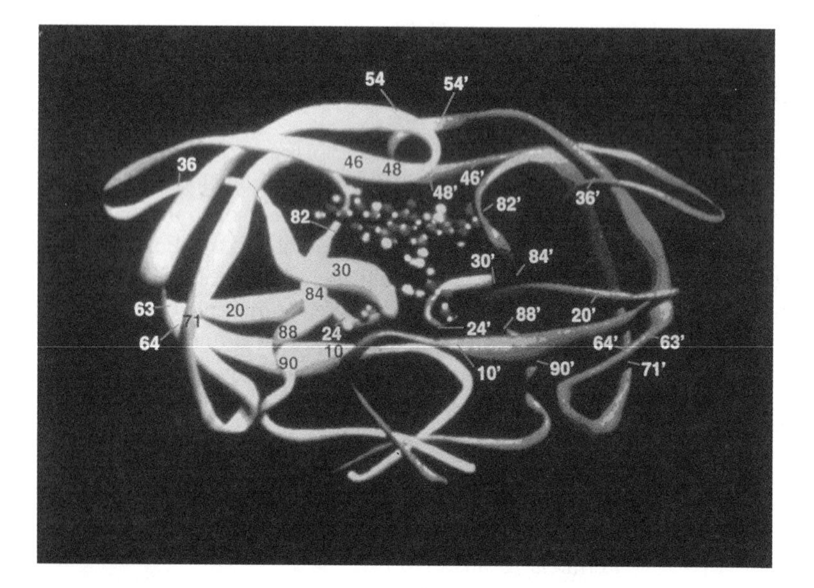

Figure 1 Ribbon diagram of HIV protease. The molecule is a homodimer, represented by ribbons. Numbers indicate the locations of key amino acid residues. A model protease inhibitor is positioned in the active site.

III. MECHANISMS OF RESISTANCE

An understanding of enzyme kinetics is essential for understanding the possible mechanisms of resistance to protease inhibitors. Several detailed kinetic models for analysis of protease activity and resistance to inhibitors have been presented (14–16). A key concept in resistance is that of enzyme specificity, the relation of the relative rate of enzyme (e.g., HIV protease) interaction with the inhibitor (e.g., protease inhibitor) and the natural substrate (e.g., Gag–Pol precursor). This is accomplished by comparing the rate or velocity (v) of the interaction of the enzyme with both the inhibitor and the substrate. Note that the velocity is dependent on the way that enzyme interacts with free, or nonbound ligand (represented by the term k_{cat}/K_m) and the concentration of the ligand. Consequently, the specificity of inhibitor and substrate in protease reactions can be related as follows:

$$\frac{v_{inhibitor}}{v_{substrate}} = \frac{(k_{cat}/K_m)_{inhibitor} \times [\text{inhibitor}]}{(k_{cat}/K_m \times [\text{substrate}]} \quad (1)$$

where v = velocity, k_{cat} is the catalytic constant, K_m the Michaelis constant, is the apparent dissociation constant and k_{cat}/K_m, the specificity constant, is the apparent second-order rate constant that relates to the reaction of free substrate and free enzyme (17). The rate of reaction with protease is related to the ratio of the interactions with the natural substrate and the protease inhibitor. The overall reaction rate is determined by the relative ratio of k_{cat}/K_m for substrate and inhibitor and not just K_m. Hence, in the presence of protease inhibitors, the rate of the reaction of the Gag–Pol precursor with HIV protease is dictated by the ratio of k_{cat}/K_m of each substrate cleavage site to the protease inhibitor.

$$\begin{array}{rcl} \text{Enzyme + substrate} & \xrightarrow{(k_{on})} & \text{enzyme–substrate complex} \\ \text{(or inhibitor)} & \xleftarrow[(k_{off})]{} & \text{(or enzyme–inhibitor complex)} \end{array} \quad (2)$$

$$K_m = \frac{k_{on}}{k_{off}} = \frac{\text{association rate}}{\text{dissociation rate}} \quad (3)$$

K_m, the apparent dissociation constant of the enzyme–substrate complex, is the composite of the second-order association rate k_{on} and the dissociation rate k_{off}. Factors or mutations that increase the dissociation rate for inhibitor have the effect of decreasing the level of inhibition. Conversely, if the dissociation rate for natural substrate is increased, the enzyme is less susceptible to inhibition.

Factors that decrease $v_{inhibitor}/v_{substrate}$ [see Eq. (1)] favor the native reaction with substrate, whereas factors that increase this ratio increase the degree of enzyme inhibition. Changes that alter any of these parameters result in enzymes with altered susceptibility to the inhibitor. The mechanistic consequences of possible changes are discussed in the following.

A. Reduced Inhibitor Binding

When the binding interaction between the enzyme and the inhibitor is impaired, higher concentrations of inhibitor will be required to achieve the same level of enzyme inhibition. This mechanism of resistance may be inhibitor-specific. However, to the extent that the chemical backbone of currently available protease inhibitors is similar, some degree of cross-resistance is anticipated. Biochemical analysis of purified HIV-1 protease from variant viruses harboring mutations in the substrate-binding site at Val-82 show a 2- to 25-fold reduction in k_{cat}/K_m for both saquinavir and amprenavir (18). Crystallographic analysis of HIV-1 protease with a codon-82 mutation shows that structural alterations remove a water molecule from the catalytic site (19). This water molecule is thought to play an important role in stabilizing the enzyme–inhibitor interaction. Elimination of the water molecule has the effect of decreasing the free energy of inhibitor binding to the enzyme. A similar study of G48V/L90M mutant protease showed dramatic increases in the dissociation constant for saquinavir, presumably mediated through effects at position 90 (20).

B. Improved Enzyme Activity

Mutations that improve the catalytic efficiency of protease could also be associated with relative resistance to inhibitors. These could result in the improvement of the relative binding affinity of substrate to inhibitor [increases in K_m(substrate)/K_m(inhibitor)], increases in catalytic chemistry or turnover, or increases in the steady-state level of enzyme (either by increasing enzyme synthesis or stability). This class of mutation is generally inhibitor-nonspecific and may also occur as secondary mutations that improve the enzymatic ''performance'' (k_{cat}) of the protease with primary mutations that have other mechanistic effects.

Previous studies have shown that evolutionary refinements of catalytic function can result from mutations that occur away from the active site of the enzyme (21). An interpretation of these data suggests that evolution may be facilitated by the ability of a protein to more readily accommodate conformational rearrangements (22). It is significant that crystallographic structural investigations of inhibitor–protease cocrystals reveal that conformational rearrangements in the backbone structures of both inhibitor and protease may account for some of the effects of secondary mutations in the enzyme (20,23,24).

C. Reduced Drug Concentration

Factors, such as suboptimal drug dose or poor bioavailability of a protease inhibitor, may have the effect of decreasing the effective concentration of the inhibitor. The consequence of decreased inhibitor concentration on Eq. (1) is to decrease the overall ratio, because the numerator is decreased. The effect of lower protease inhibitor doses on hastening the emergence of drug-resistant virus has been observed in several studies (25–27).

D. Substrate Alterations

HIV-1 protease cleaves the Gag–Pol polyprotein precursor at seven or more different sites (Fig. 2) (28). Cleavage of the Gag polyprotein results in the production of the matrix (p17), capsid (p24), nucleocapsid (p7), and several smaller proteins (p1, p2, p6), whereas cleavage of the Gag–Pol polyprotein results in the production of essential viral enzymes, including protease, reverse transcriptase, and integrase.

Alterations in the Gag–Pol substrate may act to increase the $(k_{cat}/K_m)_{substrate}$ in the context of other primary drug-resistance mutations. This form of mutation would have the effect of decreasing the overall value of $[v_{inhibitor}/v_{substrate}]$, because the denominator is increased. A detailed kinetic analysis of cleavage of the Gag–Pol polyprotein has been presented (29). These relations show how changes in the way the protease interacts with the native substrate can contribute to resistance

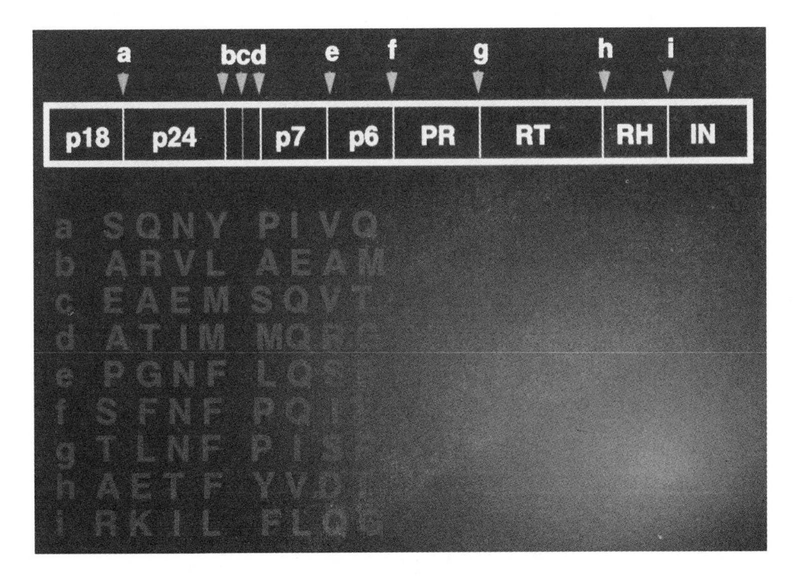

Figure 2 HIV polyprotein cleavage sites: HIV polyprotein is denoted by box. Individual component proteins are indicated: PR, protease; RT, reverse transcriptase; RH, RNaseH; IN, integrase. Lower-case letters indicate cleavage sites. Amino acid sequences adjacent to the cleavage sites are shown below.

to the inhibitor. Kinetic studies of purified protease, which contains mutations at codon 82, with a panel of substrate analogues have been performed. Results show that the K_i for inhibitor depends on both the amino acid sequence (i.e., structure) of protease, as well as the substrate (30).

In vitro studies of the protease inhibitors BILA 1906 and BILA 2185 found mutations at the p1/p6 and p7/p1 cleavage sites in addition to mutations within the PR gene of resistant viruses (31). By contrast, cleavage site mutations were not observed during in vitro selection for saquinavir resistance (32). Variant viruses with altered cleavage sites were impaired in their growth characteristics, implying an important, rate-determining function of polyprotein processing in viral growth.

gag–pol Mutations were also observed after in vitro selection with lopinavir (ABT-378) (33,34). In this study mutations were invariably seen at the p1/p6 site after 11 rounds of passage. Mutations at the p7/p1 site occurred after additional in vitro passage. The presence of these mutations allowed the replication of virus that contained several mutations in the protease gene.

Recent clinical studies provide additional evidence for a potential role of mutations in the Gag–*pol* precursor in clinical resistance to protease inhibitors. In a study of six patients failing indinavir-containing therapy, mutations were observed not only in the PR gene, but also a mutation in the *gag* p7/p1 cleavage site was observed (13). This mutation was found as early as 6 weeks after the initiation of protease inhibitor therapy. Reconstruction experiments showed that presence of the p7/p1 mutation partially compensated for the effects on viral replication on mutations at protease codons 46, 54, or 82. A recent study also found a higher proportion of viral isolates from patients lacking response to indinavir-containing combination antiretroviral therapy had genetic mutations at the p7/p1 cleavage site (35).

Mutations at the p1/p6F cleavage site were observed in patients failing combination therapy that included amprenavir (36). Resistant isolates selected in vitro by passage in the presence of amprenavir also have mutations in the p1/p6 and p17/p24 polyprotein cleavage sites (37). In the latter study, the p1/p6 mutation arose in conjunction with protease gene mutations at codons 10 and 84, whereas the p17/p24 mutation appeared in the presence of mutations at codons 10, 46, 47, and 50.

Together, these data illustrate the capacity of HIV-1 to adapt to selective pressure posed by drug therapy and the importance of both the target enzyme and substrate as targets of evolution.

IV. BASELINE POLYMORPHISMS

Naturally occurring polymorphisms are found throughout the genome of HIV-1. Polymorphisms at many positions in the protease gene are common among

HIV-1 isolates from protease inhibitor-naive patients. In one study, 47.5% of PR codons were variable when compared with the Clade B consensus sequence (38). Seven mutations previously associated with resistance to protease inhibitors were identified in the study population; viral sequence from 26% of subjects had at least two of these polymorphisms. Another analysis of 246 PR gene sequences obtained from 12 protease inhibitor-naive individuals confirmed the high degree of polymorphism in this gene (39).

Sequence analysis of viruses isolated from 28 patients with primary HIV-1 infection in Italy showed a high degree of genetic polymorphism in PR, with only 14% of isolates having sequences that were identical with the Clade B consensus sequence (40). The most commonly observed substitution occurred at codon 63 (17/28).

Naturally occurring sequence polymorphisms have also been described for protease cleavage sites within the Gag–Pol precursor (28,41). Five of six cleavage sites were highly conserved, whereas the p2/p7 cleavage site was variable. Only 19% of isolates had the consensus sequence. Sequence variability at the p7/p1 and p6/p1 sites was not observed.

Data on the clinical significance of these polymorphisms is still accumulating. Nevertheless, a reasonable hypothesis is that some key mutations might be responsible for the early emergence of drug-resistant virus and clinical failure, whereas other substitutions are functionally synonymous with those of the wild-type.

V. RESISTANCE TO INDIVIDUAL AGENTS

Extensive testing in vitro as well as in clinical trials has revealed a pattern of mutations in PR that confer resistance to the drug. Some of these mutations appear early in the emergence of drug resistance and are important determinants of resistance to a specific agent. Such mutations have been called "primary" mutations. Many of these mutations map to the substrate-binding site of the enzyme. Other mutations occur later in the selection process and are referred to as "secondary" mutations. The presence of these mutations is often shared among the mutational patterns associated with resistance to multiple inhibitors. Mutations of this class frequently map to the flap domain or to other sites away from the substrate-binding site.

Initial data on the genetic determinants of protease inhibitor resistance were based on in vitro studies that selected the ability of HIV-1 to replicate in the presence of the drug. In general, the mutational patterns that emerged from in vitro experimentation have been corroborated by analysis of viral isolates obtained from patients failing monotherapy trials of protease inhibitors (Fig. 3).

Although only a limited number of isolates from patients in clinical trials of combination antiviral have yet been analyzed, an accumulating body of evidence suggests that genetic patterns that underlie protease inhibitor resistance will be

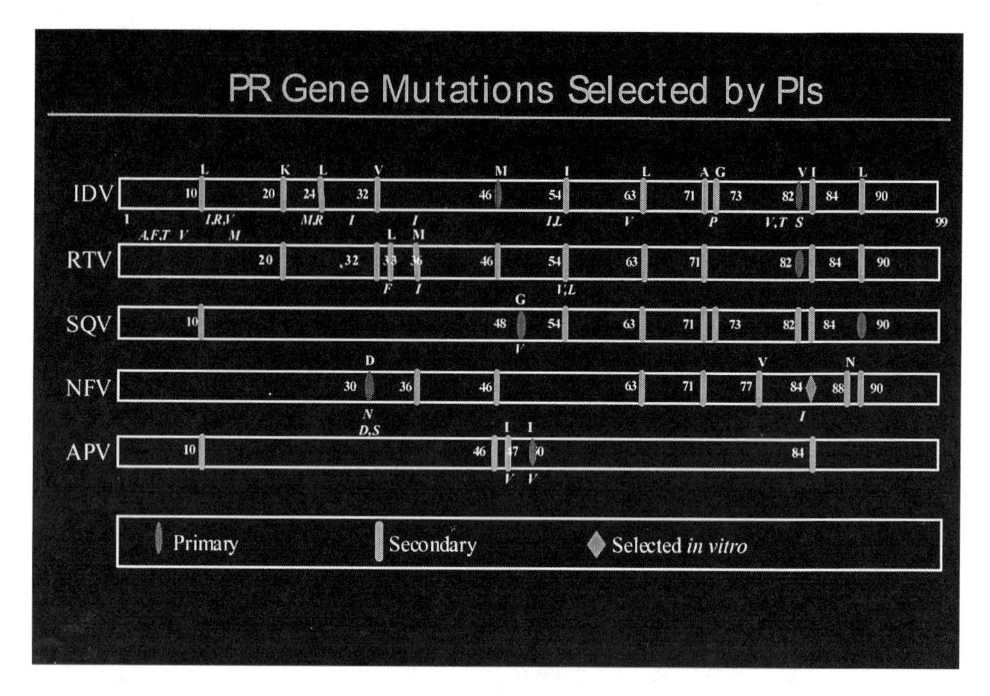

Figure 3 Protease gene mutations selected by HIV protease inhibitors: HIV PR gene is denoted by boxes. Individual mutations arising in the PR gene upon selection by HIV protease inhibitors are indicated by vertical bars. Numbers refer to protease codon. Letters refer to the amino acid residue at the position. Substituted residues are indicated by italics. (From Ref. 97.)

the same as those in monotherapy. What follows is a discussion of the genetic determinates of viral resistance to each of the currently available protease inhibitors. We will not review the important issue of the clinical usefulness of drug susceptibility testing, but refer the reader to recent reviews of the subject (42).

A. Saquinavir

1. Resistance In Vitro

Resistance to saquinavir emerges during passage of HIV-1 in tissue culture. Saquinavir resistance is associated with the appearance of G48V and L90M substitutions (32,43). The mutations at codons 48 and 90 cause a two- to sixfold reduction in susceptibility to saquinavir. Viruses carrying the double substitution G48V/L90M have an 8- to 11-fold increase in the 50% inhibitory concentration

(IC_{50}) to saquinavir. Mutations at codons 36, 54, 71, and 84 also have been reported in association with resistance to saquinavir in vitro (44,45).

The role of mutations at codons 48 and 90 has been studied in biochemically purified protease (45). In general, there is a good correlation between the viral susceptibility to, and the affinity of, purified enzyme for the inhibitor. This behavior, related to increases in the dissociation constant (compared with wild-type protease), was observed for the mutant enzyme with the L90M/G48V double substitution. Mutant enzymes retained near wild-type inhibitory constants for amprenavir. The observed increases in dissociation constant of saquinavir are related to conformational equilibria within the flap domain of the enzyme (10). Position 48 of the protease has been implicated in determining substrate specificity for the enzyme (23). Analysis of the catalytic constant (k_{cat}) of the mutant and wild-type enzymes showed no significant differences, confirming the important enzymatic role of the dissociation constant of inhibitor in the development of resistance.

An additional analysis of in vitro-selected saquinavir-resistant viral isolates with multiple mutations showed that the k_{cat}/K_m of the mutant enzymes was reduced compared with that of the wild-type (15).

2. Clinical Studies

Mutations associated with resistance to saquinavir in patients generally correlate with those observed in vitro (25,46,47). Analyses of virus obtained from patients failing saquinavir monotherapy have established mutations at codons 48 and 90 as key to the development of resistance. The most frequently observed substitution, in isolates from patients who received saquinavir monotherapy or saquinavir in combination with zidovudine, was L90M. In a study of 144 treatment-experienced patients receiving combination therapy that included saquinavir (ACTG 229), 36% of viral isolates from patients receiving saquinavir (1800 mg/day) had mutations at either codon 48 or 90. The L90M substitution was observed in isolates from 57 subjects, whereas the G48V substitution was not noted (25,48). The G48V substitution occurred most frequently in isolates from patients treated with higher doses of saquinavir (2700–3600 mg/day) (25,49). Variant viruses containing the double mutation at codons 48 and 90 were rarely detected in these clinical studies. Other mutations associated with saquinavir resistance in vivo are found at codons 10, 36, 54, 63, 71, and 84 (46,47,49). The contribution of these mutations to saquinavir resistance is less well understood than for mutations at codons 48 and 90. Polymorphisms at PR codons 10, 63, and 71 are not correlated with virological response to saquinavir in treatment-naïve individuals (50).

Early clinical studies demonstrated that higher doses of saquinavir delayed the emergence of drug resistance (25). Resistant viral isolates have the substitutions at codons 10, 54, 63, 71, but mutations at codon 82 were not observed.

Analysis of virus isolated from 16 patients failing saquinavir therapy revealed that only 8 had the L90M substitution and 3 had the G48V substitution, suggestion that no one genotypic pattern is uniquely predictive of clinical failure (51). Changes in the formulation of saquinavir from hard capsule to soft-gel capsule have not resulted in any significant alteration in the mutational patterns associated with saquinavir resistance (52).

B. Ritonavir

1. Resistance In Vitro

HIV-1 variants resistant to ritonavir are readily selected in vitro (53). Mutations that emerge with resistance include M46I, L63P, A71V, V82F, and I84V. The mutations at codons 82 and 84 are associated with five- and tenfold reduction in susceptibility to ritonavir, respectively, whereas the mutations at codons 63, 46, and 71 have no effect on susceptibility when tested in recombinant viruses as single mutations.

Crystallographic modeling of the V82A mutant enzyme complexed to inhibitor shows a decreased van der Waals interaction of ritonavir in the substrate-binding pocket (54). It has been suggested that the M46I substitution stabilizes the flaps in a closed conformation (10). A possible consequence of this structural alteration is to change the specificity of protease for the natural substrate and inhibitor [see Eq. (1)].

2. Clinical Data

Resistance to ritonavir occurs in most patients after 12–18 months of monotherapy (55). Phenotypic resistance to ritonavir requires the stepwise accumulation of multiple mutations (26). The first mutation detected generally appears at codon 82, resulting in a V82 A or F substitution. Mutations at codons 54 and 71 are also commonly observed and may be associated with substitutions at codons 8, 33, 34, 36, 57, 63, 84, or 90 (26,55).

Mutations at codons 82 or 84 in HIV-1 isolates from patients receiving ritonavir monotherapy are associated with a 20-fold resistance to ritonavir (56). Viruses with the double substitution M36I/I54V have 50-fold resistance to ritonavir. A triple mutant M36I/I54V/A71V has even greater (70-fold) resistance to ritonavir. The replicative efficiency of the mutant viruses is decreased compared with wild-type isolates, but is restored on the selection of secondary mutations (57).

Only limited data are available on the genetic determints of resistance to ritonavir in the setting of combination therapy. An analysis of HIV-1 isolates from patients receiving treatment with ZDV/ddC/ritonavir-combined therapy found no significant differences in resistance mutations compared with isolates from patients receiving ritonavir monotherapy (58).

C. Indinavir

1. Resistance In Vitro

In vitro passage of HIV-1 in the presence of indinavir produced resistant virus with the substitutions V32I, M46L, and V82A (43). Viruses harboring the triple-substitution V32I/M46L/V82A were only threefold resistant to indinavir in vitro.

The structure of indinavir cocrystallized with either wild-type or a quadruple mutant protease (M46I/L63P/V82T/I84I) has been solved (11,59). The mutant enzyme has a 70-fold increase in K_i for indinavir. Crystallographic data show that the codon 82 substitution introduces an unfavorable (for binding) hydrophobic moiety in the binding site, whereas the codon 84 mutation reduces van der Waals contacts between drug and enzyme. These studies also show a role for mutations at codons 46 and 63 in generating smaller conformational changes in the flap domain of the enzyme.

2. Clinical Data

Prolonged therapy with suboptimal doses of indinavir leads to the emergence of viral variants with reduced susceptibility to the protease inhibitor (60,61). Development of resistance is associated with variable patterns of multiple amino acid substitutions in at least 11 protease codons. In a regression analysis, mutations at codons 10, 20, 24, 46, 54, 63, 71, 82, 84, and 90 were significantly correlated with a fourfold or greater loss of susceptibility to indinavir. No particular amino acid substitution was present in all resistant isolates, implying that resistance can emerge through multiple genetic pathways. Measurable resistance (an increase of IC_{90} greater than fourfold) required the presence of at least three mutations. Continued viral replication in the presence of indinavir leads to the accumulation of additional mutations and further decreases viral susceptibility to the inhibitor. Mutations at codon 82 that result in V82 A, T, or F substitutions are most commonly correlated with clinical failure to indinavir monotherapy, and are associated with a four- to eightfold reduction in drug susceptibility (60).

D. Nelfinavir

1. Resistance In Vitro

Culture of HIV-1 in the presence of nelfinavir for 22-passages resulted in selection of a viral variant with a sevenfold decrease in nelfinavir susceptibility. This variant remained fully susceptible to indinavir, saquinavir, and ritonavir. Nucleotide sequence analysis revealed a novel mutation at PR codon 30 that resulted in a D → N substitution. Analyses of other HIV-1 variants isolated after in vitro selection by nelfinavir suggest a role for substitutions M46I, G48V, I84V, and L90M in resistance (62). Introduction of these mutations into recombinant viruses resulted in less than fivefold decreases in susceptibility (62).

2. Clinical Studies

Analysis of virus isolated from patients on nelfinavir monotherapy mirrored the results of in vitro selection experiments. The most commonly observed substitution was D30N; other substitutions observed included M36I, L36P, V77I, and I84V (63). In one study, substitutions associated with resistance to other protease inhibitors (G48V; V82 A, F, T; and I84I) were not seen, and the L90M mutation was only rarely observed (63). By contrast, another study, the G48V and L90M substitutions were reported in viruses isolated from patients after prolonged treatment with nelfinavir (63,64). The combination of ritonavir–saquinavir has been used as ''salvage therapy'' for patients failing nelfinavir (65,66). At the time of therapeutic switch, mutations at codon 30 were seen in isolates from 17/25 subjects, and the L90M mutation was observed in 5/25 (65). Presence or absence of these mutations did not predict response to the salvage regimen.

E. Amprenavir

In vitro passage of HIV-1 in the presence of amprenavir results in the selection of substitutions M46I/L, I47V, and I50V (43,67,68). Other substitutions observed included L10F and I84V (68). The mutation at codon 50 appears to be unique to amprenavir, a sulfonamide protease inhibitor. When introduced into the PR genes of recombinant viruses, the I50V mutation produced a two- to threefold reduction in susceptibility to the inhibitor (68). By contrast, the triple-mutant virus (M46I/I47V/I50V) has an approximately 20-fold reduction in susceptibility. Enzymes carrying the single and triple mutations show reductions in affinity for amprenavir of 80- and 270-fold, respectively (68). Taken together, these data reinforce the concept that increasing numbers of mutations generate higher levels of viral resistance to the protease inhibitors.

Kinetic and crystallographic analyses of an I50V mutant protease have been performed (67,69,70). These data reveal an 80-fold decreased affinity for amprenavir, but only slightly reduced affinity of the enzyme to indinavir or saquinavir. This effect is due to the loss of hydrophobic binding of the inhibitor, but not to the natural substrate. This analysis also showed that purified enzyme containing the I50V substitution has a 25-fold decreased k_{cat}/K_m for wild-type substrate analogues. Enzyme containing the triple mutation M46I/I47V/I50V has a twofold increase in k_{cat}/K_m, suggesting a compensatory role for the mutations at codons 46 and 47.

The role of resistance mutations in therapeutic failures of amprenavir has recently been studied (36). In this analysis, the I50V substitution was observed frequently in viral isolates from patients failing monotherapy with amprenavir. Other substitutions noted included M46I, I47V/M, I84L, V77I, and L101. The Lp1/p6F Gag cleavage-site mutation was also observed. Variant viruses containing the I50V mutation emerged within weeks in patients failing amprenavir monotherapy (71).

VI. CROSS-RESISTANCE AMONG PROTEASE INHIBITORS

The data on cross-resistance between individual protease inhibitors is incomplete, but suggests that significant cross-resistance is likely with prolonged exposure to the partially suppressive regimen (43,60). These observations suggest that convergent evolution occurs in response to the selective pressure exerted by different protease inhibitors. HIV-1 variants that are cross-resistant to different protease inhibitors are readily selected by in vitro passage, or by drug therapy, in patients (72). In vitro selection experiments have shown that saquinavir, ritonavir, and indinavir are similar in their capacity to select for multiple overlapping mutations in the protease gene. High levels of resistance to the selecting agent are usually accompanied by similar degrees of resistance to the other protease inhibitors (72). Indeed, in a genotypic analysis of HIV-1 isolates from heavily pretreated patients, a shared pattern of mutations in reverse transcriptase and protease genes was observed (73). Patients in this study had received between 4 and 9 years of antiretroviral therapy, and each subject had been treated with at least four nucleoside reverse transcriptase inhibitors and three protease inhibitors (saquinavir, indinavir, and ritonavir). The protease substitutions L10I, G48V, I54V/T, L63P/H/Q, A71V/L, and V82A were common to viral sequences from all patients, despite their differing treatment histories. This overlap in the genetic determinants of resistance to individual protease inhibitors is consistent with the shared drug target and generally similar structure and mechanism of action of the currently available inhibitors.

Phenotypic analysis of viral isolates from 96 patients experiencing virological failure of protease inhibitor-containing therapy has recently been presented (74). In this study, a longer duration of protease inhibitor therapy was associated with an increasing number of protease inhibitors with reduced susceptibility. The mean fold-change in susceptibility to amprenavir and saquinavir was lower than for other protease inhibitors. Viruses from patients treated with nelfinavir developed nelfinavir resistance as expected, but fewer showed cross-resistance than viruses from patients who had received indinavir or other protease inhibitors, even after accounting for duration of treatment.

Viral variants that are resistant to indinavir are usually resistant to ritonavir and, conversely, virus selected for resistance to ritonavir are cross-resistant to indinavir (75). In one study, HIV-1 isolated from patients after 48 weeks of ritonavir therapy was 170-fold resistant to ritonavir, 30-fold resistant to indinavir, and 5-fold resistant to saquinavir (55). HIV-1 isolates obtained from patients participating in clinical trials of indinavir monotherapy show significant cross-resistance to other protease inhibitors. All isolates resistant to indinavir were also resistant to ritonavir; 63 and 81% of indinavir-resistant isolates were resistant to saquinavir or amprenavir, respectively (61).

Biochemically purified enzymes with mutations at codon 82 (correlated with resistance to ritonavir, discussed in the foregoing) have 10- to 50-fold reductions in sensitivity to ritonavir but only 2- to 25-fold reductions in catalytic efficiency (k_{cat}/K_m) (18). As suggested by virological studies (26), the enzymes containing only single substitutions retain activity against other protease inhibitors (saquinavir, amprenavir).

Viral variants selected in vitro for resistance to amprenavir remain sensitive to indinavir and saquinavir (68). When the amprenavir-resistant protease was biochemically purified (with the substitutions M46I, I47I, I50V) it had a 270-fold increase in the inhibitory constant (K_i) for amprenavir. A 41-fold and 29-fold increase in K_i was found for indinavir and saquinavir, respectively, despite the lack of evidence for viral resistance. HIV-1 strains selected for amprenavir resistance in vitro remain susceptible to other protease inhibitors (68). After ten passages in the presence of amprenavir, virus becomes 170-fold resistant to amprenavir, but only 6-fold resistant to indinavir and has no change in susceptibility to saquinavir.

The evolutionary path that HIV-1 takes to acquire drug resistance may depend on the order in which protease inhibitors are administered. For example, in patients failing nelfinavir therapy, the D30N substitution occurs commonly in viral isolates (63,76). By contrast, when nelfinavir was used after failure of a previous protease inhibitor regimen the D30N mutation was identified in only 2/16 (12.5%) viral isolates from patients with virological failure (51). These observations have led to the suggestion that the presence of the D30N mutation may confer a beneficial effect toward subsequent protease inhibitor-containing therapy (77). This conclusion has been questioned by a retrospective analysis in which patients failing nelfinavir showed similar response rates to subsequent ritonavir–saquinavir therapy, irrespective of the presence or absence of the D30N mutation (66).

It is possible that the presence of certain mutations is incompatible with the emergence of other mutations required to confer resistance to a second inhibitor. When isolates selected for resistance to saquinavir were subsequently passaged in the presence of amprenavir, loss of the I47V substitutions may have resensitized the viral population to saquinavir (43). A recent analysis of viral isolates from patients failing indinavir or nelfinavir showed that the presence of the N88S mutation was associated with a 2.5- to 10-fold increase in susceptibility to amprenavir (78). These data provide support for the use of dual protease inhibitor therapy, as some viruses resistant to one inhibitor may show enhanced susceptibility to other protease inhibitors.

An accumulating body of data suggests that mutations thought to be primarily associated with resistance to one protease inhibitor may play a significant role in conferring resistance to other inhibitors. For example, the V82A substitution, which is associated with resistance to ritonavir, can emerge during treatment with saquinavir or during indinavir or nelfinavir therapy after failure with ritonavir (51). Conversely, the L90M substitution (associated with saquinavir resistance) can emerge during treatment with nelfinavir. Isolates from 23/66 (35%) patients failing saquinavir-containing regimens had mutations primarily associated with resistance to other protease inhibitors (79).

The frequent finding of the L90M mutation in isolates from unselected, mostly indinavir-experienced patients with treatment failure (80) and in five of ten patients failing nelfinavir therapy (81) suggests an important role of this substitution for in vivo resistance to indinavir, ritonavir, or nelfinavir. This conclusion is supported by data that demonstrate a significant deleterious effect of the L90M mutation on virological response to subsequent protease inhibitor-containing therapy (77). These data suggest that saquinavir resistance may limit the response to subsequent therapy with the other available protease inhibitors.

A. Nelfinavir Cross-Resistance

Resistance to nelfinavir is associated with the presence of the unique mutation D30N (63). In one study of nelfinavir, the D30N substitution was rarely associated with the presence of other protease inhibitor-resistance mutations (G48V, V82F/T, I84V, or L90M). The D30N mutation was observed in isolates from two of three patients failing combination therapy that included nelfinavir (76). In the same study, the L90M mutation was observed in virus isolated from one of three patients failing therapy. Phenotypic analysis of viral isolates with high-level resistance to nelfinavir showed full susceptibility to indinavir, saquinavir, ritonavir, and amprenavir. A recent analysis of viral phenotypes from patients failing protease inhibitor-containing therapy also showed preservation of susceptibility to other protease inhibitors among patients failing nelfinavir, but not indinavir (74). These findings provide support for the concept that the evolution of drug resistance to nelfinavir does not necessarily impart cross-resistance to other protease inhibitors.

An important issue related to nelfinavir use is whether nelfinavir resistance predisposes the viral population to resistance to other protease inhibitors. In studies of patients failing nelfinavir-containing combination therapy who were switched to saquinavir–ritonavir, the D30N substitution was found in 17/25 (68%) isolates. Significantly, some isolates from 5/25 (25%) patients also had the L90M mutation (65). Plasma HIV-RNA levels fell below 500 copies/mL in most patients who switched to ritonavir–saquinavir, but interpretation of these data is limited by the relatively short duration of follow-up (16 weeks). In another study the D30N mutation was observed in samples from 17/29 patients with nelfinavir failure, whereas L90M was observed in 11/29 samples. Virological response to subsequent therapy with an indinavir regimen was significantly more frequent among patients with D30N viruses compared with those with L90M (64). The possibility of using nelfinavir as second-line therapy in patients who have failed treatment with other protease inhibitors has examined. Phenotypic analysis of a large number of clinical isolates resistant to indinavir, ritonavir, and saquinavir showed that 40% of these isolates remained susceptible to nelfinavir (75). Nevertheless, the clinical response to nelfinavir in patients carrying such isolates has been disappointing.

VII. SECOND-GENERATION PROTEASE INHIBITORS AND CROSS-RESISTANCE

The development of ''second-generation'' protease inhibitors with chemical structures that differ from the currently available inhibitors offers the hope of more limited cross-resistance between agents, and for improved response to therapy for patients failing their first protease inhibitor-containing regimen.

A. Lopinavir (ABT-378)

ABT-378 (lopinavir) is a second-generation peptidic protease inhibitor currently in clinical development. This drug was discovered through efforts to identify compounds with activity against ritonavir-resistant HIV-1 (82). The compound was designed specifically to minimize molecular interactions at Val 82, the amino acid residue important in the generation of ritonavir resistance (83). Serial in vitro passage of HIV-1 in the presence of ABT-378 results in the sequential appearance of the substitutions L10F, V32I, M46I, I47V, I84V, and T91S (34). After prolonged passage in the presence of ABT-378, a seventh mutation (I47A) was observed, as well as reversions of 32I to wild-type. Mutations in the p1/p6 cleavage site were also observed (33,34). Virus containing all six mutations was 50-fold less susceptible to ABT-378 (82). ABT-378-resistant isolates selected in vitro retain susceptibility to ritonavir, saquinavir, and indinavir (34). The compound retains significant activity against biochemically purified, ritonavir-resistant protease (containing mutations V82A, V82F, V82S, or V82T), with picomolar inhibitory constants (18).

ABT-378 is being developed in a coformulation with low-dose ritonavir (ABT-378/r), which offers significant pharmacokinetic advantages over ABT-378 alone (84). In a study of 70 patients having failure of a first protease inhibitor-containing regimen, HIV-1 RNA levels lower than 400 copies/mL were achieved in 78% of subjects who received ABT-378/r in combination with nevirapine, lamivudine, and two new nucleoside RT inhibitors (85).

B. Tipranavir (PNU 140690)

Tipranavir is a member of a new class of nonpeptidic, sulfonamide-containing HIV-1 protease inhibitor. Similar to ABT-378, this drug retains activity against HIV-1 isolates that are moderately resistant to ritonavir and indinavir (86). Mutations associated with resistance to PNU140690 have not yet been reported.

Preclinical testing of tipranavir demonstrates activity against clinical isolates resistant to peptidic protease inhibitors, with submicromolar IC_{90}s against highly resistant laboratory strains of HIV-1 and ritonavir-resistant clinical isolates (86). A laboratory-adapted strain of HIV-1 that was 47- to more than 125-fold resistant to the currently available petidomimetic inhibitors had only a sixfold decrease in susceptibility to tipranavir. The drug is currently in clinical testing.

VIII. DUAL PROTEASE THERAPY

There has been much interest in the concomitant use of two or more protease inhibitors in the treatment of HIV-1 infection. The pharmacokinetic rationale for dual protease therapy stems from the increased drug levels achieved through the coadministration of two or more drugs (87). Moreover, the replication of dually protease inhibitor-resistant virus in vitro is significantly impaired as compared with wild-type virus (88,89).

Clinical trials of dual protease inhibitor therapy have shown significant virological responses that equal or exceed those achieved by combination therapy with two RT inhibitors and a single protease inhibitor (90). The response to dual protease inhibitors (ritonavir–saquinavir) is less encouraging in treatment-experienced patients with advanced disease (27). Viruses isolated from subjects failing ritonavir–saquinavir have substitutions at codons 10, 16, 36, 48, 63, 64, 82, and 90. Drug resistance mutations appeared as soon as 5 weeks after initiation of therapy in virus isolated from subjects who did not respond to therapy. A retrospective cohort analysis of ritonavir–saquinavir therapy in patients failing their first protease inhibitor-containing regimen has shown more encouraging results, with 51% of subjects with plasma viral loads less than the limit of quantification at 6 months (66). A multivariate analysis of this data showed that CD4 count and viral loads at the time of treatment switch were independent predictors of clinical response.

In a study of nelfinavir–saquinavir (soft gel) dual therapy, mutations associated with saquinavir resistance emerged at codons 48 and 90 in four of ten subjects after a median of 11 months, but the D30N mutation was not seen (91,92). In this study, some patients added nucleoside analogues to the protease inhibitors at 12 weeks. Despite genotypic evidence of resistance to saquinavir, nine of ten subjects had plasma HIV-1 RNA levels higher than 500 copies/mL, including three patients who had not added nucleosides to their regimen.

IX. SALVAGE THERAPY

Given that resistance mutations in the protease gene accumulate over time in isolates from patients failing a protease inhibitor, the length of time that a patient remains on a failing regimen may be an important parameter in determining the success of subsequent protease inhibitor regimens. Indeed, the length of time on a failing protease inhibitor-containing regimen is positively correlated with the degree of cross-resistance to other protease inhibitors (74). Furthermore, a higher plasma viral load at time of therapy switch is an independent predictor of poor response to subsequent therapy (66). A possible implication of these observations is that frequent monitoring of plasma HIV-1 RNA levels may be necessary along with a rapid response to rising viral loads. These data provide a theoretical justification for early therapeutic intervention in this setting to prevent ever greater levels of cross-resistance.

X. TREATMENT FAILURE WITHOUT DRUG RESISTANCE

Several studies have pointed out that clinical failure can occur in the absence of drug-resistance mutations, or with a minimal number of mutations. Analysis from viral sequences from patients failing protease inhibitor therapy found a significant number of isolates with wild-type protease genes (80,93–96). In one study, the median number of mutations observed in PR was three and 34% of isolates had two or fewer mutations (80). Genotypic evidence of resistance to each component of a multidrug regiment was not required for therapeutic failure (80). Additionally, two of nine patients failing combination therapy that included amprenavir had wild-type population sequences for both PR and Gag (36). Taken together, these data reinforce the contribution of pharmacological factors and patient adherence toward the therapeutic success of antiretroviral therapy and highlight the differences in the results of controlled clinical trials as compared with unselected clinic populations.

XI. CONCLUSIONS

The discovery and clinical application of HIV-1 protease inhibitors is a triumph of rational drug design. The clinical use of protease inhibitors has dramatically improved the prognosis of infected individuals, and offers hope for long-term prevention of disease progression. Nevertheless, in the presence of evolutionary selective pressure in the form of drug therapy and incomplete suppression of viral replication, drug-resistant viral variants are likely to emerge. The genetic and biochemical determinants of resistance to the currently available HIV-1 protease inhibitors have been studied in great detail and provide a useful model for understanding the biochemical mechanisms of evolution. The degree of cross-resistance among protease inhibitors may limit the efficacy of ''salvage'' regimens that includes a second protease inhibitor. To what extent this phenomenon applies to the ''second-generation'' protease inhibitors remains to be seen.

ACKNOWLEDGMENTS

We thank Karen Kohler and Kristin Doherty for editorial assistance. Benjamin Young was supported in part by an institutional AIDS training grant from the NIH (T32 AI-07447); Daniel Kuritzkes was supported in part by AI-42567 and by a Virology Advanced Technology Laboratory Award from the Adult AIDS Clinical Trials Group (subcontract of AI-38858).

REFERENCES

1. Carpenter CJ, Fischl MA, Hammer SM, et al. Antiretroviral therapy for HIV Infection in 1998. JAMA 1998; 280:78–86.
2. Hammer SM, Squires KE, Hughes MD, et al. A controlled trial of two nucleoside analogues plus indinavir in persons with human immunodeficiency virus infection and CD4 cell counts of 200 per cubic millimeter or less. N Engl J Med 1997; 337: 725–733.
3. Gulick RM, Mellors JW, Havlir D, et al. Treatment with indinavir, zidovudine and lamivudine in adults with human immunodeficiency virus infection and prior antiretroviral therapy. N Engl J Med 1997; 337:734–740.
4. Mouton Y, Alfandari S, Valette M, et al. Impact of protease inhibitors on AIDS-defining events and hospitalizations in 10 French AIDS reference centres. AIDS 1997; 11:F101–F105.
5. Palella FJ, Delaney KM, Moorman AC, et al. Declining morbidity and mortality among patients with advanced human immunodeficiency virus infection. HIV outpatient study investigators. N Engl J Med 1998; 338:853–860.
6. Schinazi RF, Larder BA, Mellors JW. Mutations in retroviral genes associated with drug resistance. Int Antiviral News 1997; 5:129–142.

7. Erickson JW, Burt SK. Structural mechanisms of HIV drug resistance. Annu Rev Pharmacol Toxicol 1996; 36:545–571.
8. Schock HB, Garsky VM, Kuo LC. Mutational anatomy of an HIV-1 protease variant conferring cross-resistance to protease inhibitors in clinical trials. J Biol Chem 1996; 27:31957–31963.
9. Cameron CE, Grinde B, Jacques P, et al. Comparison of the substrate-binding pockets of the Rous sarcoma virus and human immunodeficiency virus type 1 proteases. J Biol Chem 1993; 268:11711–11720.
10. Collins JR, Burt SK, Erickson JW. Flap opening in HIV-1 protease simulated by ''activated'' molecular dynamics. Struct Biol 1995; 2:334–338.
11. Chen Z, Li Y, Schock HB, Hall D, Chen E, Kuo LC. Three-dimensional structure of a mutant HIV-1 protease displaying cross-resistance to all protease inhibitors in clinical trials. J Biol Chem 1995; 270:21433–21436.
12. Smidt ML, Potts KE, Tucker SP, et al. A mutation in human immunodeficiency virus type 1 protease at position 88, located outside the active site, confers resistance to the hydroxyethylurea inhibitor SC-5539A. Antimicrob Agents Chemother 1997; 41:515–522.
13. Zhang YM, Imamichi H, Imamichi T, et al. Drug resistance during indinavir therapy is caused by mutations in the protease gene and in its Gag substrate cleavage sites. J Virol 1997; 71:6662–6670.
14. Tang J, Hartsuck JA. A kinetic model for comparing proteolytic processing activity and inhibitor resistance potential of mutant HIV-1 proteases. FEBS Lett 1995; 367: 112–116.
15. Gulnik SV, Suvorov LI, Liu B, et al. Kinetic characterization and cross-resistance patterns of HIV-1 protease mutants selected under drug pressure. Biochemistry 1995; 34:9282–928-77.
16. Ermolieff J, Lin X, Tang J. Kinetic properties of saquinavir-resistant mutants of human immunodeficiency virus type 1 protease and their implications in drug resistance in vitro. Biochemistry 1997; 36:12364–12370.
17. Fersht A. Enzyme Structure and Function. New York: WH Freeman & Co, 1985.
18. Chen C, Niu P, Kati W, et al. Activity of ABT-378 against HIV protease containing mutations conferring resistance to ritonavir. 4th Conference on Retroviruses and Opportunistic Infections, 1997; 208:103[abstr].
19. Sussman F, Villaverde MC, Davis A. Solvation effects are responsible for the reduced inhibitor affinity of some HIV-1 PR mutants. Protein Sci 1997; 6:1024–1030.
20. Shao W, Everitt L, Manchester M, Loeb DD, Hutchinson DA, Swanstrom R. Sequence requirements of the HIV-1 protease flap region determined by saturation mutagenesis and kinetic analysis of flap mutants. Proc Natl Acad Sci USA 1997; 94:2243–2248.
21. Wedemayer GJ, Patten PA, Wang LH, Schultz LH, Stevens RC. Structural insights into the evolution of an antibody combining site [published erratum in Science 1997 Sept 5, 227:1423]. Science 1997; 276:1665–1669.
22. Joyce GF. Evolutionary chemistry: getting there from here. Struct Biol 1997; 276: 1658–1659.

23. Moody MD, Pettit SC, Shao W, et al. A side chain at position 48 of the human immunodeficiency virus type-1 protease flap provides an additional specificity determinant. Virology 1995; 207:475–485.
24. Bhat TN, Randad RS, Lee AY, et al. Structural studies of inhibitor complexes of HIV-1 protease and of its drug resistance mutants. Abstracts Meeting Groups Study of Structure AIDS Related Systems and Their Applied Target Drug Design, June 25–27, 1996 [abstr].
25. Schapiro JM, Winters MA, Stewart F, et al. The effect of high-dose saquinavir on viral load and CD4+ T-cell counts in HIV-infected patients. Ann Intern Med 1996; 124:1039–1050.
26. Molla A, Korneyeva M, Gao Q, et al. Ordered accumulation of mutations in HIV protease confers resistance to ritonavir. Nat Med 1996; 2:760–766.
27. Lorenzi P, Yerly S, Abderrakim K, et al. Toxicity, efficacy, plasma drug concentrations and protease mutations in patients with advanced HIV infection treated with ritonavir plus saquinavir. AIDS 1997; 11:F95–F99.
28. Barrie KA, Perez EE, Lamers SL, et al. Natural variation in HIV-1 protease, gag p7 and p6, and protease cleavage sites within Gag/Pol polyproteins: amino acid substitutions in the absence of protease inhibitors in mothers and children infected by human immunodeficiency virus type 1. Virology 1996; 219:407–416.
29. Rasnick D. Kinetics analysis of consecutive HIV proteolytic cleavages of the Gag–Pol polyprotein. J Biol Chem 1997; 272:6348–6353.
30. Lin Y, Lin X, Hong L, et al. Effect of point mutations on the kinetics and the inhibition of human immunodeficiency virus type 1 protease: relationship to drug resistance. Biochemistry 1995; 34:1143–1152.
31. Fantuzzi L, Gessani S, Borghi P, et al. Induction of interleukin 12 (IL-12) by recombinant glycoprotein gp 120 of human immunodeficiency virus type 1 in human monocytes/macrophages: requirement of gamma interferon for IL-12 secretion. J Virol 1996; 70:4121–4124.
32. Jacobsen H, Yasargil K, Winslow DL, et al. Charactarization of human immunodeficiency virus type 1 mutants with decreased sensitivity to proteinase inhibitor Ro 31-8959. Virology 1995; 206:527–534.
33. Carrillo A, Sham H, Norbeck D, et al. Mutation in the proteolytic sites is required for the growth of mutant HIV selected by ABT-378, a new HIV protease inhibitor. 37th Interscience Conference on Antimicrobial Agents and Chemotherapy, Toronto, Ontario, Canada, 1997; I-116:264.
34. Carrillo A, Sham H, Norbeck D, et al. Selection and analysis of HIV-1 variants with increased resistance to ABT-378, a novel protease inhibitor. 4th Conference on Retroviruses and Opportunistic Infections, 1997 [abstr].
35. Delphin N, Schneider V, Nicolas JC, Rozenbaum W. Variations in the p7/p1 gag cleavage region and virological response to a HIV protease inhibitor [abstr 85]. Antiviral Ther 1999; 4:58.
36. DePasquale MP, Murphy R, Gulick R, et al. Mutations selected in HIV RNA during 141W94 therapy. 5th Conference on Retroviruses and Opportunistic Infections, 1998 [abstr 406a].
37. Alford JL, McQuaid TJ, Partaledis JA, Markland W, Byrn R. Amprenavir-resistant

mutants show reduced viral fitness: correlation of protease mutations, cleavage site mutations, enzyme kinetics, viral polyprotein processing and viral growth kinetics [abstr 44]. Antiviral Ther 1999; 4:30.

38. Kozal M, Shah N, Shen N, et al. Extensive polymorphisms observed in HIV-1 Clade B protease gene using high-density oligonucleotide arrays. Nat Med 1996; 2:753–759.
39. Lech WJ, Wang G, Yang YL, et al. In vivo sequence diversity of the protease of human immunodeficiency virus type 1: presence of protease inhibitor-resistant variants in untreated subjects. J Virol 1996; 70:2038–2043.
40. Berlusconi A, Violin M, Columbo MC, et al. Genotypic prevelence of ZDV-resistant HIV-1 strains and preexistent mutations in protease coding region of recently infected subjects. 5th Conference on Retroviruses and Opportunistic Infections, 1998; 675:206[abstr].
41. Perez E, Lamers S, Heath–Chiozzi M, et al. Emergence of resistant protease alleles and vaiant *gag* sequences in HIV-1 infected children enrolled in protease inhibitor phase I/II clinical trials. International Workshop on HIV Drug Resistance, St Petersburg, FL, 1997; 75:[abstr].
42. Kuritzkes D. Drug resistance testing: time to be used in clinical practice. AIDS Rev 1999; 1:45–50.
43. Tisdale M, Myers RE, Maschera B, Parry NR, Oliver NM, Blair ED. Cross-resistance analysis of human immunodeficiency virus type 1 variants individually selected for resistance to five different protease inhibitors. Antimicrob Agents Chemother 1995; 39:1704–1710.
44. Eberle J, Bechowsky B, Rose D, et al. Resistance of HIV type 1 to proteinase inhibitor Ro 31-8959. AIDS Res Hum Retroviruses 1995; 11:671–676.
45. Maschera B, Darby G, Palu G, et al. Human immunodeficiency virus. J Biol Chem 1996; 271:33231–33235.
46. Jacobsen H, Haenggi M, Ott M, et al. Reduced sensitivity to saquinavir: an update on genotyping from phase I/II trials. Antiviral Res 1996; 29:95–97.
47. Jacobsen H, Hanggi M, Ott M, et al. In vivo resistance to a human immunodeficiency virus type 1 proteinase inhibitor: mutations, kinetics and frequencies. J Infect Dis 1995; 173:1379–1387.
48. Schapiro JM, Lawrence J, Speck R, et al. HIV RNA and resistance mutations to saquinavir and zidovudine in patients receiving dual versus triple combination therapy. 5th Conference on Retroviruses and Opportunistic Infections, 1998; 401: 154[abstr].
49. Ives KJ, Jacobsen H, Galpin SA, et al. Emergence of resistant variants of HIV in vivo during monotherapy with the proteinase inhibitor saquinavir. J Antimicrob Chemother 1997; 39:771–779.
50. O'Sullivan E, Cammack N, Craig C. Responsiveness to saquinavir is not affected by baseline HIV protease genotype in protease-inhibitor naive patients. 5th Conference on Retroviruses and Opportunistic Infections, 1998; 399:154[abstr].
51. Lawrence J, Schapiro J, Pesano R, et al. Clinical response and genotypic resistance patterns of sequential therapy with nelfinavir followed by indinavir plus nevirapine in saquinavir/reverse transcriptase inhibitor-experienced patients. International Workshop on HIV Drug Resistance, St Petersburg, FL, 1997; 64:42[abstr].

52. Craig C, O'Sullivan E, Cammack N. Increased exposure to the HIV protease inhibitor saquinavir (SQV) does not alter the nature of key resistance mutations. 5th Conference on Retroviruses and Opportunistic Infections, 1998; 398:154[abstr].
53. Markowitz M, Mo H, Kempf DJ, et al. Selection and analysis of human immunodeficiency virus type 1 variants with increased resistance to ABT-538, a novel protease inhibitor. J Virol 1995; 69:701–706.
54. Baldwin ET, Bhat TN, Liu B, Pattabiraman N, Erickson JW. Structural basis of drug resistance for the V82A mutant of HIV-1 proteinase. Struct Biol 1995; 2:244–249.
55. Schmit JC, Ruiz L, Clotet B, et al. Resistance-related mutations in the HIV-1 protease gene of patients treated for 1 year with the protease inhibitor ritonavir (ABT-538). AIDS 1996; 10:995–999.
56. Nijhuis M, Back N, deJong D, et al. Host cell dependent replication efficacy of 3TC resistant HIV-1 variants. 5th International Workshop on HIV Drug Resistance, Whistler, BC, Canada, 1998; 86:86[abstr].
57. Nijhuis M, Schuurman R, Schipper P, et al. Reduced replication potential of HIV-1 variants initially selected during ritonavir therapy is restored upon selection of additional substitutions. 4th Conference on Retroviruses and Opportunistic Infections, 1997; 596:174[abstr].
58. Clavel F, Paulos S, Mathez D, Leibowitch L. HIV protease sequences selected during ZDV–ddC–ritonavir triple combination. 4th Conference on Retroviruses and Opportunistic Infections, 1997; 236:108[abstr].
59. Chen Z, Li Y, Chen E, et al. Crystal structures at 1.9-A resolution of human immunodeficiency virus (HIV) II protease complexed with L-735, 524, an orally bioavailable inhibitor of the HIV proteases. J Biol Chem 1994; 269:26344–26348.
60. Condra JH, Schleif WA, Blahy OM, et al. In vivo emergence of HIV-1 variants resistant to multiple protease inhibitors. Nature 1995; 374:569–571.
61. Condra JH, Holder DJ, Schleif WA, et al. Genetic correlates of in vivo viral resistance to indinavir, a human immunodeficiency virus type 1 protease inhibitor. J Virol 1996; 70:8270–8276.
62. Patick AK, Mo H, Markowitz M, et al. Antiviral and resistance studies of AG1343, an orally bioavailable inhibitor of human immunodeficiency virus protease. Antimicrob, Agents Chemother 1996; 40:292–297.
63. Patick AK, Kuritzkes D, Johnson VA, et al. Genotypic and phenotypic analyses of HIV-1 variants isolated from patients treated with nelfinavir and other HIV-1 protease inhibitors [abstr 18]. Antiviral Ther 1997; 2(suppl 5):26.
64. Condra J, Holder D, Schleif WA, et al. Genetic correlates of virological response to an indinavir-containing salvage regimen in patients with nelfinavir failure [abstr 63]. Antiviral Ther 1999; 4:44.
65. Tebas P, Patick AK, Kane EM, et al. Virologic responses to ritonavir–saquinavir-containing regimen in patients who had previously failed nelfinavir. AIDS 1999; 13:F23–F28.
66. Zolopa AR, Tebas P, Gallant J, et al. The efficacy of ritonavir (RTV)/saquinavir (SQV) antiretroviral therapy (ART) in patients who failed nelfinavir (NFV): a multicenter clinical cohort study [abstr 2065]. 39th Interscience Conference on Antimicrobial Agents and Chemotherapy, San Francisco, CA, Sept 26–29, 1999.

67. Rao BG, Dwyer MD, Thomson JA, et al. Structural and modeling analysis of the basis of viral resistance to VX-478. 5th International Workshop on HIV Drug Resistance, Whistler, BC, Canada, 1998; 22:22[abstr].
68. Partaledis JA, Yamaguchi K, Tisdale M, et al. In vitro selection and characterization of human immunodeficiency virus type 1 (HIV-1) isolates with reduced sensitivity to hydroxyethylamino sulfonamide inhibitors of HIV-1 aspartyl protease. J Virol 1995; 69:5228–5235.
69. Pazhanisamy S, Stuver CM, Cullinan AB, Margolin N, Rao BG, Livingston DJ. Kinetic characterization of human immunodeficiency virus type-1 protease-resistant variants. J Biol Chem 1996; 271:17979–17985.
70. Rao BG, Kim EE. Calculation of solvation and binding free energy differences between VX-478 and its analogs by free energy perturbation and AMSOL methods. J Comp Aided Mol Design 1996; 10:23–30.
71. Murphy R, DeGruttola V, Gulick R, et al. 141W94 with or without zidovudine/3TC in patients with no prior protease inhibitor or 3TC therapy—ACTG 347. 5th Conference on Retroviruses and Opportunistic Infections, Chicago, IL, 1998;[abstr. 512].
72. Smith T, Swanstrom R. Biological cross-resistance to HIV-1 protease inhibitors. International Workshop on HIV Drug Resistance, St. Petersburg, FL, 1997; 15: 10[abstr].
73. Shafer RW, Winters MA, Merigan TC. Multiple concurrent RT and protease mutations and multidrug resistance in heavily treated HIV-1 infected patients. International Workshop on HIV Drug Resistance, St Petersburg, FL, 1997; 39:25 [abstr].
74. Haubrich R, Kemper C, Witt M, et al. Differences in protease inhibitor (PI) phenotypic susceptibility after failure of the first PI-containing regimen. 39th Interscience Conference on Antimicrobial Agents and Chemotherapy, San Francisco, CA, Sept 26–29, 1999; abstr 1167.
75. Hertogs K, Mellors JW, Schel P, et al. Patterns of cross-resistance among protease inhibitors in 483 clinical isolates. 5th Conference on Retroviruses and Opportunistic Infections, 1998; 395:153[abstr].
76. Markowitz M, Cao Y, Hurley A, et al. Virologic and immunologic response to triple therapy with nelfinavir in combination with ZAT/3TC in 12 antiretroviral naive HIV-infected subjects at 20 months. 5th Conference on Retroviruses and Opportunistic Infections, 1998; 371:148[abstr].
77. Mayers DL, Baxter JD, Wentworth DN, Neaton JD, Merigan T, the CPCRA 046 Study Team for the Terry Beirn Community Programs for Clinical Research on AIDS (CPCRA). The impact of drug resistance mutations in plasma virus of patients failing on protease inhibitor-containing HAART regimens on subsequent virological response to the next HAART regiment: results of CPCRA 046 (GART) [abstr 74]. Antiviral Ther 1999; 4:51.
78. Ziermann R, Limoli K, Petropoulos CJ, Parkin NT. The N88S mutation in HIV-1 protease is associated with increased susceptibility to amprenavir [abstr 93]. Antiviral Ther 1999; 4:62
79. Winters MA, Schapiro JM, Lawrence J, Merigan TC. Genotypic and phenotypic analysis of the protease gene in HIV-1-infected patients that failed long-term saqui-

navir therapy and switched to other protease inhibitors. International Workshop on HIV Drug Resistance, St Petersburg, FL. 1997; 17:11[abstr].

80. Young B, Johnson S, Bakhtiari M, et al. Genotypic analysis of HIV-1 protease from patients failing highly active anti-retroviral therapy: preliminary analysis. International Workshop on HIV Drug Resistance, St Petersburg, FL, 1997; 65:[abstr].
81. Henry K, Kane E, Melroe H, Simpson J, Patick A, Winslow D. Experience with a ritonavir/saquinavir based regimen for the treatment of HIV-infection in subjects developing increased viral loads while receiving nelfinavir. 37th Interscience Conference on Antimicrobial Agents and Chemotherapy, Toronto, Ontario, Canada. 1997; I-204:282[abstr].
82. Korneyeva M, Chernyvskiy T, Norbeck D, et al. Virological evaluation of ritonavir-resistant HIV to the HIV protease inhibitor ABT-378. 4th Conference on Retroviruses and Opportunistic Infections, 1997; 212:[abstr].
83. Sham H, Kempf D, Molla A, et al. Design, synthesis and biological properties of ABT-378, a highly potent HIV protease inhibitor. 4th Conference on Retroviruses and Opportunistic Infections, 1997; 14:[abstr].
84. Kempf D, Mo H, Brun S, et al. Analysis of virological response to ABT-378/ritonavir therapy in protease inhibitor-experienced patients with respect to baseline viral phenotype and genotype. Antiviral Ther 1999; 4:[abstr].
85. Eron J, King M, Xu Y, et al. ABT-378/ritonavir (ABT-387/r) suppresses HIV RNA to >400 copies/mL in 95% of treatment-naive patients and in 78% of PI-experienced patients at 36 weeks [abstract LB-20]. 39th Interscience Conference on Antimicrobial Agents and Chemotherapy, San Francisco, CA, Sept 26–29, 1999.
86. Poppe SM, Slade DE, Chong KT, et al. Antiviral activity of the dihydropyrone PNU-140690, a new nonpeptidic human immunodeficiency virus protease inhibitor. Antimicrob Agents Chemother 1997; 41:1058–1063.
87. Kempf DJ, Marsh KC, Kumar G, et al. Pharmacokinetic enhancement of inhibitors of the human immunodeficiency virus protease by coadministration with ritonavir. Antimicrob Agents Chemother 1997; 41:654–660.
88. Rose RE, Gong YF, Greytok JA, et al. Human immunodeficiency virus type 1 viral background plays a major role in development of resistance to protease inhibitors. Proc Natl Acad Sci USA 1996; 93:1648–1653.
89. Croteau G, Doyon L, Thibeault D, McKercher G, Pilote L, Lamarre D. Impaired fitness of human immunodeficiency virus type 1 variants with high-level resistance to protease inhibitors. J Virol 1997; 71:1089–1096.
90. Farthing C, Japour A, Cohen C, et al. Cerebrospinal fluid (CSF) and plasma HIV RNA suppression with ritonavir (RIT)–saquinavir (SQV) in protease inhibitor naive patients. 37th Interscience Conference on Antimicrobial Agents and Chemotherapy, Toronto, ON, Canada, 1997; LB-3:Adden 8[abstr].
91. Kravcik S, Farnsworth A, Patick A, et al. Long term follow-up of combination protease inhibitor therapy with nelfinavir and saquinavir (soft gel) in HIV infection. 5th Conference on Retroviruses and Opportunistic Infections, 1998; 394c:153 [abstr].
92. Kravcik S, Sahai J, Kerr B, et al. Protease gene mutations and long term follow-up of HIV-infected patients treated with nelfinavir mesylate (NFV) plus saquinavir-soft gel capsule. 37th Interscience Conference on Antimicrobial Agents and Chemotherapy, Toronto, ON, Canada, 1997; I-191:278.

93. Mayers DL, Gallahan DL, Martin GL, et al. Drug resistance genotypes from plasma virus of HIV-infected patients failing combination drug therapy. International Workshop on HIV Drug Resistance, St Petersburg, FL, 1997; 80[abstr].
94. Havlir D, Hellman N, Petropoulos CJ, et al. Viral rebound in the presence of indinavir without protease inhibitor resistance [abstr LB12]. 6th Conference on Retroviruses and Opportunistic Infections, Chicago, IL, Jan 31–Feb 4, 1999.
95. Descamps D, Peytavin G, Calvez V, et al. Virologic failure, resistance and plasma drug measurements in induction maintenance therapy trial (Anrs 072, Trilege) [abstr 493]. 6th Conference on Retroviruses and Opportunistic Infections, Chicago, IL, Jan 31–Feb 4, 1999.
96. Holder DJ, Condra JH, Schleif WA, Chodakewitz J, Emini E, and Merck Research Labs. Virologic failure during combination therapy with Crixivan and RT inhibitors is often associated with expression of resistance-associated mutations in RT only [abstr 492]. 6th Conference on Retroviruses and Opportunistic Infections, Chicago, IL, Jan 31–Feb 4, 1999.
97. Hirsch M, Conway B, D'Aquila RT, et al. Antiretroviral Drug Resistance Testing in Adults with HIV Infection. JAMA 1998; 279:1984–1991.

13
Expedited Drug Approval and HIV Protease Inhibitors

Jeffrey Murray
U.S. Food and Drug Administration, Rockville, Maryland

I. INTRODUCTION

The availability of potent antiretrovirals such as HIV protease inhibitors (PRI), particularly when used in combination with other antiretrovirals, has resulted not only in decreasing morbidity and mortality (1,2), but has also improved understanding of HIV pathogenesis (3,4), and has served as a model for expediting antiretroviral drug development and approval. In some cases, the approach for evaluating HIV treatments has served as an example for the development and approval of other drugs for serious diseases.

II. EXPEDITING ANTIRETROVIRAL DRUG APPROVAL AND ACCESS

The U.S. Food and Drug Administration's (FDA) response to the HIV epidemic has included the formation of a mechanism for accelerated approval, expansion of preapproval access or ''compassionate-use'' programs, ''fast-track'' procedures for interactions between the agency and pharmaceutical sponsors, and the acceptance of plasma HIV RNA measurements as a primary endpoint in phase 3 clinical efficacy trials. These expedited drug approval mechanisms, most of which were used in the development and approval of HIV protease inhibitors, are defined briefly in the following. Some of the regulatory terms are often confused and may seem irrelevant to clinicians. However, because a routine part of

HIV clinical practice has involved using either recently approved drugs or at least one unapproved drug, a better understanding of a few regulatory terms is worthwhile.

A. Definitions of Regulatory Terms

Accelerated approvals (5) are permitted for drugs that ". . . have been studied for their safety and efficacy in treating serious and/or life-threatening illnesses and that provide meaningful therapeutic benefit to patients over existing treatments (e.g., ability to treat patients unresponsive to, or intolerant of, available therapy, or improved patient response over available therapy)." Accelerated approvals are based on surrogate endpoint data (e.g., CD4, HIV RNA, and such) or clinical endpoint data consisting of something less than irreversible morbidity or mortality. Under an accelerated approval, marketing of a new drug is subject to certain conditions outlined in the regulations, such as a requirement for confirmatory data, FDA review of all advertising materials, and an expedited drug withdrawal process if further clinical studies fail to confirm efficacy. Once clinical efficacy is confirmed in clinical trials, a drug under accelerated approval may be granted *traditional approval*, which is the usual regulatory approval for drugs used to treat most medical illnesses. Traditional approvals are not subject to the same marketing conditions outlined in the accelerated approval regulations.

It is noteworthy that didanosine, a nucleoside reverse transcriptase inhibitor (NRTI), served as the model for the accelerated approval regulations, which apply not only to antiretroviral therapies, but also to other drugs designed to treat serious illnesses. Didanosine was the first HIV drug to come to market supported by changes in surrogate markers (CD4 cell counts, p24 antigen). Subsequently, clinical endpoint studies confirmed that the didanosine-induced changes in surrogate markers conferred clinical benefit.

Subpart E drugs (6) are defined in the regulations pertaining to Investigational New Drugs (INDs). They are drugs intended to treat life-threatening and severely debilitating diseases, such as HIV/AIDS, especially where no satisfactory alternative therapy exists. In a nutshell, these regulations emphasize that FDA may exercise the "broadest flexibility" in applying the safety and efficacy standards for approval of these drugs. The Subpart E regulations also establish some general procedures designed to expedite the development, evaluation, and marketing of new therapies intended to treat life-threatening or severely debilitating illnesses.

Fast track is a new designation arising out of the Food and Drug Modernization Act (FDAMA) of 1997. It allows increased face-to-face interactions between a drug sponsor and the agency and also allows the "rolling" submission of data in a New Drug Application (NDA). Fast track also allows consideration, but not necessarily a guarantee, of a *priority review*, which means FDA is required to take a review action within 6 months, rather than within 10 months, for a standard review. The provisions of fast-track pattern the successful approach that the agency had already been using for antiretroviral drugs, specifically the protease inhibitors.

Expanded access refers to several different mechanisms that allow drugs to be used for treatment before their approval. With expanded access, a drug is made available to patients who have limited therapeutic options, through protocols that are usually uncontrolled, open-label, and streamlined for collection of safety and efficacy data. Such protocols may be conducted using a Treatment IND, Parallel Track (which is an option only for HIV drugs), or general open-label treatment protocols submitted by the sponsor. The particular mechanism for expanded access protocols depends on the stage of drug development (e.g., parallel track is an expanded access mechanism for HIV drugs in earlier stages of development).

Sometimes expanded access protocols are designed to collect data that may answer certain scientific questions relative to safety or choice of dose; however, the primary intent of these protocols is to provide early access of drugs to patients in need. Expanded access protocols should not be burdensome for clinicians that need these drugs to treat their patients, but they should also not interfere with the ability of sponsors to enroll and complete randomized, controlled, clinical studies to support drug approval.

III. APPROVAL TIMES FOR HIV PROTEASE INHIBITORS

Approval times for the first four HIV protease inhibitors, saquinavir (Invirase), ritonavir (Norvir), indinavir (Crixivan), and nelfinavir (Viracept) ranged from 42 to 90 days. These approval times were among the shortest in regulatory history and also shorter than those of the antiretroviral drugs that preceded them. Most of the regulatory procedures mentioned in the foregoing, such as accelerated approval, expanded access, and increased FDA–sponsor interactions, were applied during development of these drugs; and protease inhibitor approvals served as a model for updated regulatory procedures (FDAMA), in much the same way that the didanosine approval served as the model for accelerated approval. Table 1 shows review times for all antiretroviral drugs including protease inhibitors.

The HIV protease inhibitors were not the only antiretrovirals that have been developed and approved quickly. In general, FDA has provided short approval times for first drugs of a new class, for drugs that demonstrate a clear advantage over existing therapies in terms of activity, tolerability, or ease of administration, or for drugs that have shown efficacy in patients who have exhausted other approved treatment options. For example, zidovudine (Retrovir; ZVD) was rapidly reviewed not only because it was the first antiretroviral therapy but also because

Table 1 Review Times for All Retroviral Drugs

Drug	Approval date	Review time
Nucleoside analogues		
Retrovir (zidovudine)	March 1987	3.5 mo
Videx (didanosine)	October 1991	6 mo
Hivid (zalcitabine)	June 1992	8.5 mo
Zerit (stavudine)	June 1994	6 mo
Epivir (lamivudine)	November 1995	6 mo
Ziagen (abacavir)	December 1998	6 mo
Protease inhibitors		
Invirase (saquinavir)	December 1995	3 mo
Norvir (ritonavir)	March 1996	72 days
Crixivan (indinavir)	March 1996	42 days
Viracept (nelfinavir)	March 1997	84 days
Fortovase (saquinavir)	November 1997	6 mo
Agenerase (amprenavir)	April 1999	6 mo
Non-nucleoside reverse transcriptase inhibitors		
Viramune (nevirapine)	June 1996	3 mo
Rescriptor (delavirdine)	April 1997	9 mo
Sustiva (efavirenz)	September 1998	3 mo

it had shown an unambiguous survival benefit in a placebo-controlled clinical endpoint study. In this study of patients with advanced HIV there was one death among those randomized to ZDV compared with 19 deaths among those randomized to placebo (7). Likewise, the non-nucleoside reverse transcriptase inhibitor (NNRTI) efavirenz (Sustiva) also had a relatively short review time for several reasons. First, it was the first antiretroviral showing clinical activity with once-a-day administration. Second, clinical studies showed that efavirenz was at least as active as a protease inhibitor in the setting of combination drug therapy (8).

When the first three HIV protease inhibitors were approved, there was an enormous need for new treatment options coupled with dramatic study data and anecdotal reports that surpassed any previously approved antiretroviral treatments. Phase 2 studies had demonstrated that ritonavir and indinavir, even when given suboptimally as monotherapy, could produce marked, albeit time-limited, decreases in HIV RNA levels associated with substantial increases in CD4 cell counts (9–11). In studies of indinavir in combination with zidvoudine and lamivudine, remarkable decreases in plasma viral load were demonstrated, with decreases to the limits of assay detection (12), a concept that revolutionized the approach to treating HIV. In addition there were several reports of resolution of clinical HIV-related illnesses, such as molluscum contagiosum, Kaposi's sarcoma (KS), and even progressive multifocal leukoencephalopathy (PML) in patients treated with protease inhibitors (13–19). No previously approved antiretroviral had been associated with such a dramatic resolution of HIV-related signs and symptoms, such as KS lesions. At about the same time, study M94-247, comparing ritonavir versus placebo over a background of approved therapies showed a delay in clinical disease progression and death for patients randomized to ritonavir, compared with placebo (20). All of the forgoing data findings and observations justified mobilization of both pharmaceutical and FDA resources to expedite these drugs through the approval process with unprecedented speed.

It is noteworthy that ritonavir was the first drug since zidovudine to receive its initial approval based on data demonstrating a decrease in clinical disease progression and survival (7,20). All other antiretroviral drugs were initially approved based on surrogate marker data with clinical confirmation following completion of studies months to years after marketing under accelerated approval. Remarkably, the double-blind phase of clinical endpoint study M94-247 was enrolled and completed in fewer than 9 months. This was due to the advanced stage of the participants (median CD4 cell count at entry was approximately 20/mL), and the potent activity of ritonavir in patients who had never received protease inhibitor therapy. This illustrates the point that efficacy studies, particularly clinical endpoint studies, can be conducted efficiently in patients with advanced HIV, not only because this population is most at risk for experiencing endpoints, but also because they may respond well to new treatment regimens.

Saquinavir (Invirase), the first protease inhibitor on the market (December 1995), was also approved rapidly; however, data at the time of approval showed only modest decreases in plasma viremia (21,22). Somewhat better activity was observed when saquinavir was administered in combination with nucleoside analogues to patients who were treatment-naive (23). Saquinavir's less impressive activity was related to poor bioavailability, approximately 4% (24) for the initially approved saquinavir formulation (Invirase). Had the NDA for saquinavir followed that of ritonavir and indinavir, approval of saquinavir may not have proceeded as rapidly. Currently, with the availability of the new formulation of saquinavir (Fortovase), which permits improved bioavailability (25), the first formulation (Invirase) is no longer recommended, except for use in combination with ritonavir (26). Fortunately, even the modest surrogate marker changes produced by saquinavir (Invirase) eventually showed clinical benefit in two clinical endpoint studies (22,27). After accelerated approval, clinical benefit was also established for indinavir in two clinical endpoint studies (28,29).

Overall, the influence of protease inhibitor therapy and combination antiretroviral therapy on the health of HIV-infected patients has been significant and dramatic, with obvious declines in morbidity and mortality. In spite of several recently recognized adverse events that have been associated with the protease inhibitor class, as will be discussed in the following section, their benefits continue to outweigh their risks.

IV. POTENTIAL RISKS OF RAPID APPROVALS

A. Drug Withdrawals

Subsequent to five withdrawals of nonantiviral drugs from the U.S. market over a 12-month period during 1997–1998, the rapidity of drug development, review, and approvals has been criticized (30). Criticism generally has not been directed toward rapid approvals of drugs used to treat serious or life-threatening illnesses; however, after associations between metabolic problems (hyperglycemia, fat redistribution) and HIV protease inhibitors were postulated, the rapidity of approval of antiretroviral drugs may also be questioned.

Friedman et al. (31) investigated the five most recent drug removals from the market and compared them with drug removals in the last two decades. There was no suggestion that rapid review times were responsible for eventual market withdrawals. In fact, review times for these five drugs ranged from 15 to 75 months, review times that were quite long compared with those for antiretroviral drugs. In addition, although the withdrawals were clustered in a 12-month period, these products were introduced to the market over a wide range of years, 1973–1997. Thus, the withdrawals did not seem to be a product of recent changes in review practices in response to the Prescription Drug User Fee Act (PDUFA).

In that drug withdrawals are often based on toxicities that are considered rare, it is not practical to expect that such events will routinely be identified in premarketing studies. For example, bromfenac, a nonsteroidal anti-inflammatory drug (NSAID), was withdrawn from the market because of rare, but serious, hepatotoxic adverse events estimated to occur in approximately 1:20,000 patients who took the drug for longer than 10 days. To reliably detect such an adverse event, a clinical trial database would have to include upward of 100,000 participants. Clearly, trials of this magnitude would not be feasible or even desirable for evaluation of antiretroviral drugs. Should larger clinical trials have been required premarketing for the protease inhibitors, life-saving treatments would have been significantly delayed.

B. Possible Protease Inhibitor Class Adverse Events

After approval of the first four HIV protease inhibitors, several metabolic or endocrinological complications were observed in postmarketing reports (32). One type of metabolic adverse event that was observed with the use of protease inhibitors in combination with other antiretrovirals was hyperglycemia (33–35). The observed hyperglycemia ranged in severity and presentation from mild hyperglycemia or exacerbation of existing diabetes, to frank new diabetes mellitus with ketoacidosis. These adverse events were not recognized before marketing. An increased risk of hyperglycemia also was not apparent in a retrospective look at the clinical trial data submitted in the original NDAs for these drugs, which usually included a total drug exposure in approximately 1000 patients. This suggests that clinically recognizable diabetes or hyperglycemia in association with protease inhibitors probably occurs relatively infrequently, approximating 1 case per 1000 patients or less. Approximately 3000 patients would be required to reliably find one event of this frequency. However, as this complication has been scrutinized further, more subtle degrees of hyperglycemia or insulin resistance may be more common.

Fat redistribution is a syndrome consisting of changes in body morphology among patients receiving HAART regimens (36–39). It involves lipoatrophy or lipoaccumulation that includes wasting of face, buttocks, or limb fat, or accumulation of central fat in the abdomen, neck, or breasts, or combinations thereof. It may be associated with abnormalities in metabolic laboratory parameters, such as insulin and lipid levels (40). Long-term consequences of the morphological aspects of this syndrome are unknown, but there is concern that these changes could possibly lead to increased cardiovascular risk (41). As with hyperglycemia, these changes were not recognized before approval of the first three protease inhibitors. However, in contrast with hyperglycemia, the lack of preapproval recognition of fat redistribution was not related to a low frequency of its signs and symptoms. In fact, the frequency of fat redistribution may be common, reported to occur in approximately two-thirds of individuals treated long-term in one series (36). Rather, the delay in recognition of such a complex syndrome was probably related to the time course for this unusual constellation of signs and symptoms to develop. Specifically, fat redistribution problems appear to be gradual in onset, with signs and symptoms that are subtle initially and that have sometimes been attributed to aging, HIV progression, or weight gain associated with therapeutic success.

Had either or both hyperglycemia or fat redistribution been clearly recognized or attributed to protease inhibitors before their approval, the risk/benefit ratio would still have weighed heavily in favor of approval of these drugs.

C. HIV Resistance

Other than the potential safety concerns for rapid approval of antiretrovirals, concerns have been raised about the effect of rapid approvals on the emergence of HIV drug resistance and cross-resistance. In the past, many HIV-infected patients received sequential monotherapy as each new drug entered the marketplace. This practice created a population of patients harboring viruses that were resistant and cross-resistant to available treatment options and sometimes even future options. This unfortunate situation was not so much a result of rapid approvals, per se, but rather a result of lack of knowledge about how HIV should be properly treated and monitored to ensure durable virological suppression. Scientific data clearly support the principle that HIV should be treated with combination therapy to maintain durable virological suppression (28,42,43). Whenever possible, all patients and trial participants should have the opportunity to receive combination therapy with several, and preferably three, potentially active drugs (26,44). Therefore, one might argue that rapid development and approvals would allow the availability of more drugs to be successfully used in combination which, in turn, could offer a higher therapeutic barrier for the development of resistance.

V. LEGACIES OF PROTEASE INHIBITOR DEVELOPMENT

The development of HIV protease inhibitors provided several landmark studies and scientific investigations that formed the basis for current HIV treatment and is likely to form the foundation for future research. Some of these important milestones are shown in Figure 1.

It was the antiviral potency of drugs, such as indinavir and ritonavir, that permitted the study of viral decay and half-life as investigated by Shaw, Ho, and others. In addition, when protease inhibitors were studied in triple combination, the resulting profound and durable suppression of plasma HIV became the foundation for the therapeutic principles of current HIV treatment guidelines (26,44). Consequently, new therapies combined with application of these treatment princi-

Understanding of HIV viral dynamics

Maximal virologic suppression as goal of therapy

Obvious impact on HIV morbidity and mortality

Innovations in clinical trial design

Use of pharmacologic interactions to enhance activity/ease of dosing

Focus on therapeutic barriers for preventing viral resistance

Figure 1 Protease inhibitor legacies.

ples are credited as producing an obvious and widespread effect on HIV morbidity and mortality in the United States (1,2).

The development of protease inhibitors included innovative trial designs that allowed more flexibility for study participants. Although these specific trial designs might not be acceptable today, they helped expand ideas of how clinical studies of HIV treatments may be conducted. For example, study M94-247 that compared ritonavir versus placebo over a background of therapy, allowed patient choice of background treatment (20). Despite a choice of various background therapies, a treatment effect was still discernible. Before this study, protocols generally required patients to remain on certain narrow, fixed drug regimens owing to concerns of confounding study results. Although it is recognized that adding a new agent to a failing regimen is presently not an optimal way to treat HIV, study designs such as M94-247 opened the door for other studies that allowed participant choices, particularly for treatment combinations. In current protocols, study participants are often allowed one of several possibilities of accompanying antiretroviral therapy based on their prior treatment experience or possibly resistance testing; this allows more individualization and optimization of therapy in the setting of a clinical investigation. Another example of innovation in study design emanating from protease inhibitor development was the allowance for treatment switches for loss of virological response. Agouron studies 506 and 511 (45,46), evaluating nelfinavir in combination with nucleosides, were among the first trials that permitted treatment switches based on increases in HIV RNA levels. Many prior studies had required patients to remain on fixed regimens for prolonged periods.

The ability of the currently approved protease inhibitors to inhibit other drugs metabolized by cytochrome P (CYP) 4503a was initially viewed as a negative quality of the drug class. However, metabolic inhibition has also been used in a positive way to increase concentrations of drugs with poor bioavailability, as with the ritonavir–saquinavir combination (47). Such combinations may also promote adherence or prevent the emergence of resistance by increasing the ease of administration of drugs. This is being studied for indinavir and ritonavir combinations, which holds potential for converting indinavir from a thrice-daily regimen with food restrictions to a twice-daily regimen without food restrictions (48). In addition, a new protease inhibitor in development, lopinavir (ABT-378), produces high and sustained plasma concentrations by coformulation of this new protease inhibitor with low doses of ritonavir (49).

VI. HIV RNA AS AN ENDPOINT IN CLINICAL STUDIES

The influence of HIV protease inhibitor therapy on successful treatment of HIV was a major factor in the agency's acceptance of a new primary endpoint for clinical efficacy trials for antiretrovirals. Before July 1997, the FDA required clinical endpoint studies to support traditional approval of antiretroviral drugs. Although accelerated approvals were routinely based on changes in surrogate endpoints, such as CD4 cell counts and plasma HIV RNA levels, clinical endpoint trials assessing effects on mortality or disease progression had been a requirement for traditional approvals before this time point. However, owing to the effect of combination therapy on the decline of HIV-related clinical disease progression, it was clear that a requirement for clinical endpoint studies for every drug approval was no longer feasible. A decreasing frequency of clinical endpoints as a result of treatment successes would require increasingly larger and longer clinical endpoint studies. In addition, treatment guidelines that recommended monitoring of plasma viremia and maximal suppression of HIV RNA as the treatment goal greatly affected the ability to conduct clinical endpoint studies that would not be confounded by treatment switches. With the emphasis of HIV RNA as a treatment goal, it would not be feasible to expect a patient to be maintained on a randomized regimen until a clinical endpoint occurred. Treatment switches based on knowledge of HIV RNA levels would be expected and would reduce the ability to discriminate subsequent differences in clinical progression rates between treatment arms.

In the setting of potent combination regimens, study size and duration for a clinical endpoint study would need to be quite large. For example, to adequately power a study to detect at least a 50% difference in the time to development of a clinical endpoint (AIDS-defining illness or death) between two treatment arms, in which triple-combination therapy is the control, the study sample size would have to be at least 1600 patients (800 per arm for a two-arm study). This sample size is based on the assumption that patients are relatively advanced at baseline (<200 cells/mm^3) and that at least 2 years of follow-up would be possible. Detecting smaller differences in event rates in less-advanced patients, in which the event rate is less frequent, would require even larger and longer studies. The sample size for HIV RNA endpoint studies comparing differences in time to virological failure can be accomplished in less than a third of the sample size required of a clinical endpoint study. In addition, virological endpoint studies can be conducted over a shorter time period and in either advanced or less-advanced patients. These savings in sample size and study time would be predicted to have a positive effect on expediting antiretroviral drug development and perhaps in stimulating pharmaceutical interest in antiretroviral drug development.

Fortunately, changes in HIV RNA levels appear to be reliably predictive of clinical disease. In five separate analyses of more than 5000 patients, there was a clear association between initial decreases in plasma HIV RNA (within the first 24 weeks) and a reduction in the risk of disease progression and death (50). This relation was also observed across a range of patient characteristics, including pretreatment CD4 counts and RNA levels, prior drug experience, and treatment regimen. This relation has also been observed when initial RNA reductions were evaluated using nadir, average change over time, or change from baseline at a specified time point. Evidence for this clinical association is strengthened by a dose–response relation between plasma HIV RNA reduction and disease progression. In addition, a longer duration of initial HIV RNA reduction appears to correlate with a decreased risk of clinical progression.

VII. CONCLUSIONS

Few new classes of drugs have produced such a dramatic and immediate influence on morbidity and mortality as have HIV protease inhibitors. In this alone, these products have been amazingly successful, despite their demanding dosing schedules, drug interactions, and toxicities. The activity of protease inhibitors also permitted a leap in the understanding of HIV pathogenesis that helped in the formation of new treatment guidelines. In addition, the regulatory experience of ushering in such an important drug class served as a model for expediting drug development. However, expedited drug development should involve more than just getting a drug to the market sooner. Expedited drug development should include methods for rapid accumulation and dissemination of more and improved clinical information to physicians and patients so that antiretroviral therapy can be used rationally and safely. If safety and efficacy standards can be achieved more efficiently, additional clinical resources could be directed toward better characterizing the clinical use of drugs. This would include the availability of more drug–drug interaction data; clinical resistance and cross-resistance data; data in special populations, such as children or patients with hepatic or renal disease; and data in patients with a wide range of treatment experience, ranging from treatment-naive to heavily pretreated. Such information could make any drug entering the market more useful in the setting of the current state of HIV polypharmacy.

Despite recent successes of HIV therapies, it is certain HIV will continue to evolve. The challenge for both the scientific and regulatory communities is to keep pace with viral evolution so that the clinical gains made in the last decade are not lost. Perhaps further emphasis on including additional technological tools during drug development, such as resistance testing and therapeutic concentration monitoring, could aid in even more efficient and rational drug development in the future.

REFERENCES

1. Palella FJ, Delaney KM, Moorman AC, Loveless MO, Fuhrer J, Satten GA, Aschman DJ, Holmberg SD. Declining morbidity and mortality among patients with

advanced human immunodeficiency virus infection. N Engl J Med 1998; 338:853–860.
2. Hogg RS, Yip B, Kully C. Improved survival among HIV-infected patients after initiation of triple-drug antiretroviral regimens. Can Med Assoc J 1999; 160:659–665.
3. Ho DD, Neumann AU, Perelson AS, Chen W, Leonard JM, Makowitz M. Rapid turnover of plasma virions and CD4+ lymphocytes in HIV-1 infection. Nature 1995; 373:117–122.
4. Wei X, Ghosh SK, Taylor ME, Johnson VA, Emini EA, Deutsch P, Lifson JD, Bonhoeffer S, Nowak MA, Hahn BH. Viral dynamics in human immunodeficiency virus type 1 infection. Nature 1995; 373:117–122.
5. Code of Federal Regulations. 21; 314.500.
6. Code of Federal Regulations. 21; 314.200.
7. Fischl MA, Richman DD, Grieco MH, Gottlieb MS, Volberding PA, Laskin OL, Leedom JM, Groopman JE, Mildvan D, Schooley RT. The efficacy of azidothymidine (AZT) in the treatment of patients with AIDS and AIDS-related complex. A double-blind, placebo controlled trial. N Engl J Med 1987; 317:185–191.
8. Tashima K, Staszewski S, Stryker R, Johnson P, Nelson M, Morales–Ramirez J, Manion DJ, Farina D, Labriola D, Ruiz N, The Study 006 Investigator Team. A phase III, multicenter, randomized, open-label, study to compare the antiretroviral activity and tolerability of efavirenz (EFV) + indinavir (IDV), versus EFV + zidovudine (ZDV) + lamivudine (3TC), versus IDV + ZDV + 3TC at 48 weeks (study DMP-266-006). 6th Conference on Retroviruses and Opportunistic Infections, Chicago IL, Jan 31–Feb 4, 1999.
9. Markowitz M, Saag M, Powderly WG, Hurley AM, Hsu A, Valdes JM, Henry D, Sattler F, La Marca A, Leonard JM. A preliminary study of ritonavir, an inhibitor of HIV-1 protease, to treat HIV-1 infection. N Engl J Med 1995; 333:1534–1539.
10. Stein DS, Fish DG, Bilello JA, Preston SL, Martineau GL, Drusano GL. A 24-week open-label phase I/II evaluation of the HIV protease inhibitor MK-639 (indinavir). AIDS 1996; 10:485–492.
11. Danner SA, Carr A, Leonard JM, Lehman LM, Gudiol F, Gonzales J, Raventos A, Rubio R, Bouza E, Pintado V, the Europe and Australian Collaborative Ritonavir Study Group. A short-term study of the safety, pharmacokinetics, and efficacy of ritonavir, an inhibitor of HIV-1 protease. N Engl J Med 1995; 333:1528–1533.
12. Gulick RM, Mellors JW, Havlir D, Eron JJ, Gonzalez C, McMahon D, Richman DD, Valentine FT, Jonas L, Meibohm A, Emini EA, Chodakewitz JA. Treatment with indinavir, zidovudine, and lamivudine in adults with human immunodeficiency virus infection and prior antiretroviral therapy. N Engl J Med 1997; 337:734–739.
13. Murphy M, Armstrong D, Sepkowitz KA, Ahkami RN, Myskowski PL. Regression of AIDS-related Kaposi's sarcoma following treatment with an HIV-1 protease inhibitor. AIDS 1997; 11:261–262.
14. Grube H, Ramratnam B, Ley C, Flanigan TP. Resolution of AIDS associated cryptosporidiosis after treatment with indinavir. Am J Gastroenterol 1997; 92:726.
15. Elliot B, Aromin I, Gold R, Flanigan T, Mileno M. 2.5 year remission of AIDS-associated progressive multifocal leukoencephalopathy with combined antiretroviral therapy. Lancet 1997; 349:850.
16. Lebbe C, Blum L, Pellet C, Blanchard G, Verola O, Morel P, Danne O, Calvo F.

Clinical and biological impact of antiretroviral therapy with protease inhibitors on HIV-related Kaposi's sarcoma. AIDS 1998; 12: F45–F49.
17. Conant MA, Opp KM, Poretz D, Mills RG. Reduction of Kaposi's sarcoma lesions following treatment of AIDS with ritonavir. AIDS 1997; 11:1300–1301.
18. Burdick AE, Carmichael C, Rady PL, Tyring SK, Badiavas E. Resolution of Kaposi's sarcoma associated with undetectable level of human herpesvirus 8 DNA in a patient with AIDS after protease inhibitor therapy. J Am Acad Dermatol 1997; 37:648–649.
19. Hicks CB, Myers SA, Giner J. Resolution of intractable molluscum contagiosum in a human immunodeficiency virus-infected patient after institution of antiretroviral therapy with ritonavir. Clin Infect Dis 1997; 24:1023–1025.
20. Cameron DW, Heath–Chiozzi M, Danner S, Cohen C, Kravcik S, Maurath C, Sun E, Henry D, Rode R, Potthoff A, Leonard J. Randomised placebo-controlled trial of ritonavir in advanced HIV-1 disease. The Advanced HIV Disease Ritonavir Study Group. Lancet 1998; 351:543–549.
21. Collier AC, Coombs RW, Schoenfeld DA, Bassett RL, Timpone J, Baruch A, Jones M, Facey K, Whitacre C, McAuliffe VJ, Friedman HM, Merigan TC, Reichman RC, Hooper C, Corey L. Treatment of human immunodeficiency virus infection with saquinavir, zidovudine, and zalcitabine. N Engl J Med 1996; 334:1011–1018.
22. Invirase (saquinavir mesylate) package insert. Nutley, NJ: Roche Laboratories; Hoffmann–La Roche, Inc.
23. Vella S, Galluzzo C, Giannini G, Pirillo MF, Duncan I, Jacobsen H, Andreoni M, Sarmati L, Ercoli L. Saquinavir/zidovudine combination in patients with advanced HIV infection and no prior antiretroviral therapy: CD4+ lymphocyte/plasma RNA changes, and emergence of HIV strains with reduced phenotypic sensitivity. Antiviral Res 1996; 29:91–93.
24. Invirase (saquinavir mesylate) package insert. Nutley, NJ: Roche Laboratories; Hoffmann–La Roche, Inc.
25. Fortovase (saquinavir) package insert. Nutley, NJ: Roche Laboratories; Hoffmann–La Roche, Inc.
26. Guidelines for the use of antiretroviral agents in HIV infected adults and adolescents. MMWR 1998; 47:43–82.
27. Roche Media Release. Largest ever HIV drug study shows invirase dramatically reduces the onset of AIDS. Basel, Switzerland, June 20, 1997.
28. Hammer SM, Squires KE, Hughes MD, et al. A controlled trial of two nucleosides plus indinavir in persons with human immunodeficiency virus infection and CD4 cell counts of 200 per cubic millimeter or less. N Engl J Med 1997; 337:725–733.
29. Crixivan (indinavir sulfate). West Point, PA: Merck & Co.
30. Rubin R. Policy speeds approvals, but some say it's too risky. USA Today. July 10–12, 1998.
31. Friedman MA, Woodcock J, Lumpkin MM, Shuren JE, Hass AE, Thompson LJ. The safety of newly approved medicines: do recent market removals mean there is a problem? JAMA 1999; 281:1728–1734.
32. Dube MP. Metabolic complications of antiretroviral therapies. AID Clin Care 1998; 10:41–44.
33. Lumpkin MM. Reports of diabetes and hyperglycemia in patients receiving protease

inhibitors for the treatment of human immunodeficiency virus. Washington, DC: FDA Public Health Advisory, June 11, 1997.
34. Dube MP, Johnson DL, Currier JS, Leedom JM. Protease inhibitor-associated hyperglycemia. Lancet 1997; 350:713–714.
35. Eastone JA, Decker CF. New onset diabetes mellitus associated with the use of protease inhibitors. Ann Intern Med 1997; 948:127.
36. Carr A, Samaras K, Burton S, Law M, Freund J, Chisholm DJ, Cooper DA. A syndrome of peripheral lipodystrophy, hyperlipidemia and insulin resistance in patients receiving HIV protease inhibitors. AIDS 1998; 12:F51–F58.
37. Mann M, Piazza–Hepp T, Koller E, Struble KA, Murray JS. Unusual distributions of body fat in AIDS patients: a review of adverse events reported to the Food and Drug Administration. AIDS Patient Care STDS 1999; 5:287–295.
38. Lo JC, Mulligan K, Tai VW, Algren H, Schambelan M. ''Buffalo hump'' in men with HIV-1 infection. Lancet 1998; 351:867–870.
39. Miller KD, Jones E, Yanovski JA, Shankar R, Feuerstein I, Falloon J. Visceral fat accumulation associated with use of indinavir. Lancet 1998; 351:871–875.
40. Carr A, Samaras K, Chisholm DJ, Cooper DA. Pathogenesis of HIV-1 protease inhibitor-associated peripheral lipodystrophy, hyperlipidemia, and insulin resistance. Lancet 1998; 351:1881–1883.
41. Henry K, Melroe H, Huebsch J, Hermundson J, Levine C, Swensen L, Daley J. Severe premature coronary artery disease with protease inhibitors. Lancet 1998; 351: 1328.
42. Kempf DJ, Rode RA, Xu Y, Sun E, Heath–Chiozzi ME, Valdes J, Japour AJ, Danner S, Boucher C, Molla A, Leonard JM. The duration of viral suppression during protease inhibitor therapy for HIV-1 infection is predicted by plasma HIV-1 RNA at the nadir. AIDS 1998; 12:F9–F14.
43. Coffin JM. HIV population dynamics in vivo: implications for genetic variation, pathogenesis, and therapy. Science 1995; 267:483–489.
44. Carpenter CC, Fischl MA, Hammer SM, Hirsch MS, Jacobsen DM, Katzenstein DA, Montaner JS, Richman DD, Saag MS, Schooley RT, Thompson MA, Vella S, Yeni PG, Volberding PA. Antiretroviral therapy for HIV infection in 1998, updated recommendations for the international AIDS society–USA panel. JAMA 1998; 280:78–86.
45. Viracept (nelfinavir mesylate) package insert. La Jolla, CA: Agouron Pharmaceuticals.
46. Saag M, Gersten M, Chang Y. Long term virological and immunological effect of the HIV protease inhibitor Viracep (nelfinavir mesylate) in combination with zidovudine (AZT) and lamivudine (3TC). Infectious Diseases Society of America 35th Annual Meeting, San Francisco, CA, Sept 13–16, 1997.
47. Kempf DJ, Marsh KC, Kumar G, Rodrigues AD, Denissen JF, McDonald E, Kukulka MJ, Hsu A, Granneman GR, Baroldi PA, Sun E, Pizzuti D, Plattner JJ, Norbeck DW, Leonard JM. Pharmacokinetic enhancement of inhibitors of the human immunodeficiency virus protease by coadministration with ritonavir. Antimicrob Agents Chemother 1997; 41:654–660.
48. Burger DM, Hugen PW, Prins JM, Van De Ende ME, Reiss P, Lange JM. Pharmacokinetics of an indinavir/ritonavir 800/100 mg bid regimen. 6th Conference on Retroviruses and Opportunistic Infections. Chicago, IL, Jan 31–Feb 4, 1999.

49. Sham HL, Kempf DJ, Molla A, Marsh KC, Kumar GN, Chen CM, Kati W, Stewart K, Lal R, Hsu A, Beterbenner D, Korneyeva M, Vasavanonda S, McDonald E, Saldivar A, Wideburg N, Chen X, Niu P, Park C, Jayanti V, Grabowski B, Granneman GR, Sun E, Japour AJ, Norbeck DW. ABT-378, a highly potent inhibitor of the human immunodeficiency virus protease. Antimicrob Agents Chemother 1998; 42: 3218–3224.
50. Murray JS, Elashoff MR, Iacono–Connors LC, Cvetkovich TA, Struble KA. The use of plasma HIV RNA as a study endpoint in efficacy trials of antiretroviral drugs. AIDS 1999; 13:797–804.

Postscript

Use of Protease Inhibitors in the Management of HIV/AIDS

Richard C. Ogden
Agouron Pharmaceuticals, San Diego, California

Charles W. Flexner
The Johns Hopkins University School of Medicine and School of Hygiene and Public Health, Baltimore, Maryland

In the 5 years since inhibitors of HIV protease first became available to people living with AIDS, we continue to witness the increasingly sharp contrast between the impact that AIDS is having in the developed, as opposed to the developing, world. In the developed world, where access to healthcare, including the use of protease inhibitor–based combination therapy, is straightforward, mortality and morbidity associated with HIV infection have declined for most (1). In the developing world, where there are medical, social, educational, and financial barriers to healthcare, infection rates remain high and the future prosperity of entire countries seems in jeopardy. For many of us, the Thirteenth World AIDS Conference, held in Durban, South Africa in July 2000, served as a direct reminder of this dilemma and provides a basis for renewed dialogue between investigators, public health officials, the pharmaceutical industry, and the affected communities. The UNAIDS report on the global HIV/AIDS epidemic provides a comprehensive summary (2). From such a global perspective, a book discussing the discovery, development, and therapeutic benefits of one class of antiretroviral agents may seem presumptuous. However, as is the case with many diseases, these and other drugs help scientists and clinicians better understand HIV pathogenesis. In addition to providing therapeutic options for some, protease inhibitors can be clinical research tools to assist the search for long-term solutions for all.

The emergence of HIV/AIDS in the developed world in the early 1980s coincided with the widespread availability of new scientific methodologies arising from the discovery of in vitro recombinant DNA technology. It was no great surprise that following identification and isolation of the virus, therapeutic targets could be identified and drug discovery initiated very rapidly. HIV protease was first identified as a potential viral protein by nucleic acid sequence homology to known aspartyl proteases. Following the establishment of its critical role in viral replication, it rapidly became the target for many groups employing the most current tools of modern drug discovery. As outlined in the early chapters of this book, discovery of the protease inhibitors is a textbook example of how structural biology in conjunction with medicinal chemistry can accelerate lead compound and drug discovery.

The appearance of protease inhibitors on the market drove the realization that viral suppression in infected individuals would require the simultaneous chronic administration of three or more drugs. The ability of protease inhibitors, in combination with reverse transcriptase inhibitors, to suppress viremia gave researchers the means to understand the dynamic nature of virus production and immune depletion. They also revealed the magnitude of the task of eradicating virus from sanctuary sites and finding a cure. As cornerstone drugs of the first HAART regimens, their promise as key components of long-term viral suppression and immune restoration was, and remains, high.

The realities of chronic combination drug therapy, however, began to become apparent when substantial differences emerged between response rates to therapy in clinical practice compared to clinical research studies. The reasons behind this difference are varied, but often related to patient adherence. One challenge for the pharmaceutical industry in recent years has been to deliver the same or better antiviral efficacy with fewer pills and less frequent dosing—important factors in improving patient adherence to medication. Nowadays all marketed HIV protease inhibitors are prescribed for the most part in simpler regimens than those originally investigated and approved by the regulatory authorities. Indeed, the HIV protease inhibitor ritonavir is currently at the forefront of clinicians' efforts to enhance the plasma levels of this class of protease inhibitors. The intent is thereby to reduce dosing intervals and remove meal or concomitant medication restrictions that complicate the daily routine of patients.

As with any chronic therapy, and as the threat of serious morbidities and death declined in the developed world, the appearance of toxicities and the incidence of adverse events not anticipated from the accelerated approval process for certain protease inhibitors and other antiretroviral drugs have become a focus. Differentiating the contributions of individual agents, of classes of agents, or simply of controlled but persistent HIV infection to these toxicities is not straightforward and is the subject of ongoing research. Clearly, chronic toxicities are a barrier to adherence. With HIV infection, the problems of missed doses and subtherapeutic drug levels are of particular concern because of the rapid selection of drug-resistant virus under those conditions. Treatment of drug-resistant HIV in many patients can be a challenging task for even the most experienced clinician. In contrast to some classes of antiretroviral agents, there is encouraging early clinical data that supports the sequential use of protease inhibitors in patients who experience virological failure. This intrinsic limited cross-resistance between some members of the class will ensure them a continuing role as components of secondary or tertiary regimens.

Research and development involving the class of protease inhibitors continues at an intense pace. New agents are in all stages of clinical development and discovery research programs exist in industry and academia. Unexpected observations have been made by clinicians, immunologists, and virologists regarding the in vitro and clinical use of HIV protease inhibitors. These preliminary studies cover such new areas as investigating immune reconstitution differences very early or late in the course of infection, and exploring the relevance of the inhibition of certain cellular proteases by HIV protease inhibitors. For example, a certain proportion of patients experiencing virological failure on combination drug therapy, often in a ''salvage'' regimen, experience prolonged rises in CD4 T-cells. They do not, over the short term, appear to be experiencing immune deficiency. T-cells in these patients have a longer half-life. Virus infecting these cells may be impaired in its replication and these observations may be correlated to a maintenance of certain mutations within the virus by continued drug selection pressure. The link between rising CD4 T-cells, impaired replicative capacity, and reduced viral pathogenicity remains an intriguing idea to be tested in controlled studies with protease inhibitors and other antiretroviral drugs.

As this book goes to press, the prospects for controlling HIV/AIDS for many years in infected individuals with access to drug therapy are good. The prospects for limiting infection and finding prophylactic or therapeutic vaccines remain challenging and the notion of a ''cure'' for a chronically infected individual remains a distant goal. The dedicated efforts of all whose work impacts this disease and whose life is impacted by this disease, represented in small part by the science and medicine described here, will continue.

References

1. Palella FJ, Delaney KM, Moorman AC, et al. Declining morbidity and mortality among patients with advanced human immunodeficiency virus infection. N Engl J Med 1998; 338:853–860.
2. Report on the global HIV/AIDS epidemic. UNAIDS, June 2000.

Index